To Harriet, Julia, and David

Contents

PREFACE	xi
ACKNOWLEDGMENTS	xiii
AUTHOR'S PREFACE TO THE SECOND EDITION	xv
EDITOR'S PREFACE TO THE SECOND EDITION	xvi

Introduction

Objectives	1
The Problem of Optimal Search	2
Example	3
Outline of Chapters	8

Chapter I Search Model

1.1	The Target Distribution	17
1.2	Detection	22
1.3	Basic Search Problem	32
	Notes	34

Chapter II Uniformly Optimal Search Plans

2.1	Optimization by the Use of Lagrange Multipliers	35
2.2	Uniformly Optimal Search Plans for Regular Detection Functions	41
2.3	Optimal Search with Uncertain Sweep Width	59
2.4	Uniformly Optimal Search Plans for the General Case	70
	Notes	81

Chapter III Properties of Optimal Search Plans

3.1	An Interpretation of Lagrange Multipliers	84
3.2	Maximum Probability Search	87
3.3	Incremental Optimization	91
3.4	Multiple Constraints	96
	Notes	100

Chapter IV Search with Discrete Effort

4.1	Model for Search with Discrete Effort	102
4.2	Uniformly Optimal Search Plans	104
4.3	Minimizing Expected Cost to Find the Target	110
4.4	Optimal Whereabouts Search	114
	Notes	116

Chapter V Optimal Search and Stop

5.1	Optimal Search and Stop with Discrete Effort	119
5.2	Optimal Search and Stop with Continuous Effort	130
	Notes	135

Chapter VI Search in the Presence of False Targets

6.1	False Target Models	137
6.2	Criterion for Optimality	145
6.3	Optimal Nonadaptive Plans	146
6.4	Optimization Theorems	155
6.5	Examples	162
6.6	Semiadaptive Plans	165
	Notes	177

Chapter VII Approximation of Optimal Plans

7.1	Approximations That Are Optimal in a Restricted Class	179
7.2	Incremental Approximations	189
	Notes	196

Chapter VIII Conditionally Deterministic Target Motion

8.1	Description of Search Problem	198
8.2	Uniformly Optimal Plans when the Target Motion Is Factorable	202
8.3	Optimal Plans within $\psi(m_1, \infty, K)$	213
8.4	Necessary and Sufficient Conditions for t-Optimal Search Plans	217
	Notes	219

Chapter IX Markovian Target Motion

9.1	Two-Cell Discrete-Time Markovian Motion	222
9.2	Two-Cell Continuous-Time Markovian Motion	229
9.3	Continuous-Time Markovian Motion in Euclidean N-Space	231
	Notes	233

Contents ix

Appendix A **Reference Theorems** 235

Appendix B **Necessary Conditions for Constrained Optimization of Separable Functionals**

B.1 Necessary Conditions for the Discrete Space J 238
B.2 Necessary Conditions for the Continuous Space X 241
Notes 244

References 245

AUTHOR INDEX 251

SUBJECT INDEX 255

NOTATION INDEX 259

Appendix C **Recent Results in Optimal Search for a Moving Target**

C.1 Optimal Search Density for a Moving Target 262
C.2 Optimal Searcher Path for a Moving Target 273
C.3 References 275

Appendix D **Corrections** 278

Preface

This book deals with the problem of optimal allocation of effort to detect a target. A Bayesian approach is taken in which it is assumed that there is a prior distribution for the target's location which is known to the searcher as well as a function which relates the conditional probability of detecting a target given it is located at a point (or in a cell) to the effort applied there. Problems involving computer search for the maximum of a function do not, for the most part, conform to this framework and are not addressed.

The allocation problems considered are all one-sided in the sense that only the searcher chooses how to allocate effort. For example, the target is allowed to move but not to evade. Thus, pursuit and evasion problems are not considered.

The primary focus of the book is on problems in which the target is stationary. For this case, we use a generalized Lagrange multiplier technique, which allows inequality constraints and does not require differentiability assumptions, to provide a unified method for finding optimal search allocations. This method is also extended to find optimal plans in some cases involving false and moving targets. For the case of Markovian target motion, results are, for the most part, presented without proof, and the reader is referred to the appropriate papers for proofs.

There are a large number of search-related problems, and this book does not claim to cover them all. Instead, the following basic classes of search problems are considered: search for a stationary target, search in the presence of false targets, optimal search and stop, and search for targets with conditionally deterministic motion and Markovian motion. Approximation of optimal plans is also briefly considered. Work on many of the interesting variations of these problems is referenced in the notes at the end of chapters.

Since this book gives the first unified presentation of the basic results in search theory, the notes also attempt to credit these results to the people who originally published them. The search-related problem of modeling the detection process is not considered in this book.

The choice of material included in this book was strongly influenced by experience in major search operations and in constructing computerized search planning systems. For example, the material on uncertain sweep width in Chapter II and that on false targets in Chapter VI was developed in response to problems encountered in the 1968 search for the missing submarine Scorpion. The problems in Chapter VIII on conditionally deterministic motion were first investigated when the problem of searching for a drifting target was considered for a computer-assisted search planning program developed for the U.S. Coast Guard.

The book is written for a person interested in operations research who has a strong mathematics background. The presentation of material in the early chapters is leisurely with proofs and motivation given in detail. In the later chapters the presentation is more condensed. The introductory chapter contains a chapter-by-chapter summary of the book and a guide to the logical dependence of chapters.

LAWRENCE D. STONE

June 1975

Acknowledgments

This book was written during the academic year 1973–1974 which I was invited to spend at the Naval War College in Newport, Rhode Island, for this purpose. The opportunity to do this was initiated by the suggestion of Warren F. Rogers, Chairman of the Department of Management at the Naval War College. The work in search theory by myself and my colleagues at Daniel H. Wagner, Associates, which provided the knowledge and experience required to write this book, has been primarily supported by the Naval Analysis Programs of the Office of Naval Research under the sponsorship of Robert J. Miller and J. Randolph Simpson. Substantial portions of this work are referenced in or adapted to the book.

I wish to acknowledge the debt that this book owes to the opportunity that my colleagues and I have had to participate in various major search operations. In fact, significant portions of the theory presented in this book have been motivated by and in part applied to these searches. This involvement in major search operations began with the participation of Henry R. Richardson in the 1966 search for the H-bomb lost in the Mediterranean near Palomares, Spain. Further experience in search was obtained under contract to the Navy's Deep Submergence Systems Project, headed by John P. Craven. This experience was deepened by participation in the 1968 search for the missing United States nuclear submarine Scorpion and the other projects discussed below. As a result of the knowledge gained in these operations, we prepared a manual for the operations analysis of search for submerged stationary objects (Richardson *et al.*, 1971) in conjunction with Frank A. Andrews for the Navy's Supervisor of Salvage. At this time, the Office of Naval Research began to support work for the purpose of solving some of the theoretical problems presented by the Scorpion search, notably those involving uncertain sweep width and false targets. Subsequent work on search problems has included the development of computer programs to assist the U.S. Coast Guard in their search and rescue operations. This work has led to a broader understanding of search problems and to some of the theoretical developments on moving-target problems presented in Chapter VIII. Recently analysis support was provided to the U.S. Navy in order to help them assist the Egyptians in clearing ordnance from the Suez Canal.

I would like to thank Henry R. Richardson for writing a first draft of the section of Chapter II which deals with uncertain sweep width. The book benefited greatly from comments made by the many people who read all or part of the manuscript. Among these are Warren F. Rogers, George F. Brown, Richmond M. Lloyd, and Chantee Lewis of the Naval War College; Daniel H. Wagner, Henry R. Richardson, and Bernard J. McCabe of Wagner Associates; and Joseph P. Kadane of Carnegie-Mellon University.

I greatly appreciate the patient and good natured efforts of Grace P. McCrane, who typed the major portion of the original draft. I also wish to thank Jane James, Linda D. Johnson, Colleen V. Penlington, Lisa L. DiLullo, and Christine M. McAllister for their assistance in preparing the manuscript.

The preparation of this book was accomplished under contracts N00140–74–C–6089, N00140–75–M–7015, N00014–74–C–0329, and N00014–70–C–0232 which were funded by the Naval War College and the Office of Naval Research.

Author's Preface to the Second Edition

This book was awarded the 1975 Lanchester Prize by the Operations Research Society of America. Since then there have been major developments in search theory and its applications. On the theoretical side, there has been significant progress made in solving the problem of optimal search for a moving target. Appendix C reviews this progress and provides a bibliography of references to work in this area. In the area of optimal search for stationary targets, the results published in the first edition have stood the test of time in the sense that more recent results have tended to be modest generalizations of existing ones, often proved with great effort.

On the applied side, the development of cheap and powerful microcomputers with high resolution color graphics has allowed the development of real-time interactive search planning systems. Because of these computer developments, the trend in search theory is toward algorithms for computers and away from the theorem-proof style of presentation given in this book.

Appendix D contains some corrections to the first edition.

LAWRENCE D. STONE

February 1989

Editor's Preface to the Second Edition

The search process is inherently a nervous one. Either you will find the "target" or you won't. This involves more stress than the continuous penalties or payoffs associated with dullness or brilliance in dealing with problems such as scheduling or logistics. This discontinuity makes search a little like litigation. During an actual "case" there is a sense of urgency and emergency. This stress can trigger a major, sometimes frantic, effort. Experts can be mobilized. Armies (or navies) can be sent scurrying around. A nervous principal or client can make intuitive decisions that are painfully wrong. In short, search theory is a delightful challenge for operations research.

Stone's book, first published in 1975, was and is an elegant response to that challenge. As Stone himself says, B. O. Koopman pioneered the application of coherent mathematical process to the search problem during World War II. Stone's book exploits the advances that we have witnessed in pure mathematics since then. The origins of these advances -- from men like Lebesgue and Borel -- clearly antedated the War; but it was not until the 1940's that the mathematical community began to experience the energetic and pervasive consequences of the pre-war concepts.

Stone wasn't kidding when he stated in his 1975 introduction that the book was written for those with "a strong mathematics background". Even those with first-class graduate credentials in operations research will find themselves stretched to appreciate clearly the underlying assumptions for his results, and equally stretched to digest the results themselves. From this community, it will be the truly remarkable exception that will comprehend -- much less enjoy -- his proofs and existence theorems. Even at that, the stretching will be good for us, and the results constitute an invaluable definitive set of solutions to central problems in search theory.

Oddly enough, a collateral target of this book should be pure mathematicians themselves. We have already noted that it takes a real mathematician to appreciate and enjoy the subtleties and rigor that the book presents. There is a real dearth of monographs that reach this far into sophisticated mathematics in support of real problems in the world of operations research. In our day, mathematicians had a dismaying tendency to excise any intuitive or applications-oriented origins of what were presented to the mathematical community as elegant, abstract ideas. As a result, many mathematics students were deprived of the very stimulations that fuel some of the best mathematical advances. This book should provide a welcome meal for the reality-starved pure mathematician.

JOHN D. KETTELLE

February 1989

Introduction

We begin by discussing the objectives of this book and its intended audience. A brief description of the setting of search theory is given and an example is presented to illustrate some of the concepts and methods of the theory. Then, the results of the book are outlined in a chapter-by-chapter fashion. Finally, there is a guide to the logical dependence of the chapters.

Search theory is one of the oldest areas of operations research. The initial developments were made by Bernard Koopman and his colleagues in the Anti-Submarine Warfare Operations Research Group of the US Navy during World War II to provide efficient methods of detecting submarines. Search theory is still widely used for this purpose. In addition, the theory has been used by the US Navy to plan searches for objects such as the H-bomb lost in the ocean near Palomares, Spain, in 1966 and the submarine Scorpion lost in 1968, as well as numerous, less well-known objects. The US Coast Guard is also using search theory to plan some of its more complicated search and rescue efforts.

Other possible areas of application of search theory include prospecting and allocating marketing or law enforcement effort (see Larson, 1972). Searches for persons lost on land also fit into the framework of the theory.

OBJECTIVES

In spite of its history and usefulness, no unified and comprehensive presentation of the results in search theory has previously been published. One objective of this book is to make such a presentation of the major results in the theory of optimal search. Prior to the writing of this book, the results of search theory have appeared scattered throughout the literature of operations research, applied mathematics, statistics, and optimization theory. The methods used in one article appear, on the surface, to be unrelated to the methods in another. In this book, a basic optimization technique, i.e., use of Lagrange multipliers and maximization of Lagrangians, is presented and used to solve most of the problems of optimal allocation encountered here.

Another objective is to give a presentation of search theory appropriate to someone interested in applying the basic theory rather than becoming a specialist. This is done in Chapters I, II, IV, and VII, which contain results that allow the reader to find optimal search plans for the most basic search problem, that of finding a stationary target when no false targets are present.

The scope of the book is not encyclopedic. It contains what the author considers to be the core of the subject as it now stands. In selecting results to present, the author has tried to include those applicable to a wide range of search problems or important to the development of search theory. In addition, there is an emphasis on results that allow one to compute search plans. However, certain results have not been presented because their development would be too technical and tedious for this book. This is particularly true for searches including false or moving targets. Often these results will be described and referenced for the interested reader.

It is important to emphasize that the theory of optimal search is still developing, particularly in the areas involving moving or false targets. It is hoped that this book will outline what is presently known and provide impetus for researchers to extend the theory.

The presentation of the book is designed for an operations researcher with a strong mathematics background. In particular, it is assumed that the reader has had a graduate level course in probability theory and a solid background in real analysis. Knowledge of Lebesgue integration is desirable but not essential. In the early chapters the presentation is leisurely, with proofs and motivation given in detail. In the later chapters the presentation is more condensed. Parts of the book that are more technical than the general level of the chapter in which they appear are indicated by an asterisk. These may be skipped without impairing the reader's ability to follow most of the subsequent material.

THE PROBLEM OF OPTIMAL SEARCH

The problem of optimal search begins with an object of interest, the target, which the searcher wishes to find. The target is assumed to be located either

Example 3

at a point in Euclidean n-space, X, or in one of a possibly infinite collection of cells, J. The search space X is called continuous and the space J discrete. While the target's exact position is unknown, it is assumed that there is a probability distribution, known to the searcher, for the target's position at time 0. For most of the search problems considered in this book, the target is assumed to be stationary.

There is a detection function b that relates effort applied in a region or cell to the probability of detecting the target given that it is in that region or cell. In addition, there is a constraint on effort. The basic search problem is to find an allocation of effort in space that maximizes the probability of detecting the target subject to the given constraint on effort.

In some cases optimal search for moving targets will be investigated. However, situations in which the target is evading are not considered. Mathematically speaking, only one-sided optimization problems are considered; that is, the only choice to be made is the allocation of the searcher's effort. If the target moves, it does so independently of the searcher's actions. Thus, the target may move randomly according to distributions assumed known to the searcher, but it is not allowed to take evasive action related to the searcher's action.

In addition to maximizing the probability of detection within a fixed constraint on effort, other measures of optimization are considered. Attention is given to search plans that yield the maximum probability of detection at each time instant during an interval of time or that minimize the mean time to find the target.

In order to illustrate the concepts just discussed in a more specific manner, a simple example of a search problem and its solution are given.

EXAMPLE

The following simple example illustrates some of the basic concepts and techniques of search theory. The example also provides a convenient background for the discussion of results given later in this chapter.

Consider an idealized search with two cells, and suppose the target has probability $p(1) = \frac{2}{3}$ of being in cell 1 and probability $p(2) = \frac{1}{3}$ of being in cell 2 (see Fig. 1). One could think of the cells as drawers filled with coins and the searcher as looking for a particular coin, which is very valuable. He feels the valuable coin is more likely to be in drawer 1 than in drawer 2.

Suppose there is a fixed amount of time K in which to find the target. How should time be divided between the two cells in order to maximize the probability of detecting the target (i.e., finding the coin)? To answer this question, it is necessary to assume there is a detection function b that relates time spent

$p(1) = 2/3$	$p(2) = 1/3$
Cell 1	Cell 2

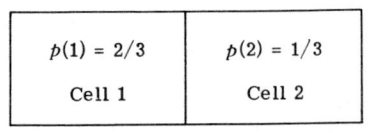

Fig. 1. Two-cell target distribution.

looking in a cell to probability of detecting the target given that it is in the cell. Assume that the detection function is exponential, i.e.,

$$b(z) = 1 - e^{-z} \quad \text{for } z \geq 0, \tag{1}$$

where z is measured in hours. Thus, if z hours are spent looking in cell 1, there is probability $b(z)$ of finding the valuable coin given that it is in cell 1.

Without justifying the exact form of the detection function b, we note that it has two intuitively plausible features. First, if the target is in a given cell and some time is spent looking in that cell, then there is always positive probability that the target will be overlooked. In the case of the coins, this could result from the possibility that the searcher may not recognize the valuable coin even if he looks right at it. The second is a saturation effect or law of diminishing rate of returns. This is evidenced by the decreasing nature of the derivative $b'(z) = e^{-z}$ of the detection function. The result is that the longer one looks in a given cell, the slower the rate of increase of probability of detection.

Suppose that z_1 time is spent looking in cell 1 and z_2 time in cell 2. The probability of detecting the target with this allocation of time is

$$p(1)(1 - e^{-z_1}) + p(2)(1 - e^{-z_2})$$

and the total time (or cost) required by this allocation is $z_1 + z_2$.

Let f be an allocation, i.e., $f(j)$ tells the amount of time spent looking in cell j for $j = 1, 2$. Define

$$P[f] = p(1)b(f(1)) + p(2)b(f(2)), \tag{2}$$

$$C[f] = f(1) + f(2). \tag{3}$$

Then $P[f]$ gives the probability of detection and $C[f]$ the cost (or total time) associated with the allocation f. Of course, $f(j) \geq 0$ for $j = 1, 2$. In mathematical terms, this simple search problem becomes that of finding an allocation f^* such that $C[f^*] \leq K$ and

$$P[f^*] = \text{maximum of } P[f] \text{ over all allocations } f \text{ such that } C[f] \leq K. \tag{4}$$

Such an f^* is called *optimal* for cost K.

Suppose the searcher has spent z_1 time looking in cell 1 and z_2 time looking in cell 2 and is considering spending a small increment of time h looking in cell 1 again. The increase in probability resulting from this increment is approximately

$$p(1)b'(z_1)h.$$

If the increment were added to cell 2, the increase would be approximately

$$p(2)b'(z_2)h.$$

In the short term the searcher would benefit most from placing the increment in the cell having the highest value of $p(j)b'(z_j)$.

Let r be the probability that the target has not been detected after spending z_1 time looking in cell 1 and z_2 looking in cell 2. Since $b'(z) = e^{-z} = 1 - b(z)$, it follows that

$$p(j)b'(z)/r = p(j)[1 - b(z_j)]/r$$

is the posterior probability that the target is in cell j given that the target has not been detected. Thus, for an exponential detection function, searching in the cell with the highest value of $p(j)b'(z_j)$ is equivalent to searching in the cell with the highest posterior probability. This results from the following property of the exponential detection function:

$$\text{Pr\{detection in time } z + h \mid \text{failure by time } z\} = \frac{b(z+h) - b(z)}{1 - b(z)} = 1 - e^{-h},$$

i.e., the probability of detecting in the next increment of time h given failure to detect previously is independent of the amount of time spent searching. In this sense the exponential distribution does not "remember" how much effort has been expended searching in a given cell. As the search progresses and the target is not found, only the posterior target location probabilities change. Thus it is clear that the optimal short-term policy for an exponential detection function should be to place the next small increment of effort in the cell or cells with the highest posterior probabilities.

As is well known, the exponential detection function is the only one having this lack of memory. For other detection functions one would expect that it is necessary to take account of both the posterior target probability and the posterior detection function given failure to detect by time z_j. In fact, the probability of detecting the target with an increment h of time spent in cell j given failure to detect the target previously is

$$\frac{1}{r} p(j)[b(z_j + h) - b(z_j)].$$

As h becomes small, the above expression is approximately equal to

$$p(j)b'(z_j)h.$$

Thus, taking into account both the posterior target probabilities and the posterior detection function, one obtains the criterion of searching where $p(j)b'(z_j)$ is highest.

It still remains to show that the policy that yields the maximum short-term gain also produces an optimal long-term policy. To do this, we define

$$\rho(j, z) = p(j)b'(z) \quad \text{for} \quad j = 1, 2, \quad z \geq 0.$$

Then ρ is called the *rate of return function*. Consider a search policy that always places the next "small" increment of effort in the cell having the highest rate of return, considering the time previously spent looking in that cell. That is, the next increment goes into the cell j^* such that

$$\rho(j^*, z_{j^*}) = \max_{j=1,2} \rho(j, z_j),$$

where z_j gives the amount of time previously spent looking in cell j. Such a search policy is called *locally optimal*.

At time 0 when no search has been made in either cell,

$$\rho(1, 0) = p(1)b'(0) = \tfrac{2}{3}, \qquad \rho(2, 0) = p(2)b'(0) = \tfrac{1}{3}.$$

Thus, a locally optimal policy calls for looking solely in cell 1 until time s, such that

$$\rho(1, s) = \tfrac{2}{3} e^{-s} = \tfrac{1}{3} = \rho(2, 0),$$

i.e.,

$$s = \ln 2.$$

In order to continue the locally optimal policy, the additional times d_1 and d_2 spent looking in cell 1 and 2, respectively, must be split so that

$$\rho(1, s + d_1) = \tfrac{2}{3} e^{-(s + d_1)} = \tfrac{1}{3} e^{-d_2} = \rho(2, d_2),$$

i.e.,

$$d_1 = d_2.$$

This search policy may be described as follows: Let $\varphi^*(j, t)$ be the amount of time out of the first t hours spent looking in cell j for $j = 1, 2$. Then

$$\varphi^*(1, t) = \begin{cases} t & \text{for} \quad 0 \leq t \leq \ln 2, \\ \tfrac{1}{2}(t + \ln 2) & \text{for} \quad \ln 2 < t < \infty, \end{cases}$$

$$\varphi^*(2, t) = \begin{cases} 0 & \text{for} \quad 0 \leq t \leq \ln 2, \\ \tfrac{1}{2}(t - \ln 2) & \text{for} \quad \ln 2 < t < \infty. \end{cases} \tag{5}$$

Suppose that $K = 4$ hr is the amount of time available for the search. Let

$$f^*(1) = \varphi^*(1, 4) = 2 + \tfrac{1}{2} \ln 2, \qquad f^*(2) = \varphi^*(2, 4) = 2 - \tfrac{1}{2} \ln 2,$$

Example

so that f^* gives the allocation resulting from following the plan φ^* for 4 hr. It is claimed that f^* is optimal for cost $K = 4$ hr. To see this, consider the function ℓ defined by

$$\ell(j, \lambda, z) = p(j)b(z) - \lambda z \qquad \text{for} \quad j = 1, 2, \quad \lambda \geq 0, \quad z \geq 0.$$

This function is called the *pointwise Lagrangian*, and λ is called a *Lagrange multiplier*. Let

$$\lambda = p(1, f^*(1)) = p(2, f^*(2)) = \frac{\sqrt{2}}{3} e^{-2}.$$

Then one may check that, for $j = 1, 2$,

$$\ell(j, \lambda, f^*(j)) = \text{maximum of } \ell(j, \lambda, z) \qquad \text{over} \quad z \geq 0 \qquad (6)$$

by taking the derivative ℓ' of ℓ with respect to z and observing that for $j = 1, 2$,

$$\begin{aligned} \ell'(j, \lambda, z) &\geq 0 \qquad \text{for} \qquad 0 \leq z \leq f^*(j), \\ &\leq 0 \qquad \text{for} \quad f^*(j) < z < \infty. \end{aligned}$$

The fact that f^* satisfies (6) allows one to show that f^* is optimal for cost $K = 4$ as follows. Suppose that f is an allocation such that $C[f] \leq K = 4$ hours. Then by (6),

$$\ell(j, \lambda, f^*(j)) \geq \ell(j, \lambda, f(j)) \qquad \text{for} \quad j = 1, 2. \qquad (7)$$

Summing both sides of (7), one obtains

$$P[f^*] - \lambda C[f^*] \geq P[f] - \lambda C[f],$$

which implies

$$P[f^*] - P[f] \geq \lambda(C[f^*] - C[f]) \geq 0,$$

where the last inequality follows from $\lambda \geq 0$ and $K = C[f^*] \geq C[f]$. Thus f^* is optimal for cost $K = 4$ hr. One may check that $P[f^*] \approx 0.87$.

If λ and f^* satisfy (6) for $j = 1, 2$, then we say that (λ, f^*) *maximizes the pointwise Lagrangian*. By setting $\lambda = p(1, \varphi^*(1, t))$ and $f^*(j) = \varphi^*(j, t)$ for $j = 1, 2$, one may check that (λ, f^*) maximizes the pointwise Lagrangian and that f^* is optimal for cost $K = t$ for $t \geq 0$. Because of this, the plan φ^* is called *uniformly optimal*. Let $\mu(\varphi)$ be the mean time to find the target using plan φ and let $\varphi(\cdot, t)$ be the allocation of search effort resulting from following plan φ, for time t. Since

$$\mu(\varphi) = \int_0^\infty (1 - P[\varphi(\cdot, t)]) \, dt,$$

and φ^* is uniformly optimal (i.e., $P[\varphi^*(\cdot, t)] \geq P[\varphi(\cdot, t)]$ for $t \geq 0$), one can see that φ^* minimizes the mean time to find the target among all search plans, such that $C[\varphi(\cdot, t)] = t$ for $t \geq 0$.

OUTLINE OF CHAPTERS

Chapter I: Search Model

The basic search problem is defined and motivated here. Target distributions and methods for generating them are discussed. Detection functions are defined and examples of detection functions are derived for some special situations. The class of regular detection functions is introduced, where a regular detection function b is one such that $b(0) = 0$ and b' is continuous, positive, and strictly decreasing. (Note that the detection function in the preceding example is regular.) In Section 1.3, the basic problem of optimal search is stated in a concise and mathematical form for both the search space X, Euclidean n-space, and the search space J, a possibly infinite subset of the positive integers. Each $j \in J$ may be thought of as representing a cell where the target may be located.

Chapter II: Uniformly Optimal Search Plans

Section 2.1 presents optimization techniques, based on Lagrange multipliers, that do not require differentiability assumptions. In Theorems 2.1.2 and 2.1.3, it is shown that maximization of the pointwise Lagrangian is sufficient for constrained optimality. Theorems 2.1.4 and 2.1.5 give conditions under which maximization of a pointwise Lagrangian is necessary and sufficient for constrained optimization. There is an interesting difference here between the continuous and discrete search spaces. For the discrete space, one must assume that the detection function b is concave to guarantee that maximizing a pointwise Lagrangian is necessary for constrained optimality. For a continuous search space, no such assumption is required.

The results of Section 2.1 permit search problems to be attacked by maximizing pointwise Lagrangians. Looking at the preceding example, we see that this problem is equivalent to maximizing a real-valued function of a real variable. The techniques developed in Section 2.1 are basic to the presentation in this book and are used in various forms to solve most of the problems considered. Exceptions to this are the problems of optimal stopping considered in Chapter V and searches for targets with Markovian motion discussed in Chapter IX.

Section 2.2 finds uniformly optimal search plans when the detection function is regular by using the Lagrange multiplier techniques of Section 2.1. The uniformly optimal plan is computed and its properties are studied for the case

Outline of Chapters

of a circular normal target distribution having density function p defined on the plane and an exponential detection function b, i.e.,

$$p(x) = \frac{1}{2\pi\sigma^2} \exp\left(-\frac{x_1^2 + x_2^2}{2\sigma^2}\right) \quad \text{for} \quad x = (x_1, x_2) \in X,$$

and

$$b(z) = 1 - e^{-z} \quad \text{for} \quad z \geq 0.$$

In Example 2.2.8 an algorithm is given for optimally allocating effort to a finite number of cells when the detection function is exponential. Example 2.2.9 shows that if the detection function is not regular, a uniformly optimal plan may not exist for a discrete search space.

In Section 2.3 application is made of the results of Section 2.2 to find uniformly optimal plans when the search sensor has an uncertain detection capability. In Example 2.3.5 the uniformly optimal search plan is calculated for a search in which the target distribution is circular normal and the sweep width of the search sensor is gamma distributed. The mean time to detect the target using the optimal plan is compared to suboptimal plans, which do not take full account of the uncertainty in sweep width. It is shown that substantial penalties in mean time are possible when using these suboptimal plans.

Section 2.4 considers more general situations in which uniformly optimal plans exist. In particular, when the search space is X, it is shown that a uniformly optimal search plan exists whenever the detection function is right continuous and increasing. For discrete search space, uniformly optimal plans are shown to exist whenever the detection function is concave and continuous.

Chapter III: Properties of Optimal Search Plans

Properties of optimal search plans are investigated in this chapter. In Section 3.1 an interpretation of Lagrange multipliers is given in terms of marginal rates of return, and it is shown that locally optimal search plans are uniformly optimal when the detection function is regular. The proof is a generalization of the one given in the preceding example. In Section 3.2 it shown that always searching where the posterior target distribution is the highest produces a uniformly optimal search plan if and only if the detection function is homogeneous (i.e., not space dependent) and exponential. Section 3.3 considers search plans that are obtained in an incremental fashion by allocating a number of increments of effort so that each increment yields the maximum increase in probability of detection considering the application of the previous increments. It is shown that when the search space is X, this incrementally optimal allocation always produces an optimal allocation of the total effort involved.

In Section 3.4 the question of incremental and total optimality is considered in the context of multiple constraints. In this case, Example 3.4.4 shows that incrementally optimal allocations may fail to be totally optimal even if the search space is X and the detection and cost functions are linear. This is in strong contrast to the case of a single constraint, where the result is true in great generality for the search space X.

Chapter IV: Search with Discrete Effort

This chapter considers discrete search. In this case effort must be applied in discrete units in any cell. These units are called *looks*. Let

$\beta(j, k)$ = probability of detecting the target on but not before the kth look in cell j given that the target is in cell j,

$\gamma(j, k)$ = cost of kth look in cell j.

A search plan that always takes the next look in a cell j^* such that

$$\frac{p(j^*)\beta(j^*, k_{j^*} + 1)}{\gamma(j^*, k_{j^*} + 1)} = \max_{j \in J} \frac{p(j)\beta(j, k_j + 1)}{\gamma(j, k_j + 1)},$$

where k_j is the number of looks previously made in cell j, is called locally optimal. The ratio $p(j)\beta(j, k)/\gamma(j, k)$ is the discrete counterpart of $\rho(j, z)$. Theorem 4.2.5 shows that if γ is constant and if $p(j)\beta(j, \cdot)$ is decreasing, then a locally optimal search plan is uniformly optimal. Observe that requiring $p(j)\beta(j, \cdot)$ to be decreasing is the discrete analog of requiring the detection function b to be regular. Theorem 4.3.2 shows that if γ is bounded away from 0 and if $p(j)\beta(j, \cdot)/\gamma(j, \cdot)$ is decreasing, then a locally optimal plan minimizes the mean cost to find the target.

Section 4.4 considers whereabouts searches, in which the objective is to say which cell contains the target within a given constraint on cost. There are two ways the searcher may correctly state the target's position: He may detect the target within the cost constraint or, failing to do this, he can correctly guess the cell containing the target. Thus, a whereabouts search strategy is the combination of a detection search strategy and a cell to guess if the search fails to detect the target. In this section it is shown that an optimal whereabouts search proceeds by choosing a cell to guess and performing an optimal detection search in the remaining cells. In special cases, rules are given for choosing the cell to be guessed. In particular, if for some $\alpha > 0$,

$$\beta(j, k) = \alpha(1 - \alpha)^{k-1}$$
$$\gamma(j, k) = 1 \qquad \text{for } j \in J, \ k = 1, 2, \ldots,$$

Outline of Chapters

then one picks a cell j^* such that $p(j^*) = \max_{j \in J} p(j)$ to guess in case the detection search fails and performs a locally optimal search in the remaining cells.

Chapter V: Optimal Search and Stop

Here the problem of how to search and when to stop in order to maximize expected return is considered when there are a reward for finding the target and costs associated with searching. The combination of a search plan and a stopping time s determines a search and stop plan as follows. The search plan is followed until either the target is detected or time s is reached. When either of these events occurs, the search stops. A search and stop plan that maximizes expected return is called optimal.

Section 5.1 considers searches with discrete effort. In addition to satisfying the assumptions of the model of Chapter IV, there is a function V such that $V(j)$ gives the value of the target or reward obtained if the target is found in cell j for $j \in J$. If J is finite and γ is bounded away from 0, then Theorem 5.1.1 shows that an optimal search and stop plan exists and the optimal return function satisfies the optimality equation of dynamic programming.

Two special situations are considered first, namely, those in which $s = 0$ is optimal and those in which an optimal search and stop plan can be found in the class Ξ_∞ of plans with $s = \infty$.

If $\sum_{j \in J} p(j) V(j) < \infty$, then Theorem 5.1.3 shows that any plan that minimizes the expected cost to find the target is optimal within Ξ_∞. Theorem 5.1.6 gives conditions under which an optimal search and stop plan can be found within Ξ_∞. An immediate consequence of these theorems and Theorem 4.3.2 is that if the conditions of Theorem 5.1.6 are satisfied, the ratios $p(j)\beta(j, \cdot)/\gamma(j, \cdot)$ are decreasing for $j \in J$, and γ is bounded away from 0, then any locally optimal search plan coupled with $s = \infty$ produces an optimal search and stop plan.

Theorem 5.1.8 shows that if J is finite, γ is bounded away from 0 and

$$\frac{\beta(j, k)}{1 - b(j, k-1)} V(j) < \gamma(j, k) \qquad \text{for} \quad k = 1, 2, \ldots, \quad j \in J,$$

then $s = 0$ is an optimal search and stop plan.

The main result of Section 5.1 is given in Theorem 5.1.11, which shows that if $V(j) = V_0$ for $j \in J$, J is finite, and $p(j)\beta(j, \cdot)/\gamma(j, \cdot)$ is decreasing for $j \in J$, then an optimal search and stop plan follows a locally optimal plan as long as it continues to search. Lemma 5.1.9 implies that all locally optimal plans with the same stopping time have the same expected return. Thus, under the conditions of Theorem 5.1.11, the problem of optimal search and stop is reduced to the problem of optimal stopping when using a locally optimal search plan. Unfortunately, the optimal stopping problem is difficult to solve.

Section 5.2 considers optimal search and stop when effort is continuous. In contrast to Section 5.1, the cost of search is a function of time only and not of the search plan. For a given cumulative effort function M, search plans φ are restricted to be in the class $\Phi(M)$, i.e.,

$$\int_X \varphi(x, t)\, dx = M(t) \quad \text{for} \quad t \geq 0$$

or $\sum_{j \in J} \varphi(j, t) = M(t)$ for $t \geq 0$ if the search space is J. For a fixed stopping time, Theorem 5.2.3 shows that it is optimal to follow a plan that is uniformly optimal within $\Phi(M)$. This reduces the optimal search and stop problem to an optimal stopping problem. Theorem 5.2.5 provides sufficient conditions for a stopping time to be optimal.

Chapter VI: Search in the Presence of False Targets

This chapter introduces the problem of search in the presence of false targets. False targets are assumed to be real stationary objects that may be mistaken for the target by the primary or detection sensor. Thus, when a detection is made, an investigation is required to determine whether the contact is a false or real target. Models are developed for the location and detection of false targets, and the distinction between adaptive and nonadaptive plans is discussed.

The optimization criterion used for searches in the presence of false targets is minimization of mean time. For searches in which contact investigation is immediate and conclusive (i.e., contacts must be investigated until identified), an optimal nonadaptive plan is found. It is then shown that immediate contact identification is still optimal in a class of plans in which one is allowed to delay contact investigation.

A class of semiadaptive plans is introduced that takes advantage of the optimal nonadaptive plan and feedback obtained from the search to produce a smaller mean time to find the target than the optimal nonadaptive plan. The semiadaptive plan does not do as well as the optimal adaptive plan, but it is much easier to obtain. In fact, optimal adaptive plans have been found only in very special cases. Applications of semiadaptive plans to searching for multiple targets are also discussed.

Chapter VII: Approximation of Optimal Plans

The optimal search plans found in Chapters II and VI may be difficult to calculate without the use of a computer. Even when the optimal plan has a simple analytic form such as the one in Example 2.2.7, it may call for applying very small amounts of effort to some areas. Both of these properties can make the optimal plan difficult to realize in practice.

Outline of Chapters 13

In order to overcome the second difficulty, Section 7.1 explores approximating plans that are optimal in a restricted class. In this restricted class, effort density must be applied in multiples of a fixed number. Plans that are optimal in this restricted class are appropriate for approximating the plans of Chapter II, which do not involve false targets. The calculation of these approximating plans may still require the use of a computer.

Section 7.2 introduces a class of incremental plans called Δ-plans, which can be used when the target distribution is discrete. For a Δ-plan, one fixes an increment of effort Δ and proceeds so that at each step the increment Δ is applied to a single cell. The choice of the cell in which to place the increment is made by examining ρ, the marginal rate of return in each of the cells, and choosing the cell having the highest rate. The computation of these rates is easy even for searches involving false targets. Thus, Δ-plans overcome both the difficulties mentioned above. In Section 7.2 it is shown that a Δ-plan approximates the optimal plan in mean time to find the target, in the sense that the mean time resulting from a Δ-plan approaches the mean time of the optimal plan as Δ approaches 0.

Chapter VIII: Conditionally Deterministic Target Motion

A simple but interesting type of moving-target search problem is considered in this chapter. The target's motion takes place in Euclidean n-space and is determined by an n-dimensional stochastic parameter ξ such as the target's position at time 0. There is a target motion function η such that if $\xi = x$, then $\eta(x, t)$ gives the target's position at time t for $t \geq 0$. Thus, the target motion is deterministic when conditioned on the value of ξ. The distribution of ξ is assumed to be known to the searcher and is given by the probability density function p.

Let η_t denote the transformation that maps x into $\eta(x, t)$. Then it is assumed that η_t is one-to-one for all $t \geq 0$ and that the Jacobian $\mathbf{J}(t, x)$ of η_t evaluated at x is positive for all $x \in X$ and $t \geq 0$. A moving-target search plan is a function ψ such that $\psi(x, t)$ gives the rate at which effort density accumulates at point x at time t. There is a detection function b such that if $\xi = x$ and the search plan ψ is followed, then $b\left(\int_0^t \psi(\eta_s(x), s)\, ds\right)$ is the probability of detecting the target by time t.

Suppose the rate at which search effort may be applied at time t is bounded by $m_2(t)$ for $t \geq 0$. If the Jacobian $\mathbf{J}$ factors into space- and time-dependent parts [i.e., $\mathbf{J}(x, t) = \mathbf{j}(x)\mathbf{n}(t)$ for $x \in X$ and $t \geq 0$], then Section 8.2 presents a method of finding uniformly optimal moving-target search plans among the plans ψ such that

$$\int_X \psi(x, t)\, dx \leq m_2(t) \quad \text{for} \quad t \geq 0.$$

Section 8.3 gives sufficient conditions for finding plans that maximize probability of detection among plans ψ such that $\psi(x, t) \leq m_1(t)$ for $x \in X$ and $t \geq 0$ and

$$\int_0^\infty \int_X \psi(x, t)\, dx\, dt \leq K.$$

No factorability assumption is made in Section 8.3.

Theorem 8.4.1 gives necessary and sufficient conditions for a plan ψ^* to maximize probability of detection by time t among plans ψ that satisfy

$$\int_X \psi(x, s)\, dx \leq m_2(s) \qquad \text{for} \quad s \geq 0.$$

The conditions have the form that there exists a nonnegative function λ such that

$$p(x)b'\left(\int_0^t \psi^*(\eta_u(x), u)\, du\right) = \lambda(s)\mathbf{J}(x, s) \qquad \text{for} \quad \psi^*(\eta_s(x), s) > 0,$$

$$\leq \lambda(s)\mathbf{J}(x, s) \qquad \text{otherwise.}$$

In the case of sufficiency, b is required to be concave. Again, no factorability assumption is made. The function λ is a generalization of the Lagrange multiplier introduced in Section 2.1. In addition, the above conditions are generalizations of the necessary conditions given in Corollary 2.1.7.

Chapter IX: Markovian Target Motion

Search for targets having Markovian motion is investigated in this chapter. Section 9.1 deals with the simplest type of Markovian motion, a two-cell, discrete-time Markov process. The problem of maximizing the probability of detecting the target in n looks is shown to be a standard dynamic programming problem that is solvable, in principle, for all cases. The problem of minimizing the expected number of looks to find the target is not so straightforward even though the optimal plan satisfies the usual dynamic programming equation. For this case optimal plans are found when detection occurs with probability one if the searcher looks in the cell containing the target. The optimal plan is characterized by a threshold probability π^*, which depends on the transition matrix of the Markov process. The plan proceeds by looking in cell 1 each time the posterior probability in that cell is greater than π^* and looking in cell 2 otherwise until the target is detected.

Let a_{ij} be the probability that the target transitions to cell j at time $n + 1$ given that it is in cell i at time n for $n = 0, 1, \ldots$. If $a_{11} = a_{21}$, then from the first look on, the posterior target location probabilities given failure to detect the target are $\tilde{p}(1) = a_{21}$, $\tilde{p}(2) = 1 - a_{21}$, regardless of the sequence of looks

Outline of Chapters

or the prior target distribution. This is called the *no-learning case*. For this case, the plan that minimizes mean time to detect the target is also found.

In Section 9.2 the continuous-time version of the two-cell problem is presented. The nature of the plan that maximizes the probability of detection by time t and a method for finding the plan are discussed.

In Section 9.3 necessary conditions are given for a search plan ψ^* to maximize the probability of detection by time t when the target's position is specified by a continuous-time Markov process $\{X_t, t \geq 0\}$ in Euclidean n-space. Let

$$G(x, s, t, \psi) = \Pr\{\text{target not detected during } [s, t] \text{ using plan } \psi \mid X_s = x\}.$$

Let $g(x, s, y, t, \psi)$ be the transition density for the process representing undetected targets, that is,

$$\int_S g(x, s, y, t, \psi) \, dy$$
$$= \Pr\{X_t \in S \text{ and target not detected in } [s, t] \text{ using plan } \psi \mid X_s = x\}.$$

Let p be the probability density of the target's location at time $t = 0$ and let $m_2(s)$ constrain the rate at which search effort may be applied at time s for $s \geq 0$. If ψ^* maximizes the probability of detecting the target by time T among plans ψ that satisfy

$$\int_X \psi(x, s) \, dx \leq m_2(s) \quad \text{for} \quad s \geq 0,$$

then Theorem 9.3.1 gives the following necessary condition, which ψ^* must satisfy: There is a nonnegative function λ such that for $0 \leq t \leq T$

$$\int_X g(x, 0, y, t, \psi^*) G(y, t, T, \psi^*) p(x) \, dx = \lambda(t) \quad \text{for} \quad \psi^*(y, t) > 0,$$
$$\leq \lambda(t) \quad \text{for} \quad \psi^*(y, t) = 0.$$

Appendix A: Reference Theorems

This appendix gives the statements of some basic theorems from real analysis, which are quoted in the book.

Appendix B: Necessary Conditions for Constrained Optimization of Separable Functionals

This appendix proves that maximizing the pointwise Lagrangian is a necessary condition for a constrained optimum involving a separable nonnegative payoff functional and a vector-valued constraint functional whose coordinate functionals are nonnegative and separable. Section B.1 presents

```
        1.3
         │
         ▼
        2.1
       ╱   ╲
      ▼     ▼
    4.1     2.2
     │       │
     ▼       ▼
    4.2   2.3, 2.4,
     │    III, 5.2,
     ▼    VI, VII,
    4.3   VIII, IX
     │
     ▼
  4.4, 5.1
```

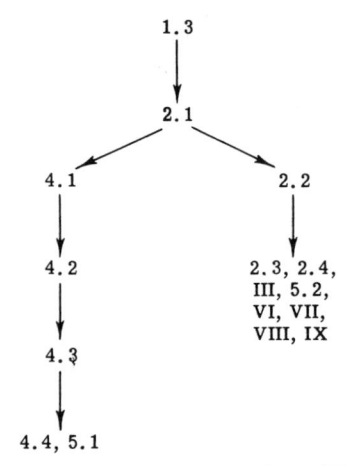

Fig. 2. Guide to dependence of chapters and sections. (*Note:* Chapters are indicated by Roman numerals and sections by arabic numbers.)

the proof for the case of a discrete search space and Section B.2 for the case of a continuous search space.

Logical Dependence of Chapters

Figure 2 gives a guide to the logical dependence of chapters and sections. Thus, if one wishes to read Section 5.1, Optimal Search and Stop with Discrete Effort, the guide shows that one may read Sections 1.3, 2.1, and 4.1–4.3 and then Section 5.1. If one wishes to read Chapter VIII, then the guide shows that one can do this by reading Sections 1.3, 2.1, and 2.2 and proceeding directly to Chapter VIII.

Chapter I

Search Model

This chapter develops and motivates a basic search model for searches involving stationary targets and sensors with perfect discrimination. The first two sections discuss and define target distributions and detection functions. The third section gives a concise statement of the basic problem of optimal search, which may be read without reading Sections 1.1 and 1.2.

1.1. THE TARGET DISTRIBUTION

The purpose of a search is to find or locate the position of an object called the target. The target's position is uncertain, but there is some knowledge about its location. For search theory it is assumed that this knowledge has been quantified in the form of a probability distribution called the *target distribution*. Except for the final chapters of this book, the target is assumed to be stationary. In the simplest situation there are two cells and the probability distribution is given by $p(1)$ and $p(2)$ as shown in Fig. 1.1.1. That is, $p(1)$ is the probability that the target is in cell 1 and $p(2)$ the corresponding probability for cell 2.

It is reasonable to ask how one obtains the probabilities $p(1)$ and $p(2)$ or the target distribution in general. Usually this is a question of quantifying, by subjective means, the available information concerning the target's position. Unfortunately, there is no general method for doing this. Suggestions for specific situations are given by Richardson *et al.* (1971) and United States Coast Guard (1959). While this problem will not be dealt with in detail in this book, we give a few examples to show how a target distribution might be obtained. The general area of subjective probabilities and methods for quantifying them are discussed by Savage (1954, 1971) and Raiffa (1968).

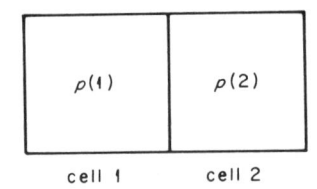

Fig. 1.1.1. Two-celled probability distribution.

As an example, a simple target distribution is constructed that would have been appropriate for the search for the lost submarine Thresher. This example is based on Richardson *et al.* (1971, Chapter 2). On April 10, 1963, the US nuclear submarine Thresher was conducting her second day of sea trials accompanied by the surface ship USS Skylark. At 9:17 A.M. the Thresher reported to Skylark by underwater telephone that she was having trouble maintaining trim. Skylark obtained a measurement of her position (latitude 41°45′N, longitude 65°00′W) by use of an electronic navigation system called Loran A. Seconds after the last communication from Thresher, Skylark heard break-up noises over the underwater telephone. On the following day at 5:00 P.M., an oil slick was sighted near the Skylark's position at the time she heard the break-up noises.

The maximum range of an underwater telephone is approximately 2.5 nautical miles. Let point A designate the position of Skylark as obtained by Loran A. If this gives the true position of Skylark, then Thresher went down within a circle of radius 2.5 mi of point A, as shown in Fig. 1.1.2. Because of uncertainty in the position of Skylark the search area was extended for "good measure" to a 10 × 10-mi square centered on point A.

A simple-minded target distribution can be found by reasoning as follows. There is high probability that the target is in the 10 × 10 mi square. Within this square the region inside the circle is more likely to contain the target. Let us assume that the target is contained inside the square with probability 0.9 and inside the circle with probability 0.6. Thus, the target distribution may be thought of as composed of two regions. The first region is inside the circle, which has probability 0.6 of containing the target. The second is the

1.1 The Target Distribution

shaded portion of Fig. 1.1.2, which has probability 0.9 − 0.6 = 0.3 of containing the target.

Notice that this probability distribution is defective, i.e., the probabilities sum to a number less than 1. This simply means that the target has some probability of being outside the region in which search is to be conducted. Usually, defective target distributions are allowed, but when they are not, this restriction will be mentioned at the beginning of the appropriate chapter or section.

A more sophisticated approach could be used to obtain a target distribution for the above case. One could analyze the uncertainties in position A and the

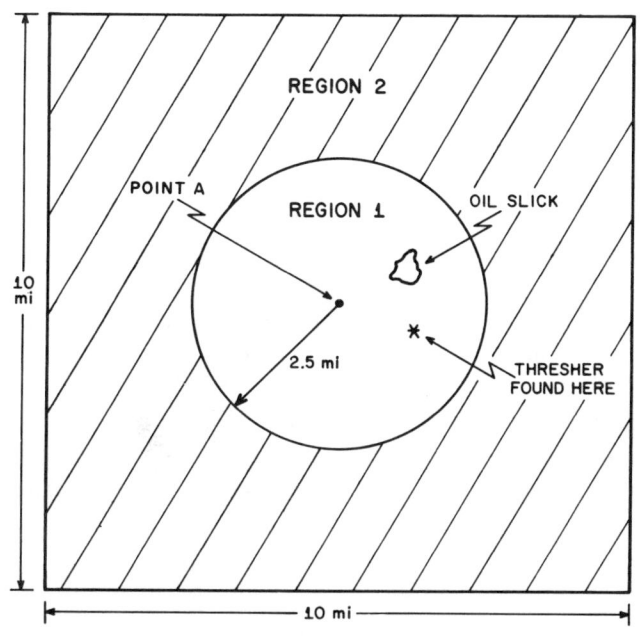

Fig. 1.1.2. Thresher loss area.

range of the underwater telephone to find probability distributions for both of these. These distributions could then be combined analytically or by Monte Carlo methods to obtain a target distribution.

Also, the oil slick might be taken into account by postulating a target distribution based on the distance and direction that the current and wind might move a quantity of diesel oil lost by a bottomed submarine. This would yield a second target distribution, which could be combined with the first by assigning weights or probabilities to each distribution and then taking their weighted average as the target distribution. The approach of postulating several target distributions and averaging them according to weights was used to obtain target distributions for the 1966 Mediterranean H-bomb search and the 1968 search for the lost submarine Scorpion (see Richardson and Stone, 1971).

As another example of a target distribution, let us consider a ship in distress that reports its position to be at a point in the ocean that will be taken as the origin of our coordinate system. According to the design of the navigation system used to obtain the ship's position, the error in the system has a bivariate normal distribution with parameters σ_1 and σ_2. Consider the surface of the ocean to be a standard Euclidean 2-space X, with points designated by ordered pairs (x_1, x_2). This is a reasonable assumption for small areas about the target's reported position. In this coordinate system, the target's reported position is $(0, 0)$. Let $(\mathbf{x}_1, \mathbf{x}_2)$ give the target's actual position. Then $\mathbf{x}_1$ is normally distributed with mean 0 and standard deviation σ_1. In addition, $\mathbf{x}_1$ is independent of $\mathbf{x}_2$, which is normally distributed with mean 0 and standard deviation σ_2.

Let

$$p(x_1, x_2) = \frac{1}{2\pi\sigma_1\sigma_2} \exp\left[-\frac{1}{2}\left(\frac{x_1^2}{\sigma_1^2} + \frac{x_2^2}{\sigma_2^2}\right)\right] \quad \text{for} \quad (x_1, x_2) \in X. \quad (1.1.1)$$

The function p is the probability density function of the *bivariate normal* distribution. Thus the distribution of error in the navigation system yields p as given in (1.1.1) for the density of the target distribution for the ship in distress. If $\sigma_1 = \sigma_2 = \sigma$, then (1.1.1) becomes

$$p(x_1, x_2) = \frac{1}{2\pi\sigma^2} \exp\left[-\frac{x_1^2 + x_2^2}{2\sigma^2}\right] \quad \text{for} \quad (x_1, x_2) \in X, \quad (1.1.2)$$

and the target distribution is called *circular normal*.

If the drift of the ship in distress is negligible compared to the parameters σ_1, σ_2, then the problem of finding the ship is essentially that of finding a stationary target with a bivariate normal distribution.

In order to define target distributions mathematically, the following notation is used: "The function $h: D \to R$" means that the function h is defined on the set D and takes its values in the set R. The set D is called the *domain* of h and the set R is called the *range* of h. The set of image points of h, i.e., $\{r \in R : r = h(y) \text{ for some } y \in D\}$ is called the *image* of h. Note that the range may contain points that are not image points, but the image of h consists solely of image points. If E is an event, then $\Pr\{E\}$ denotes the probability of the event E.

Two types of target distributions will be considered, discrete and continuous. In the discrete case the target may be in one of a set J of cells. The

1.1 The Target Distribution

set J is taken to be a subset of the positive integers, for convenience. The function p is a *target distribution on J* if $p: J \to [0, 1]$ and

$$\sum_{j \in J} p(j) \leq 1.$$

That is, $p(j)$ is the probability that the target is in cell j.

The distinguishing feature of a discrete target distribution is that the cells represent the smallest regions over which search effort can be allocated. For example, one cannot apply search effort to only half a cell.

In the continuous case, the target is located in Euclidean n-space X. The function p is a *target density on X* if $p: X \to [0, \infty)$ has the property that for any n-dimensional rectangles S[1]

$$\Pr\{\text{target contained in } S\} = \int_S p(x)\, dx. \tag{1.1.3}$$

The integration in (1.1.3) may be thought of as standard n-dimensional integration. Thinking of $x = (x_1, \ldots, x_n)$ as a vector, one could write the integral in (1.1.3) as

$$\int_S p(x_1, \ldots, x_n)\, dx_1 \cdots dx_n.$$

Figure 1.1.1 depicts a discrete target distribution with two cells while (1.1.1) and (1.1.2) define continuous target distributions. The distribution obtained for the Thresher search is incomplete in the sense that one must decide whether to treat the two regions in Fig. 1.1.2 as cells, making the target distribution discrete, or to divide the probabilities of the target's being in the two regions by their respective areas to obtain a probability density function for a continuous target distribution.

Requiring the two regions to be cells does not make sense here since one can obviously search in one part of a cell without searching in the other. So, to make a continuous target distribution one takes X to be Euclidean 2-space and divides 0.6 by the area of region 1 and 0.3 by the area of region 2 to obtain probability densities in these two regions. Setting the probability density equal to 0 outside the 10 × 10 mi square one obtains the following probability density function for the target distribution:

$$p(x) = \begin{cases} \dfrac{0.6}{6.25\pi \text{ mi}^2} & \text{for } x \text{ in region 1,} \\[6pt] \dfrac{0.3}{(100 - 6.25\pi) \text{ mi}^2} & \text{for } x \text{ in region 2,} \\[6pt] 0 & \text{otherwise.} \end{cases}$$

[1] More generally, the notation dx indicates integration with respect to n-dimensional Lebesgue measure, S is any Borel subset of X, and p is Borel measurable.

Often, as in the case of the Scorpion search, a continuous target distribution is approximated by a discrete one (see Figure 2 of Richardson and Stone, 1971). This is usually done by dividing the search region into rectangles and computing the probability that the target is located in each rectangle. The rectangles are taken as cells, and the computed probabilities form the discrete target distribution. However, when this is done, one must require that the search effort be spread uniformly within each rectangle or cell.

1.2. DETECTION

Having obtained a target distribution, the searcher requires a method of relating his search effort to the probability of detecting the target. Search is conducted with a sensor, i.e., a device capable of detecting the target. Some examples of sensors are the eye, radar, sonar, television, and cameras. In this chapter an idealized sensor will be considered that has perfect discrimination. That is, it will call a detection only when the target is present. However, detection of the target is not certain, that is, the sensor may fail to call a detection when the target is present.

This section presents examples of how one calculates probability of detection in some specific instances and uses these examples to motivate the definition of detection function given here.

Detection Models

First we develop the concepts of lateral range and sweep width. Consider a stationary target and a sensor in motion along a straight-line track as shown in Fig. 1.2.1. Let r be the minimum distance between the sensor and the target as it moves along the track. Then r is the *lateral range* of the target for that track. Suppose that the probability of detecting the target depends only on lateral range (i.e., it is independent of the track provided only that the track

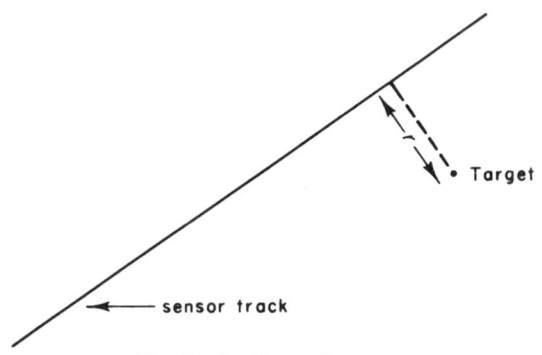

Fig. 1.2.1. Lateral range.

1.2 Detection

is a straight line) so that one can define $\hat{\alpha}(r)$ to be the probability of detecting the target along a track having lateral range r to the target. Then $\hat{\alpha}$ is called the *lateral-range function* for the sensor. Note that the lateral-range function is not a probability density function. The *sweep width* W of the sensor is defined as

$$W = 2 \int_0^\infty \hat{\alpha}(r) \, dr. \tag{1.2.1}$$

When detection is certain inside lateral range d and detection cannot occur beyond d, then

$$\hat{\alpha}(r) = \begin{cases} 1 & \text{for } 0 \leq r \leq d, \\ 0 & \text{for } r > d, \end{cases}$$

and the sensor is said to obey a *definite-range law*. In this case $W = 2d$.

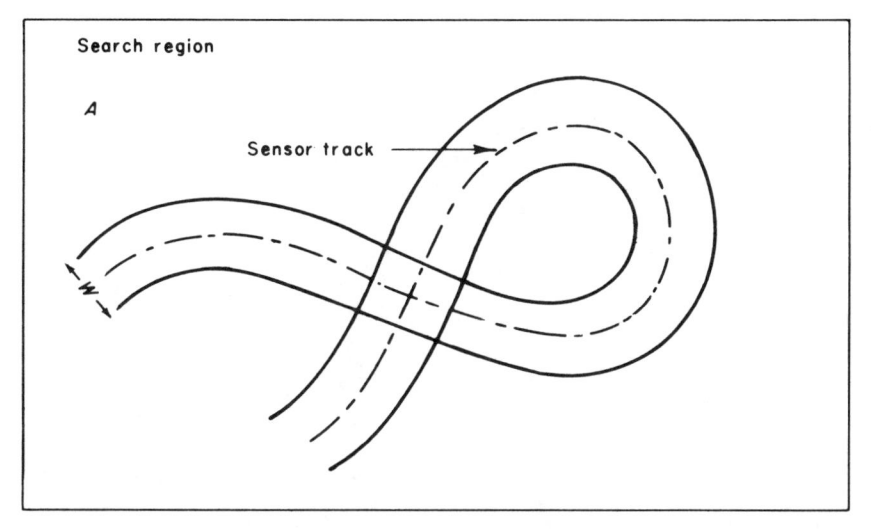

Fig. 1.2.2. Random search.

In the following paragraphs a heuristic derivation of the random-search formula is presented. Even though the derivation of this law is not precise, it remains an important tool for search theory for reasons discussed below.

Consider a sensor having a definite-range law and sweep width $W = 2d$. The search region is a rectangle of area A as shown in Fig. 1.2.2. Suppose the following assumptions hold:

(a) The target distribution is uniform over the rectangle.
(b) The sensor's track is randomly but uniformly distributed in the

rectangle. That is, disjoint sections of track are distributed uniformly and independently within the rectangle.

(c) No effort falls outside the search region.

It is hard to imagine a sensor track satisfying assumption (b). However, an irregular sensor track that wanders throughout the rectangle in such a way as to achieve a reasonably uniform track density over the rectangle would approximately satisfy this assumption. Assumption (c) will be reasonably well satisfied when the maximum detection range is much smaller than the dimensions of the search region and it is unlikely that the sensor track will wander outside the search region.

Consider $g(h)$, the conditional probability of detecting the target along a small increment h of track length given failure to detect the target previously. From Fig. 1.2.2 one can see that the area swept by the sensor in this increment is Wh, assuming no overlap in the swept area (i.e., the area of the region of width $\frac{1}{2}W$ on either side of the track). By assumption (a) and the uniformity assumption of (b), the target distribution given failure to detect is uniform over the rectangle. Thus the probability of detecting the target on this increment is Wh/A. Since the increment is placed independently of past track segments [assumption (b)], the probability of detection on this increment of track is independent of past detection, and $g(h) = Wh/A$.

Define $b(z)$ to be the probability that the target is detected by the time the sensor has traveled track length z. Then $[1 - b(z)] Wh/A$ is the probability of failing to detect the target by track length z but succeeding in the next increment h. Thus,

$$b(z + h) = b(z) + [1 - b(z)]\frac{Wh}{A},$$

and

$$b'(z) = \lim_{h \to 0} \frac{b(z + h) - b(z)}{h} = [1 - b(z)]\frac{W}{A}.$$

Since $b(0) = 0$, the above differential equation has the well-known solution

$$b(z) = 1 - e^{-zW/A} \quad \text{for} \quad z \geq 0. \tag{1.2.2}$$

This is the random-search formula of Koopman. A search satisfying assumptions (a)–(c) is called a *random search*.

In the above derivation it is assumed that the sensor has a definite-range law. This is for purposes of visualization only. The same argument can be given for any lateral-range function if the sweep width resulting from the given lateral-range function is used in place of the sweep width from the definite-range law. The result is that the random-search formula remains unchanged.

1.2 Detection

Thus, for situations in which the random-search formula holds, the performance of the search sensor may be characterized, for detection purposes, by the single number W, the sweep width.

Let us examine an important property of the random-search formula that is graphed in Fig. 1.2.3. Initially, the slope of b is W/A since at the beginning of the search there is almost no chance of overlapping with previous search.

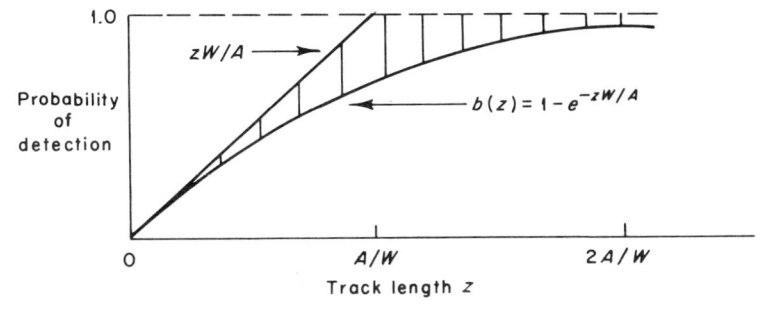

Fig. 1.2.3. Detection probability for random search.

However, as the search continues the chance of overlap with previously searched areas of the rectangle increases, which slows the rate at which the probability of detection increases. Thus, the random-search formula shows a form of the law of diminishing rate of returns. This diminishing rate or decreasing derivative of b is characteristic of the primary class of detection functions considered in this book.

The usefulness of the random-search formula arises from two of its properties: its mathematical simplicity, and its ability to provide a lower bound on the effectiveness of a systematic search of a rectangle for most search situations. As evidence of the latter, consider the following search pattern involving a sensor with definite-range law and sweep width W. A parallel-path search pattern is performed in a rectangle containing the target with the tracks nominally spaced one sweep width apart as shown in Fig. 1.2.4.

If one were able to place the paths exactly as shown in Fig. 1.2.4, the probability of detecting the target would be 1. However, this is not likely to be the case in an actual search. Instead there will be displacements or errors in the placement of the tracks. Assume that all tracks are indeed parallel to the x_2 axis as shown in Fig. 1.2.4 but that the error or difference in the x_1-coordinate of the path from its intended value is normally distributed with mean 0 and standard deviation σ. In addition, errors in the placement of the paths are mutually independent.

The result of these errors is to cause the swept areas to overlap in some places and leave gaps in others. Thus the probability of detection is less than

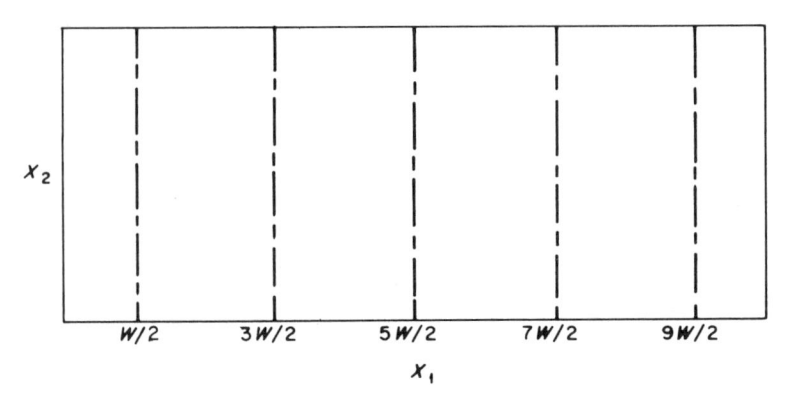

Fig. 1.2.4. Nominal track placement for parallel-path search.

1. Based on the work of Reber (1956, 1957) one can show that if edge effects are ignored (i.e., when the dimensions of the search region are large compared to W and σ), then the probability of detecting the target depends only on the ratio σ/W. This probability is shown in Fig. 1.2.5. Observe that as the ratio σ/W approaches ∞, the probability of detection approaches $1 - e^{-1}$. The track length required to complete this parallel-path search plan is A/W. Observe from (1.2.2) that the probability of detection resulting from a random search of track length A/W is $1 - e^{-1}$. Thus, the random-search

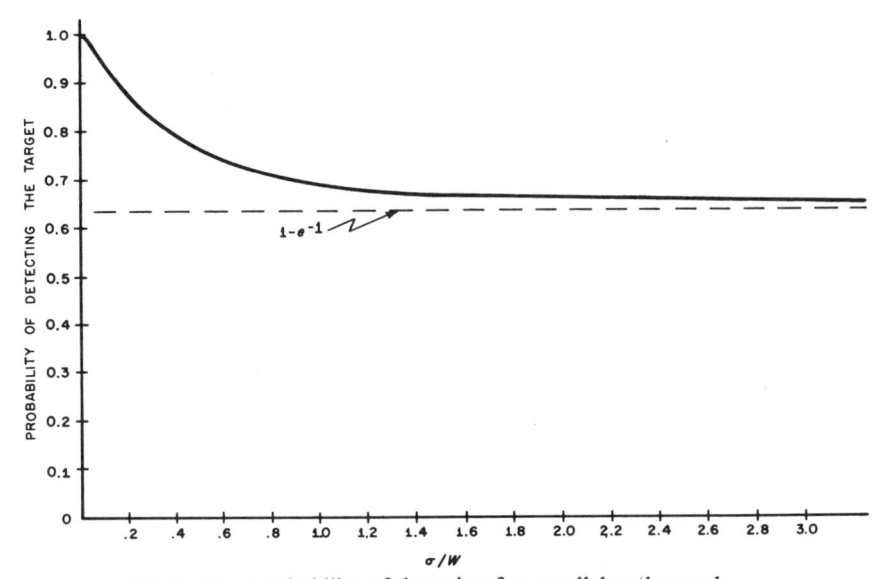

Fig. 1.2.5. Probability of detection for parallel-path search.

1.2 Detection

formula gives a lower bound on the probability of detection obtainable from a parallel-path search plan.

Suppose one fixes σ and W and allows the path spacing S for the parallel-path search plan to vary. Each value of S corresponds to a track length A/S required to complete the pattern. Each choice of S and the resulting track length A/S yields a probability of detection that may be obtained from Reber (1956, 1957). Thus, for this parallel-path search situation, one may think of a function that gives the probability of detection as a function of track length z for fixed values of σ and W. In general, the graph of this function will fall in the shaded area of Fig. 1.2.3.

In many underwater search situations, navigation errors are so large compared to the sweep width of the sensor used (often visual) that one is essentially in the limiting case of the random-search formula when trying to perform a parallel-path search plan. In many cases, one uses the random-search formula to evaluate probability of detection because it is a conservative and reasonable estimate.

In the work of Reber (1956, 1957) it is shown that the analysis of parallel-path search may be extended to a large class of lateral-range functions other than the definite-range law considered above. Most lateral-range functions that are zero beyond a given range belong to this class.

As another example, a commonly used model for visual detection is considered. Only the basic assumptions and results are outlined here. One may consult Koopman (1956b) for a detailed presentation of this model. The model for visual detection assumes that the probability of detecting a target in an interval $[0, t]$ can be expressed as

$$1 - \exp\left[-\int_0^t \hat{\gamma}(u)\,du\right], \tag{1.2.3}$$

where $\hat{\gamma}(t)$ is proportional to the solid angle subtended by the target at the observer at time t. Consider an observer in an airplane at height h above the ocean trying to sight a target on the surface at range y at time t (see Fig. 1.2.6). When the area of the target is small compared to h and y, one may show that

$$\hat{\gamma}(t) = kh/(h^2 + y^2)^{3/2},$$

where k is a constant that depends on the area of the target, visibility, and other fixed factors of the search. If in addition, h is small compared to y, then

$$\hat{\gamma}(t) = kh/y^3. \tag{1.2.4}$$

If the height of the aircraft remains constant, then $\hat{\gamma}$ depends only on y. This model is called the inverse-cube detection law. Suppose the target is stationary and the aircraft travels at speed v along an (infinitely) long straight-line track

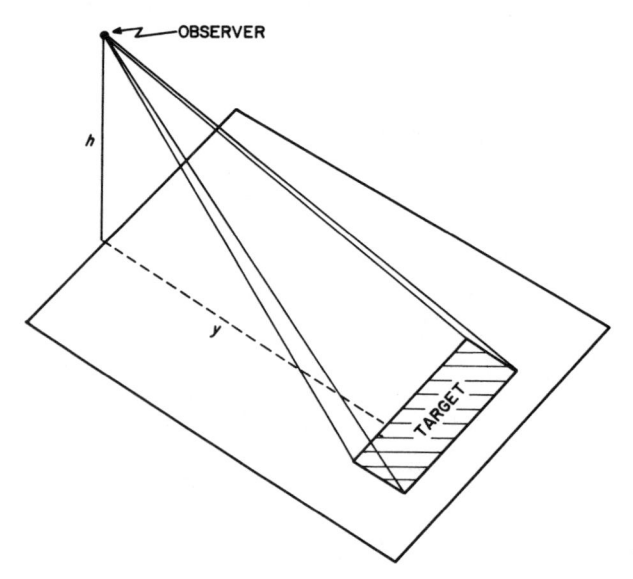

Fig. 1.2.6. Solid angle subtended by a target.

that has lateral range r to the target. Then one may use (1.2.3) and (1.2.4) to show that the probability of detecting the target is

$$\hat{\alpha}(r) = 1 - \exp(-2kh/vr^2) \quad \text{for } r > 0.$$

That is, $\hat{\alpha}$ gives the lateral range function. In addition the sweep width may be computed as

$$W = 2(2\pi kh/v)^{1/2}.$$

For a sensor with an inverse-cube detection law, consider a parallel-path search with track spacing S and no error in the placement of tracks. It is shown by Koopman (1956b) that the probability of detection for this search is

$$\text{erf}(\sqrt{\pi}W/2S),$$

where

$$\text{erf}(u) = \frac{2}{\sqrt{\pi}} \int_0^u e^{-s^2} ds \quad \text{for } -\infty < u < \infty.$$

Again a track spacing S in a rectangle of area A requires a track length $z = A/S$. Thus, the probability of detection as a function of track length for this parallel-path search is given by

$$\hat{b}(z) = \text{erf}(\sqrt{\pi}Wz/2A) \quad \text{for } z \geq 0. \tag{1.2.5}$$

The graph of this function again falls in the shaded region of Fig. 1.2.3. Let $\hat{b}'$ be the derivative of $\hat{b}$. Then

$$\hat{b}'(z) = \frac{W}{A}\exp(-\pi W^2 z^2/4A^2) \quad \text{for} \quad z \geq 0.$$

The detection function $\hat{b}$ has an important property in common with the random-search detection function given in (1.2.2), namely, the value of the derivative continually decreases as z increases. That is, $\hat{b}$ experiences a diminishing rate of return.

Detection and Cost Functions

Intuitively a detection function gives the probability of detection for a search as a function of effort. Effort may be track length, swept area, time, or whatever is appropriate for the given search. In order to define a detection function mathematically the notation $S_1 \times S_2$ is used to denote the Cartesian product of the sets S_1 and S_2, that is,

$$S_1 \times S_2 = \{(s_1, s_2): s_1 \in S_1 \text{ and } s_2 \in S_2\}.$$

For example, Euclidean 2-space, the plane, is the cartesian product of the real line with itself. If g is a function defined on $S_1 \times S_2$, then $g(s_1, \cdot)$ indicates the function defined on S_2 obtained by fixing the first variable of g at s_1. The function $g(\cdot, s_2)$ is similarly defined.

A *detection function on J* is a function $b: J \times [0, \infty) \to [0, 1]$ such that $b(j, z)$ gives the conditional probability of detecting the target with z amount of effort placed in cell j given that the target is in cell j. There is an implicit assumption here that the probability of detection in cell j depends only on the total amount of effort applied there and not on the way that the effort is applied.

A detection function gives us the means of evaluating the effectiveness of search effort in terms of probability of detecting the target. Let p be a target distribution on J. Suppose that a search is contemplated in which $f(j)$ effort is to be placed in cell $j \in J$. The probability that the target is in cell j and will be detected by $f(j)$ effort is $p(j)b(j, f(j))$. The total probability of detecting the target is

$$\sum_{j \in J} p(j)b(j, f(j)).$$

The total effort involved is

$$\sum_{j \in J} f(j).$$

As an example, let us consider a two-cell target distribution as shown in

Fig. 1.1.1. Let $J = \{1, 2\}$ and p be a target distribution on J. Suppose that cell 1 is a rectangle with area A_1 and cell 2 a rectangle with area A_2. Search in cell j is a random search with a sensor having sweep width W_j for $j = 1, 2$. Effort is measured in track length, so that by (1.2.2)

$$b(j, z) = 1 - e^{-zW_j/A_j} \quad \text{for} \quad z \geq 0, \quad j = 1, 2.$$

Naturally the units of z, W_j, and A_j must be compatible. If $f(j)$ track length is placed in cell j for $j = 1, 2$, then the probability of detection is

$$p(1)[1 - e^{-f(1)W_1/A_1}] + p(2)[1 - e^{-f(2)W_2/A_2}].$$

The total track length traveled (i.e., total effort) is

$$f(1) + f(2).$$

If the sensor moves along its track at a constant speed v, then one could use time as a measure of effort. In this case $f(j) =$ time spent in cell j and

$$b(j, z) = 1 - e^{-vzW_j/A_j} \quad \text{for} \quad z \geq 0, \quad j \in J.$$

If $W_1 = W_2$, then swept area (i.e., sweep width times track length) can be used as a measure of effort. The resulting detection function is given by

$$b(j, z) = 1 - e^{-z/A} \quad \text{for} \quad z > 0, \quad j \in J.$$

A *cost function on J* is a function $c: J \times [0, \infty) \to [0, \infty)$ such that $c(j, z)$ gives the cost of applying z effort in cell j. In many cases $c(j, z) = z$ for $j \in J$ and $z \geq 0$, which means that cost is measured in terms of effort. In other cases the measures of effort and cost are not the same. For example, effort may be measured in track length and cost in dollars. The choice of the measures of effort and cost depend on the particular search problem.

The function $f: J \to [0, \infty)$ giving the amount of effort placed in each cell is called an *allocation on J*. If we fix a target distribution p, detection function b, and cost function c on J, then for each allocation f on J, we define

$$P[f] = \sum_{j \in J} p(j) b(j, f(j)), \quad C[f] = \sum_{j \in J} c(j, f(j)).$$

Thus $P[f]$ and $C[f]$ give the probability of detection and cost resulting from the allocation f.

For a continuous search space, an *allocation on X* is a function $f: X \to [0, \infty)$ such that for any n-dimensional rectangle S

$$\int_S f(x) \, dx = [\text{amount of effort placed in the set } S].$$

Intuitively one might think of $f(x)$ as being the average effort per unit volume (or area in 2-space) in a small region about the point x.

1.2 Detection

A *detection function on* X is a function $b: X \times [0, \infty) \to [0, 1]$ such that $b(x, z)$ gives the conditional probability of detecting the target given that the target is located at x and the effort density is z at the point x. A *cost function on* X is a function $c: X \times [0, \infty) \to [0, \infty)$ such that $c(x, z)$ gives the cost density of applying effort density z at point x.

If a target distribution p, detection function b, and cost function c on X are fixed, then for each allocation f on X we define

$$P[f] = \int_X p(x) b(x, f(x))\, dx, \qquad C[f] = \int_X c(x, f(x))\, dx.$$

Again, $P[f]$ and $C[f]$ give the probability of detection and cost resulting from the allocation f.

Consider the functions defined by (1.2.2) and (1.2.5). If we think of them as applying to a single cell and having the first variable suppressed, then they become detection functions on $J = \{1\}$. The properties of these functions suggest a class of detection functions to consider in search problems.

Let h be a real-valued function of a real variable. The function h is said to be *decreasing* if $t_1 \leq t_2$ implies $h(t_1) \geq h(t_2)$ and *strictly decreasing* if $t_1 < t_2$ implies $h(t_1) > h(t_2)$. Analogous definitions apply for the terms increasing and strictly increasing. Let b be a detection function on J. For $j \in J$, let $b'(j, \cdot)$ denote the derivative of $b(j, \cdot)$. The function b is *regular* if

(i) $b(j, 0) = 0$, and
(ii) $b'(j, \cdot)$ is continuous, positive, and strictly decreasing for $j \in J$. (1.2.6)

Similarly, if b is a detection function on X, then $b'(x, \cdot)$ denotes the derivative of $b(x, \cdot)$ and b is *regular* if

(i) $b(x, 0) = 0$, and
(ii) $b'(x, \cdot)$ is continuous, positive, and strictly decreasing for $x \in X$.

(1.2.7)

Observe that the functions defined by (1.2.2) and (1.2.5) are regular. Also a regular detection function is strictly concave and has a diminishing rate of return. If b is a regular detection function on J, then $b'(j, \cdot)$ is continuous and, by Theorem 6.19 of Rudin (1953),

$$\int_0^z b'(j, y)\, dy = b(j, z) \qquad \text{for } z \geq 0, \ j \in J. \qquad (1.2.8)$$

That is, $b(j, \cdot)$ is absolutely continuous for $j \in J$. Although $b'(j, z) > 0$ for $z > 0$, (1.2.8) implies that $\lim_{z \to \infty} b'(j, z) = 0$ since $b(j, z) \leq 1$ for $z \geq 0$ and $j \in J$. The same properties hold if b is a regular detection function on X.

If $b(j, \cdot) = b(i, \cdot)$ for all $i, j \in J$, then the detection function b is called *homogeneous.* This means the detection function does not vary over the search space. If for $j \in J$,

$$b(j, z) = 1 - e^{-\alpha_j z} \quad \text{for} \quad z \geq 0,$$

where $\alpha_j > 0$, then b is an *exponential detection function.* Analogous definitions of homogeneous and exponential detection functions are understood to hold for detection functions on X.

Remark 1.2.1. Readers familiar with reliability theory will recall that if b is the distribution function of the lifetime of a piece of equipment, then $b'(z)/[1 - b(z)]$ for $z \geq 0$ gives the failure rate function for this equipment. For detection functions, one could think of $b'(z)/[(1 - b(z)]$ for $z \geq 0$ as giving the success rate for b. Detection functions having a continuous, decreasing, and positive success rate, and for which $b(0) = 0$, are clearly regular. However, Remark 5.2.7 shows that a regular detection function need not have a decreasing success rate.

1.3. BASIC SEARCH PROBLEM

This section gives a mathematical statement of the basic problem of optimal search and a concise statement of the search models developed in Section 1.2 for discrete and continuous target distributions. Usually one has only a limited resource or amount of effort available to conduct a search and wishes to maximize the probability of detecting the target within the limit or constraint imposed on cost. This, in words, is the basic problem of optimal search.

Two fundamental search situations are considered, namely, the discrete and continuous target distributions. In the discrete case the target is in one of a set J of cells that are taken to be a (possibly infinite) subset of the positive integers. In the continuous case the target is located at a point in n-dimensional Euclidean space X. In both cases the target is stationary. The sets J and X are called *search spaces.* An *allocation* f is a nonnegative (Borel-measurable)[1] function defined on the search space. When we wish to be specific about the search space involved, we will say that f is an allocation on J or X, whichever is appropriate. For a fixed search space let F be the set of allocations on that space.

For $f \in F$ let $P[f]$ be the probability of detecting the target with allocation

---

[1] For the reader unfamiliar with measure theory it is suggested that he simply ignore the words "Borel" and "Borel measurable" wherever they appear. There are important (to a mathematician) reasons for restricting discussion to Borel-measurable functions and sets, but in practice all functions and sets encountered are Borel measurable. Thus the restriction does not affect applications.

1.3 Basic Search Problem

f and $C[f]$ the cost associated with f. Suppose that K constrains the cost of the search. The basic search problem is to find $f^* \in F$ such that

$$C[f^*] \leq K, \quad \text{and} \quad P[f^*] = \max\{P[f] : f \in F \text{ and } C[f] \leq K\}. \quad (1.3.1)$$

Such an f^* is called *optimal for cost K*.

If the search space is J, then the target distribution is given by a function $p: J \to [0, 1]$ such that

$$\sum_{j \in J} p(j) \leq 1,$$

and $p(j)$ is the probability that the target is in cell j. There is a detection function $b: J \times [0, \infty) \to [0, 1]$ such that $b(j, z)$ is the conditional probability of detecting the target with z amount of effort placed in cell j given that the target is in cell j. In addition there is a cost function $c: J \times [0, \infty) \to [0, \infty)$ such that $c(j, z)$ gives the cost of applying z effort in cell j. For this case,

$$F = F(J) \equiv \text{set of functions } f: J \to [0, \infty),$$

$$P[f] = \sum_{j \in J} p(j) b(j, f(j)) \quad (1.3.2)$$

$$\text{for } f \in F(J).$$

$$C[f] = \sum_{j \in J} c(j, f(j)) \quad (1.3.3)$$

If the search space is X, then the target distribution is specified by a (possibly defective) probability density function $p: X \to [0, \infty)$, which is Borel measurable. The detection function $b: X \times [0, \infty) \to [0, 1]$ is Borel measurable and $b(x, z)$ is the conditional probability of detecting the target given that the target is located at x and the effort density is z at point x. The cost function $c: X \times [0, \infty) \to [0, \infty)$ gives the cost density of applying effort density z at point x. In this case,

$$F = F(X) \equiv \text{set of Borel functions } f: X \to [0, \infty),$$

$$P[f] = \int_X p(x) b(x, f(x)) \, dx \quad (1.3.4)$$

$$\text{for } f \in F(X).$$

$$C[f] = \int_X c(x, f(x)) \, dx \quad (1.3.5)$$

An examination of the definitions of $P[f]$ and $C[f]$ for the discrete target distribution shows that each is composed of a sum over j of functions that depend only on $f(j)$ and not on the value of f in any cell other than j. That is, the effort placed in cell j does not affect the cost or the detection probability in any other cell. In the terminology of nonlinear programming, the functionals

P and C are called *separable*, as is the basic search problem given in (1.3.1). Generalizing this notion to functionals and optimization problems defined in terms of integration, we see that the basic search problem for continuous target distributions is also separable. It is separability that will allow us to attack and in many cases solve these two problems using the Lagrange multiplier methods discussed in the next section.

Remark 1.3.1. A unified way of describing the above search models would be to let X be an arbitrary set and ν be a measure on X. By taking X to be countable and ν to be counting measure, the summations in (1.3.2) and (1.3.3) could be indicated by integration with respect to ν. For (1.3.4) and (1.3.5), one could take X to be Euclidean n-space and ν n-dimensional Lebesgue measure. Specifying a topology on X (the discrete topology in the case where X is countable) one could define F to be the set of nonnegative Borel functions defined on X and write

$$P[f] = \int_X p(x)b(x, f(x))\nu(dx)$$
$$C[f] = \int_X c(x, f(x))\nu(dx)$$

for $f \in F$.

This would unify and generalize the presentation. However, much of the intuitive feeling of the subject would be lost. Thus, at the expense of some repetition and loss of generality, the theory will be developed in terms of the continuous and discrete cases presented above. These two cases provide the motivation for the theory and also provide the framework for most applications.

NOTES

The detection models presented in this section are taken from Koopman (1956b), which is the second in a series of three articles (Koopman, 1956a, b, 1957) on search theory. Both the random search formula (1.2.2) and the detection function defined by (1.2.5) are obtained by Koopman (1956b). The three articles on search theory are based on a report (Koopman, 1946) expounding the developments in search and detection theory that were made by the Anti-Submarine Warfare Operations Research Group of the US Navy during the years 1942–1945.

Chapter II

Uniformly Optimal Search Plans

In this chapter uniformly optimal search plans are found for searches involving a stationary target and a sensor having perfect discrimination. For these searches, effort may be divided as finely as desired, in contrast to Chapter IV, where effort must be applied in discrete units.

Section 2.1 presents optimization techniques involving Lagrange multipliers. These techniques are basic to search theory and are used throughout the book to find optimal search plans. Lagrange multiplier techniques are used in Section 2.2 to find uniformly optimal search plans when the detection function is regular. In Section 2.3, uniformly optimal plans are found when the search sensor has an uncertain sweep width. The existence of uniformly optimal search plans is shown under very general conditions in Section 2.4.

2.1. OPTIMIZATION BY THE USE OF LAGRANGE MULTIPLIERS

The use of Lagrange multipliers in finding allocations f^* that solve the basic search problem given in (1.3.1) is discussed in this section. A function f^* that satisfies (1.3.1) is an example of a constrained optimum, that is, f^* maximizes the functional P subject to a constraint K on the functional C. As we shall see, one virtue of using Lagrange multipliers is that they convert a constrained optimization problem into an unconstrained one. The discussion of Lagrange multipliers given in this section differs from the standard treatment in advanced calculus texts in that no derivatives are required, and sufficient as well as necessary conditions for optimality are obtained. In addition, inequality constraints are permitted. However, the results of this section (except for Theorem 2.1.1) are restricted to separable optimization problems.

Readers interested in proceeding quickly to Section 2.2, where optimal search allocations are found, may wish to read this section only as far as Theorem 2.1.3 and return to the remaining theorems as they are needed.

Although the basic search problem is stated in terms of finding an allocation f^* in $F(X)$ or $F(J)$, it will be useful when considering discrete search (Chapter IV) and approximations to optimal search plans (Chapter VII) to be able to restrict ourselves to allocations in a subset $\hat{F}(X) \subset F(X)$ or $\hat{F}(J) \subset F(J)$.

For $j \in J$, let $Z(j)$ be an arbitrary set of nonnegative real numbers and define

$$\hat{F}(J) = \begin{bmatrix} \text{set of functions } f: J \to [0, \infty) \\ \text{such that } f(j) \in Z(j) \quad \text{for} \quad j \in J \end{bmatrix}.$$

Thus a function $f \in \hat{F}(J)$ is a member of $F(J)$ that is restricted so that the value of $f(j)$ must be in the set $Z(j)$ for $j \in J$. If $Z(j) = [0, \infty)$ for $j \in J$, then $\hat{F}(J) = F(J)$.

Similarly for $x \in X$, let $Z(x)$ be an arbitrary set of nonnegative real numbers and define

$$\hat{F}(X) = \begin{bmatrix} \text{set of Borel functions } f: X \to [0, \infty) \\ \text{such that } f(x) \in Z(x) \quad \text{for} \quad x \in X \end{bmatrix}.$$

As with F, the symbol $\hat{F}$ will be used to mean either $\hat{F}(J)$ or $\hat{F}(X)$.

The notion of optimal for cost K is also extended to the class $\hat{F}$. If $f^* \in \hat{F}$ is such that

$$C[f^*] \leq K \quad \text{and} \quad P[f^*] = \max\{P[f]: f \in \hat{F} \text{ and } C[f] \leq K\},$$

then f^* is *optimal within $\hat{F}$ for cost K*.

Sufficient Conditions for Constrained Optimality

Theorem 2.1.1. *Suppose there is a finite $\lambda \geq 0$ and an allocation $f_\lambda^* \in \hat{F}$ such that $C[f_\lambda^*] < \infty$ and*

$$P[f_\lambda^*] - \lambda C[f_\lambda^*] \geq P[f] - \lambda C[f] \quad \text{for} \quad f \in \hat{F} \quad \text{such that} \quad C[f] < \infty. \tag{2.1.1}$$

Then

$$P[f_\lambda^*] = \max\{P[f]: f \in \hat{F} \text{ and } C[f] \leq C[f_\lambda^*]\}. \tag{2.1.2}$$

2.1 Optimization by the Use of Lagrange Multipliers

Proof. Suppose $f \in \hat{F}$ and $C[f] \leq C[f_\lambda^*]$. By (2.1.1)

$$P[f_\lambda^*] - P[f] \geq \lambda(C[f_\lambda^*] - C[f]) \geq 0,$$

where the last inequality follows from $\lambda \geq 0$ and $C[f] \leq C[f_\lambda^*]$. This proves (2.1.2) and the theorem.

The conclusion of Theorem 2.1.1 says that f^* is optimal within $\hat{F}$ for cost $C[f_\lambda^*]$. Notice that the theorem allows one to convert the constrained optimization problem of finding f^* to satisfy (2.1.2) to the unconstrained problem of finding f_λ^* to maximize $P - \lambda C$. The number λ is called a *Lagrange multiplier*.

Define a function ℓ as follows: If the search space is J,

$$\ell(j, \lambda, z) = p(j)b(j, z) - \lambda c(j, z) \quad \text{for} \quad j \in J, \quad \lambda \geq 0, \quad z \geq 0. \quad (2.1.3)$$

If the search space is X,

$$\ell(x, \lambda, z) = p(x)b(x, z) - \lambda c(x, z) \quad \text{for} \quad x \in X, \quad \lambda \geq 0, \quad z \geq 0. \quad (2.1.4)$$

The function ℓ is called the *pointwise Lagrangian*. It is the crucial function for solving separable optimization problems with constraints.

Theorem 2.1.2. *Suppose there is a finite $\lambda \geq 0$ and a $f_\lambda^* \in \hat{F}(J)$ such that $C[f_\lambda^*] < \infty$ and*

$$\ell(j, \lambda, f_\lambda^*(j)) = \max\{\ell(j, \lambda, z) : z \in Z(j)\} \quad \text{for} \quad j \in J. \quad (2.1.5)$$

Then

$$P[f_\lambda^*] = \max\{P[f] : f \in \hat{F}(J) \quad \text{and} \quad C[f] \leq C[f_\lambda^*]\}. \quad (2.1.6)$$

Proof. This theorem is proved by showing that f_λ^* and λ satisfy (2.1.1). Let $f \in \hat{F}(J)$. By (2.1.5)

$$\ell(j, \lambda, f_\lambda^*(j)) \geq \ell(j, \lambda, f(j)) \quad \text{for} \quad j \in J. \quad (2.1.7)$$

If $C[f] < \infty$, then

$$\sum_{j \in J} \ell(j, \lambda, f(j)) = P[f] - \lambda C[f] \quad \text{for} \quad f \in \hat{F}(J). \quad (2.1.8)$$

Summing both sides of (2.1.7) and using (2.1.8), one obtains

$$P[f_\lambda^*] - \lambda C[f_\lambda^*] \geq P[f] - \lambda C[f] \quad \text{for} \quad f \in \hat{F}(J) \quad \text{such that} \quad C[f] < \infty.$$

Thus by Theorem 2.1.1, f_λ^* satisfies (2.1.6), and the theorem is proved.

Occasionally we shall write "a.e. $x \in S$," where S is a subset of X. The abbreviation a.e. stands for "almost every" and a.e. $x \in S$ means all $x \in S$ except for those in a set of measure 0. A set of measure 0 is one that can be contained in a subset having arbitrarily small n-dimensional volume. Typical examples of sets of measure 0 are those having only a finite or countable number of points, although there are others. For example, a straight line has measure 0 in the plane because the area (i.e., 2-dimensional volume) of a line is 0.

Theorem 2.1.3. *Suppose there is a finite $\lambda \geq 0$ and $f_\lambda^* \in \hat{F}(X)$ such that $C[f_\lambda^*] < \infty$ and*

$$\ell(x, \lambda, f_\lambda^*(x)) = \max\{\ell(x, \lambda, z) : z \in Z(x)\} \quad \text{for} \quad \text{a.e.} \; x \in X. \quad (2.1.9)$$

Then

$$P[f_\lambda^*] = \max\{P[f] : f \in \hat{F}(X) \quad \text{and} \quad C[f] \leq C[f_\lambda^*]\}.$$

Proof. The proof follows that of Theorem 2.1.2 with integration replacing summation.

Theorem 2.1.2 is very useful since it allows us to replace the problem of finding a constrained optimum for the functional P with the problem of maximizing $\ell(j, \lambda, \cdot)$ over $Z(j)$ for $j \in J$. For fixed j and λ, $\ell(j, \lambda, \cdot)$ is a function of one real variable. If this function is differentiable and $Z(j)$ is an interval, then one may apply the standard methods of calculus in the effort to find the maximum of $\ell(j, \lambda, \cdot)$. Analogous comments apply to Theorem 2.1.3.

We can now describe a method of finding allocations that are optimal for cost K. Choose a $\lambda \geq 0$. For each $j \in J$, find $f_\lambda^*(j)$ to maximize $\ell(j, \lambda, \cdot)$. The resulting allocation f_λ^* is optimal for cost $C[f_\lambda^*]$ by Theorem 2.1.2. If the cost constraint K is fixed in advance, one could try a succession of λ's until one is found for which $C[f_\lambda^*] = K$. In the next section a method is given, in the case of regular detection functions, for finding f_λ^* and for choosing λ so that $C[f_\lambda^*] = K$.

Theorems 2.1.1–2.1.3 give sufficient conditions for an allocation f^* to be a constrained optimum. That is, if f^* satisfies the hypotheses of one of these theorems, then f^* is a constrained optimum. The remaining two theorems in this section give necessary and sufficient conditions for constrained optima. By a necessary condition we mean that if f^* is a constrained optimum, it must satisfy the given condition. Necessary conditions for optimality are of interest since they allow us to examine the properties of optimal allocations.

Necessary Conditions for Constrained Optimality

Let g be a real-valued function defined on an interval of real numbers. Then g is *convex* if $g(\theta y_1 + (1 - \theta) y_2) \leq \theta g(y_1) + (1 - \theta) g(y_2)$ for $0 \leq \theta \leq 1$ and y_1, y_2 in the domain of g. The function g is *concave* if $-g$ is convex (i.e., the inequality in the definition of convex is reversed). Define the image of the

2.1 Optimization by the Use of Lagrange Multipliers

functional C to be the set of real numbers y such that $C(f) = y$ for some $f \in \hat{F}$. Often the image of C will be an interval of real numbers. In this case, to say that $C[f^*]$ is in the interior of the image of C means that $C[f^*]$ is in the interior of this interval.

If $f^* \in \hat{F}(J)$ and $0 \le \lambda < \infty$ satisfy

$$\ell(j, \lambda, f^*(j)) = \max\{\ell(j, \lambda, z) : z \in Z(j)\} \quad \text{for } j \in J, \quad (2.1.10)$$

then we say that (λ, f^*) *maximizes the pointwise Lagrangian with respect to* $\hat{F}(J)$. Similarly, if $f^* \in \hat{F}(X)$ and $0 \le \lambda < \infty$ satisfy

$$\ell(x, \lambda, f^*(x)) = \max\{\ell(x, \lambda, z) : z \in Z(x)\} \quad \text{for a.e. } x \in X, \quad (2.1.11)$$

then we say that (λ, f^*) *maximizes the pointwise Lagrangian with respect to* $\hat{F}(X)$. If $\hat{F} = F$, we say simply that (λ, f^*) maximizes the pointwise Lagrangian.

Theorem 2.1.4. *Let $b(j, \cdot)$ be concave, $c(j, \cdot)$ be convex, and $Z(j)$ be an interval for $j \in J$. Let $f^* \in \hat{F}(J)$ and $C[f^*]$ be in the interior of the image of C. Then a necessary and sufficient condition for q^* to be optimal within $\hat{F}(J)$ for cost $C[f^*]$ is that there exists a finite $\lambda \ge 0$ such that (λ, f^*) maximizes the pointwise Lagrangian with respect to $\hat{F}(J)$.*

Theorem 2.1.5. *Let $\{(x, z) : x \in X \text{ and } z \in Z(x)\}$ be a Borel set. Suppose $f^* \in \hat{F}(X)$ and $C[f^*]$ is in the interior of the image of C. Then a necessary and sufficient condition for f^* to be optimal within $\hat{F}(X)$ for cost $C[f^*]$ is that there exists a finite $\lambda \ge 0$ such that (λ, f^*) maximizes the pointwise Lagrangian, with respect to $\hat{F}(X)$.*

The sufficiency parts of Theorems 2.1.4 and 2.1.5 follow from Theorems 2.1.2 and 2.1.3, respectively. The necessity parts are proved in Appendix B.

Observe that there is an important difference between the discrete and continuous cases given in Theorems 2.1.4 and 2.1.5, respectively. In the discrete case (Theorem 2.1.4) $b(j, \cdot)$ is assumed concave and $c(j, \cdot)$ is assumed convex for $j \in J$. No such assumptions are made for a continuous case (Theorem 2.1.5). Moreover, if one removes the concavity/convexity assumptions of Theorem 2.1.4, then the existence of a $\lambda \ge 0$ such that (λ, f^*) maximizes the pointwise Lagrangian is no longer a necessary condition for f^* to be optimal for cost $C[f^*]$. This is shown in Example 3.3.5.

For the case where $\hat{F} = F$, Theorems 2.1.4 and 2.1.5 are used below to obtain necessary conditions for optimality under differentiability assumptions. Let $b'(j, \cdot)$ and $c'(j, \cdot)$ denote the derivatives of $b(j, \cdot)$ and $c(j, \cdot)$ for $j \in J$, and let $\ell'(j, \lambda, \cdot)$ denote the derivative of $\ell(j, \lambda, \cdot)$ for $\lambda \ge 0$, $j \in J$. Analogous definitions of $b'(x, \cdot)$, $c'(x, \cdot)$, and $\ell'(x, \lambda, \cdot)$ are understood to hold for $\lambda \ge 0$ and $x \in X$.

Corollary 2.1.6. Let $b'(j, \cdot)$ be decreasing and continuous and $c'(j, \cdot)$ increasing and continuous for $j \in J$. If $f \in F(J)$ is optimal for cost $C[f^*]$ in the interior of the image of C, then there exists a $\lambda \geq 0$ such that

$$p(j)b'(j, f^*(j)) - \lambda c'(j, f^*(j)) \begin{matrix} = 0 & \text{if } f^*(j) > 0, \\ \leq 0 & \text{if } f^*(j) = 0. \end{matrix} \quad (2.1.12)$$

Proof. Since $b'(j, \cdot)$ is increasing and $c'(j, \cdot)$ decreasing, $b(j, \cdot)$ is concave and $c(j, \cdot)$ convex for $j \in J$. By Theorem 2.1.4 there exists a $\lambda \geq 0$ such that (λ, f^*) maximizes the pointwise Lagrangian, that is,

$$\ell(j, \lambda, f^*(j)) = \max\{\ell(j, \lambda, z) : z \geq 0\} \quad \text{for} \quad j \in J.$$

Since $b'(j, \cdot)$ and $c'(j, \cdot)$ are continuous, $\ell'(j, \lambda, \cdot)$ is continuous. If $f^*(j) > 0$ [i.e., the maximum of $\ell(j, \lambda, \cdot)$ occurs at an interior point], then by a standard result of calculus

$$\ell'(j, \lambda, f^*(j)) = 0,$$

i.e.,

$$p(j)b'(j, f^*(j)) - \lambda c'(j, f^*(j)) = 0.$$

If $f^*(j) = 0$, then since 0 is a left-hand endpoint of the domain of $\ell(j, \lambda, \cdot)$,

$$p(j)b'(j, f^*(j)) - \lambda c'(j, f^*(j)) = \ell'(j, \lambda, f^*(j)) \leq 0.$$

This proves the corollary.

If $b(j, \cdot)$ and $c(j, \cdot)$ are concave and convex, respectively, $\ell(j, \lambda, \cdot)$ is concave for $j \in J$. If (λ, f^*) satisfies (2.1.12), then (λ, f^*) maximizes the pointwise Lagrangian and, by Theorem 2.1.4, f^* is optimal for cost $C[f^*]$.

Corollary 2.1.7. Let $b'(x, \cdot)$ and $c'(x, \cdot)$ be continuous for $x \in X$. If $f^* \in F(X)$ is optimal for cost $C[f^*]$ in the interior of the image of C, then there exists a $\lambda \geq 0$ such that

$$p(x)b'(x, f^*(x)) - \lambda c'(x, f^*(x)) \begin{matrix} = 0 & \text{for a.e. } x \text{ such that } f^*(x) > 0, \\ \leq 0 & \text{for a.e. } x \text{ such that } f^*(x) = 0. \end{matrix}$$
$$(2.1.13)$$

Proof. The corollary follows from Theorem 2.1.5 in the same manner as Corollary 2.1.6 follows from Theorem 2.1.4.

Note that no assumptions about the increasing or decreasing nature of $b'(x, \cdot)$ and $c'(x, \cdot)$ are made in Corollary 2.1.7. This is because no concavity/convexity assumptions are required for Theorem 2.1.5. On the other hand, since $\ell(x, \lambda, \cdot)$ need not be concave under the assumptions of Corollary 2.1.7, the fact that (λ, f^*) satisfies (2.1.13) does imply that (λ, f^*) maximizes the pointwise Lagrangian or that f^* is optimal for cost $C[f^*]$.

*Remark 2.1.8. Readers familiar with the Kuhn–Tucker necessary conditions will recognize the similarity between Corollary 2.1.6 and the Kuhn–Tucker theorem. However, there are some differences. In order to apply the Kuhn–Tucker theorem in its usual form, one assumes that J is a finite set of k elements and that $b'(j, \cdot)$ and $c'(j, \cdot)$ are continuous for $j \in J$. Then P and C become functions defined on the nonnegative orthant of k-space. If $f^* \in F(J)$ is optimal for cost K, then the Kuhn–Tucker theorem gurantees the existence of a finite $\lambda \geq 0$ such that λ and f^* satisfy (2.1.12) and, in addition, $\lambda(K - C[f^*]) = 0$. Note that no concavity/convexity assumptions are made, but the number of cells must be finite.

The necessity of maximizing a pointwise Lagrangian for constrained optimality when the target distribution is continuous can be compared to Pontryagin's maximum principle. However, the separable nature of the problem considered here allows one to obtain conditions that are necessary and sufficient without the differentiability or concavity/convexity assumptions usually required for the maximum principle.

*Remark 2.1.9. Theorems 2.1 and 5.3 and Corollary 5.2 of Wagner and Stone (1974) generalize the results in the above theorems to more general separable functionals and to problems with multiple constraints. It is interesting to note that the Borel assumptions on p, b, and c are crucial for the truth of Theorem 2.1.5. Remark 5.5 of Wagner and Stone (1974) shows that if these functions are allowed to be Lebesgue measurable, then maximizing a pointwise Lagrangian is no longer necessary for constrained optimality.

2.2. UNIFORMLY OPTIMAL SEARCH PLANS FOR REGULAR DETECTION FUNCTIONS

This section introduces the concept of uniformly optimal search plans and shows how to find such plans when the detection function is regular. In Example 2.2.1 a uniformly optimal search plan is found for an important search problem by the use of the Lagrange multiplier methods of Section 2.1. This example is then abstracted into the general methods given in Theorems 2.2.4 and 2.2.5.

Example 2.2.1. Consider a search in the plane, and let $x = (x_1, x_2)$ denote a point in the plane. The target distribution is circular normal, i.e.,

$$p(x_1, x_2) = \frac{1}{2\pi\sigma^2} \exp\left(-\frac{x_1^2 + x_2^2}{2\sigma^2}\right) \quad \text{for} \quad (x_1, x_2) \in X. \quad (2.2.1)$$

The detection function is exponential,

$$b(x, z) = 1 - e^{-z} \quad \text{for} \quad x \in X, \quad z \geq 0.$$

For this example, effort is measured in swept area so that the effort density z is dimensionless.

Suppose the amount of effort K is fixed, and one wishes to maximize the probability of detecting the target within effort K. In order to use Theorem 2.1.3, let $\hat{F} = F$ and

$$c(x, z) = z \quad \text{for} \quad x \in X, \quad z \geq 0.$$

Then

$$C[f] = \int_X f(x)\, dx \quad \text{for} \quad f \in F(X),$$

and we seek an f^* that is optimal for cost K. We do this by finding f_λ^* to satisfy (2.1.9) where $Z(x) = [0, \infty)$ for $x \in X$.

The pointwise Lagrangian is

$$\ell(x, \lambda, z) = p(x)(1 - e^{-z}) - \lambda z \quad \text{for} \quad z \geq 0, \quad \lambda \geq 0, \quad x \in X.$$

For the moment fix $\lambda > 0$. Let $\ell'(x, \lambda, \cdot)$ denote the derivative of $\ell(x, \lambda, \cdot)$ for $x \in X$. Then

$$\ell'(x, \lambda, z) = p(x)e^{-z} - \lambda.$$

Setting $z^* = \ln[p(x)/\lambda]$, we see that $\ell'(x, \lambda, z^*) = 0$. Observe that

$$\begin{aligned}\ell'(x, \lambda, z) &\geq 0 \quad \text{for} \quad z \leq z^*,\\ &\leq 0 \quad \text{for} \quad z \geq z^*.\end{aligned}$$

Thus, z^* maximizes $\ell(x, \lambda, \cdot)$ and we may take $f_\lambda^*(x) = z^*$ provided $z^* \geq 0$. The restriction that $z^* \geq 0$ is necessary in order that f_λ^* be a member of $F(X)$, the set of nonnegative Borel functions defined on X. Intuitively, the restriction is necessary because negative effort densities do not make sense in this problem. If $z^* < 0$, then $p(x)b'(x, 0) < \lambda$. Since $b'(x, \cdot)$ is decreasing, $\ell'(x, \lambda, z) < 0$ for $z \geq 0$, and $\ell(x, \lambda, 0) \geq \ell(x, \lambda, z)$ for $z \geq 0$, that is, $f_\lambda^*(x) = 0$ maximizes $\ell(x, \lambda, \cdot)$ over nonnegative values of the argument.

Let

$$\{s\}^+ = \begin{cases} s & \text{if } s \geq 0,\\ 0 & \text{if } s < 0.\end{cases}$$

Then we may combine the two observations just made by letting

$$f_\lambda^*(x) = \{\ln[p(x)/\lambda]\}^+, \tag{2.2.2}$$

so that

$$\ell(x, \lambda, f_\lambda^*(x)) = \max\{\ell(x, \lambda, z) : z \geq 0\}.$$

2.2 Regular Detection Functions

By Theorem 2.1.3, f_λ^* is optimal for cost $C[f_\lambda^*]$. This for any $\lambda > 0$, we are able to find a f_λ^* that is optimal for its cost. Now the problem is to select λ so that $C[f_\lambda^*] = K$.

Combining (2.2.1) and (2.2.2), we obtain

$$f_\lambda^*(x_1, x_2) = \begin{cases} -\ln(2\pi\sigma^2\lambda) - \dfrac{x_1^2 + x_2^2}{2\sigma^2} & \text{for} \quad x_1^2 + x_2^2 \leq -2\sigma^2 \ln(2\pi\sigma^2\lambda), \\ 0 & \text{for} \quad x_1^2 + x_2^2 > -2\sigma^2 \ln(2\pi\sigma^2\lambda). \end{cases}$$

At this point it is convenient to switch to polar coordinates. Thus,

$$f_\lambda^*(r, \theta) = \begin{cases} -\ln(2\pi\sigma^2\lambda) - \dfrac{r^2}{2\sigma^2} & \text{for} \quad r^2 \leq -2\sigma^2 \ln(2\pi\sigma^2\lambda), \\ 0 & \text{for} \quad r^2 > -2\sigma^2 \ln(2\pi\sigma^2\lambda). \end{cases} \quad (2.2.3)$$

Let $U(\lambda) = C[f_\lambda^*]$ for $\lambda > 0$. Then

$$U(\lambda) = \int_0^{2\pi} \int_0^\infty f_\lambda^*(r, \theta) r \, dr \, d\theta$$

$$= 2\pi \int_0^{\sigma[-2\ln(2\pi\sigma^2\lambda)]^{1/2}} \left[-\ln(2\pi\sigma^2\lambda) - \frac{r^2}{2\sigma^2} \right] r \, dr$$

$$= \pi\sigma^2 [\ln(2\pi\sigma^2\lambda)]^2.$$

Solving

$$K = \pi\sigma^2 [\ln(2\pi\sigma^2\lambda)]^2$$

for λ and taking the negative root, we obtain

$$\lambda = \frac{1}{2\pi\sigma^2} \exp\left[-\left(\frac{K}{\pi\sigma^2}\right)^{1/2}\right]. \quad (2.2.4)$$

Substituting this for λ in (2.2.3), we have

$$f_\lambda^*(r, \theta) = \begin{cases} \left(\dfrac{K}{\pi\sigma^2}\right)^{1/2} - \dfrac{r^2}{2\sigma^2} & \text{for} \quad r^2 \leq 2\sigma^2 \left(\dfrac{K}{\pi\sigma^2}\right)^{1/2}, \\ 0 & \text{for} \quad r^2 > 2\sigma^2 \left(\dfrac{K}{\pi\sigma^2}\right)^{1/2}. \end{cases} \quad (2.2.5)$$

Since $C[f_\lambda^*] = U(\lambda) = K$, the allocation f_λ^* given by (2.2.5) is optimal for cost K. This allocation was first obtained by Koopman (1946).

Suppose that one is searching at a constant speed v using a sensor with sweep width W. If there is time t available to search, then $K = Wvt$. Let

$$H = (Wv/\pi\sigma^2)^{1/2} \quad \text{and} \quad R^2(t) = 2\sigma^2 H \sqrt{t} \quad \text{for} \quad t > 0.$$

Let $\varphi^*(\cdot, t)$ be the allocation that is optimal for cost Wvt. Then for $t > 0$,

$$\varphi^*((r, \theta), t) = \begin{cases} H\sqrt{t} - \dfrac{r^2}{2\sigma^2} & \text{for } r \leq R(t), \\ 0 & \text{for } r > R(t). \end{cases} \quad (2.2.6)$$

If one could imagine a search that is performed so that at each instant in time t the resulting effort density is given by $\varphi^*(\cdot, t)$ in (2.2.6), then one would have a search that maximizes the probability of detection at each time t within the constraint Wvt on the amount of effort available by time t. Such a search would begin at the origin and spread so that effort is placed inside a circle of radius $R(t)$ for $t \geq 0$. As the search progresses, more and more effort accumulates around the origin, which has the highest prior probability density for the target's location.

The optimal allocation just obtained canot be realized in practice, but it is the basis for many search plans that attempt to approximate the optimal allocation. One example may be found in "National Search and Rescue Manual" (United States Coast Guard, 1959). Others are given in Richardson et al. (1971) and Reber (1956, 1957).

Using the preceding example as motivation, we define a search plan. A *search plan* on X is a function $\varphi: X \times [0, \infty) \to [0, \infty)$ such that

(i) $\varphi(\cdot, t) \in F(X)$ for $t \in [0, \infty)$,
(ii) $\varphi(x, \cdot)$ is increasing for $x \in X$.

Condition (ii) gurantees that the allocations forming the search plan are consistent in the sense that they do not require that effort density be removed from any region in X.

Suppose that $M(t)$ gives the effort available by time t for $t \geq 0$. Then M is called the *cumulative effort function* and is always assumed to be increasing. Let $\Phi(M)$ be the class of search plans φ that satisfy

$$\int_X \varphi(x, t) \, dx = M(t) \quad \text{for } t \geq 0.$$

Then $\varphi^* \in \Phi(M)$ is called *uniformly optimal within* $\Phi(M)$ if

$$P[\varphi^*(\cdot, t)] = \max\{P[\varphi(\cdot, t)] : \varphi \in \Phi(M)\} \quad \text{for } t \geq 0.$$

That is, φ^* maximizes the probability of detection at each time t. From the above discussion φ^* defined in (2.2.6) is uniformly optimal in $\Phi(M)$, where $M(t) = Wvt$ for $t \geq 0$.

Analogous definitions of search plans and uniform optimality hold for the case of a discrete search space.

2.2 Regular Detection Functions

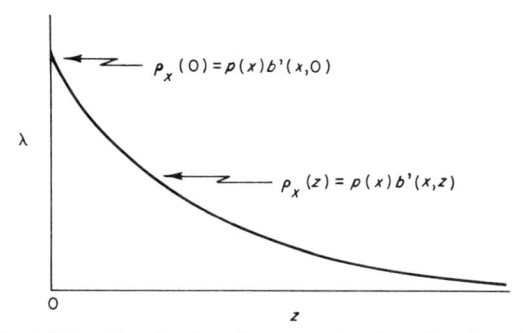

Fig. 2.2.1. Graph of ρ_x for a regular detection function.

The method of obtaining φ^* in (2.2.6) is the model for the method used to find uniformly optimal plans for regular detection functions. For this reason it is important to note several features of the method.

Recall that a regular detection function b on X is one such that

(i) $b(x, 0) = 0$, and
(ii) $b'(x, \cdot)$ is continuous, positive, and strictly decreasing for $x \in X$.

A similar definition holds for regular detection functions on J. The negative exponential detection function used in the above example is regular. In order to find z^* to maximize $\ell(x, \lambda, \cdot)$, we used the inverse of the function ρ_x defined by

$$\rho_x(z) = p(x)b'(x, z) = p(x)e^{-z} \quad \text{for} \quad z \geq 0.$$

That is, $z = \rho_x^{-1}(\lambda) = \ln[p(x)/\lambda]$ satisfies $\rho_x(z) = \lambda$. Since technically $b'(x, z)$ is defined only for $z \geq 0$, $\rho_x^{-1}(\lambda)$ is defined only for $0 < \lambda \leq p(x)b'(x, 0) = p(x)$ (see Fig. 2.2.1). By extending the definition of $\rho_x^{-1}(\lambda)$ so that $\rho_x^{-1}(\lambda) = 0$ when $\lambda > p(x)b'(x, 0)$, we find that for the negative exponential detection function

$$\rho_x^{-1}(\lambda) = \{\ln[p(x)/\lambda]\}^+ \quad \text{for} \quad x \in X, \quad \lambda > 0.$$

Thus, ρ_x^{-1} is the crucial function for finding f_λ^*.

Observe that ρ_x^{-1} is continuous and that $\lim_{\lambda \downarrow 0} \rho_x^{-1}(\lambda) = \infty$. The latter property results from the fact that, although $\lim_{z \to \infty} b'(x, z) = 0$, $b'(x, z) > 0$ for $z \geq 0$.

One other observation should be made here. In order to find the correct λ for a given cost K, we found the inverse function for U, that is, (2.2.4) really defines a function U^{-1} by

$$U^{-1}(K) = \frac{1}{2\pi\sigma^2} \exp\left[-\left(\frac{K}{\pi\sigma^2}\right)^{1/2}\right] \quad \text{for} \quad K \geq 0,$$

such that $U(U^{-1}(K)) = K$.

We now generalize the above observations to arbitrary regular detection functions and continuous target distributions. These generalizations will lead to Theorems 2.2.2 and 2.2.3. Let

$$c(x, z) = z \quad \text{for} \quad z \geq 0, \quad x \in X,$$

so that

$$C[f] = \int_X f(x)\,dx \quad \text{for} \quad f \in F(X).$$

The pointwise Lagrangian ℓ becomes

$$\ell(x, \lambda, z) = p(x)b(x, z) - \lambda z \quad \text{for} \quad x \in X, \quad \lambda \geq 0, \quad z \geq 0.$$

Assume that b is a regular detection function on X. Let $\ell'(x, \lambda, \cdot)$ be the derivative of $\ell(x, \lambda, \cdot)$ for $x \in X$, $\lambda \geq 0$, and define

$$\rho(x, z) = p(x)b'(x, z) \quad \text{for} \quad x \in X, \quad z \geq 0. \tag{2.2.7}$$

We call ρ the *rate of return function*. This terminology is motivated in Chapter III, where Lagrange multipliers are also discussed in more detail. Observe that $\rho_x = \rho(x, \cdot)$ for $x \in X$.

Fix $\lambda > 0$ and $x \in X$. To find z^* to maximize $\ell(x, \lambda, \cdot)$, we consider two cases.

Case 1: $\lambda > \rho(x, 0)$. The regularity of b guarantees that $b'(x, \cdot)$ and hence $\rho(x, \cdot)$ is decreasing. Thus

$$\ell'(x, \lambda, z) = \rho(x, z) - \lambda < 0 \quad \text{for} \quad z \geq 0, \tag{2.2.8}$$

and $z^* = 0$ satisfies

$$\ell(x, \lambda, 0) = \max\{\ell(x, \lambda, z) : z \geq 0\}.$$

Case 2: $\lambda \leq \rho(x, 0)$. The regularity of b implies that $b'(x, \cdot)$ is continous, strictly decreasing, and $\lim_{z \to \infty} b'(x, z) = 0$. Thus, there is a unique $z^* \geq 0$ such that

$$\ell'(x, \lambda, z^*) = \rho(x, z^*) - \lambda = 0. \tag{2.2.9}$$

Furthermore,

$$\ell'(x, \lambda, z) > 0 \quad \text{for} \quad 0 < z < z^*,$$
$$< 0 \quad \text{for} \quad z^* < z < \infty.$$

Thus,

$$\ell(x, \lambda, z^*) = \max\{\ell(x, \lambda, z) : z \geq 0\}.$$

Since $\rho_x = \rho(x, \cdot) = p(x)b'(x, \cdot)$ is continuous and strictly decreasing on

2.2 Regular Detection Functions

$[0, \infty)$, we may define an inverse function p_x^{-1} on $(0, p_x(0)]$. In case 2, z^* satisfies (2.2.9) so that

$$z^* = p_x^{-1}(\lambda). \tag{2.2.10}$$

Let us extend the domain of p_x^{-1} so that $p_x^{-1}(\lambda) = 0$ for $\lambda > p_x(0)$. Thus when case 1 holds, $p_x^{-1}(\lambda) = 0$ and (2.2.10) holds for both cases 1 and 2. The result is that f_λ^* defined by

$$f_\lambda^*(x) = p_x^{-1}(\lambda) \quad \text{for} \quad x \in X, \ \lambda > 0 \tag{2.2.11}$$

satisfies

$$\ell(x, \lambda, f_\lambda^*(x)) = \max\{\ell(x, \lambda, z) : z \geq 0\} \quad \text{for} \quad x \in X, \ \lambda > 0. \tag{2.2.12}$$

The function p_x^{-1} has some additional properties that will be used in the proof of Theorem 2.2.2, namely,

(a) p_x^{-1} is continuous,
(b) p_x^{-1} is decreasing on $(0, \infty)$ and strictly decreasing on $(0, p_x(0)]$.
(c) $\lim_{\lambda \downarrow 0} p_x^{-1}(\lambda) = \infty$ for x such that $p(x) > 0$.

(2.2.13)

Statement (c) may be verified by noting that the regularity of b guarantees $b'(x, z) > 0$ for $z > 0$. Thus if $p(x) > 0$, then $p(x, z) > 0$ for $z > 0$. The result is that $p(x, z)$ decreases asymptotically to 0 as $z \to \infty$, and $\lim_{\lambda \downarrow 0} p_x^{-1}(\lambda) = \infty$.

Let us generalize the definition of U given in Example 2.2.1, so that

$$U(\lambda) = \int_X p_x^{-1}(\lambda) \, dx \quad \text{for} \quad \lambda > 0.$$

Clearly, $U(\lambda) = C[f_\lambda^*]$. Let U^{-1} denote the inverse function for U.

Theorem 2.2.2. *Let $c(x, z) = z$ for $z \geq 0$ and $x \in X$, and assume b is a regular detection function on X. Fix a cost $K > 0$, and let*

$$\lambda = U^{-1}(K).$$

Then f_λ^* defined by

$$f_\lambda^*(x) = p_x^{-1}(\lambda) \quad \text{for} \quad x \in X$$

is optimal for cost K, and $C[f_\lambda^*] = K$.

Proof. By Theorem 2.1.3 and the discussion above.

$$P[f_\lambda^*] = \max\{P[f] : f \in F(X) \text{ and } C[f] \leq C[f_\lambda^*]\}.$$

Thus one needs to show only that $C[f_\lambda^*] = K$. This is accomplished by showing that U has a well-defined inverse function U^{-1} on $[0, \infty)$.

Let
$$\lambda_u = \sup\{\lambda : U(\lambda) > 0\}.$$

To show the existence of U^{-1} it is sufficient to prove that

(i) U is continuous on $(0, \infty)$ and strictly decreasing on $(0, \lambda_u)$,
(ii) $\lim_{\lambda \uparrow \lambda_u} U(\lambda) = 0$,
(iii) $\lim_{\lambda \downarrow 0} U(\lambda) = \infty$.

Fix $\lambda > 0$ and $x \in X$ such that $\rho_x^{-1}(\lambda) > 0$. Then the decreasing nature of $\rho(x, \cdot)$ yields

$$p(x)b'(x, z) = \rho(x, z) \geq \rho(x, \rho_x^{-1}(\lambda)) = \lambda \quad \text{for} \quad 0 < z < \rho_x^{-1}(\lambda).$$

Thus,

$$p(x) \geq p(x)b(x, \rho_x^{-1}(\lambda)) = \int_0^{\rho_x^{-1}(\lambda)} p(x)b'(x, z)\, dz \geq \lambda \rho_x^{-1}(\lambda),$$

that is,

$$\rho_x^{-1}(\lambda) \leq p(x)/\lambda \quad \text{for} \quad \lambda > 0.$$

From the above, it follows that

$$U(\lambda) = \int_X \rho_x^{-1}(\lambda)\, dx \leq \frac{1}{\lambda} \int_X p(x)\, dx = \frac{1}{\lambda} \quad \text{for} \quad \lambda > 0. \qquad (2.2.14)$$

To establish the continuity of U, let $\lambda_0 > 0$. Then for $\lambda > \frac{1}{2}\lambda_0$, $\rho_x^{-1}(\lambda) \leq 2p(x)/\lambda_0$ for $x \in X$. By the dominated convergence theorem (Theorem A.1 of Appendix A) and the continuity of ρ_x^{-1}

$$\lim_{\lambda \to \lambda_0} U(\lambda) = \lim_{\lambda \to \lambda_0} \int_X \rho_x^{-1}(\lambda)\, dx$$
$$= \int_X [\lim_{\lambda \to \lambda_0} \rho_x^{-1}(\lambda)]\, dx$$
$$= \int_X \rho_x^{-1}(\lambda_0)\, dx = U(\lambda_0).$$

This proves the continuity of U. The fact that ρ_x^{-1} is strictly decreasing in $(0, \rho(x, 0))$ may be used to show that U is strictly decreasing on $(0, \lambda_u)$, which proves (i). That $\lim_{\lambda \to \infty} U(\lambda) = 0$ follows from (2.2.14). Thus (ii) follows from the continuity of U and the definition of λ_u. Statement (iii) follows from property (c) in (2.2.13) and the monotone convergence theorem (see Theorem A.2). This proves the existence of U^{-1} and the theorem.

2.2 Regular Detection Functions

For the case of a discrete target distribution, define

$$p_j(z) = p(j, z) = p(j)b'(j, z) \quad \text{for} \quad z \geq 0, \quad j \in J,$$

$$p_j^{-1}(\lambda) = \begin{cases} \text{inverse of } p_j \text{ evaluated at } \lambda & \text{for} \quad 0 < \lambda \leq p_j(0), \\ 0 & \text{for} \quad \lambda > p_j(0), \end{cases}$$

$$U(\lambda) = \sum_{j \in J} p_j^{-1}(\lambda).$$

As above, U^{-1} denotes the inverse function for U.

Theorem 2.2.3. Let $c(j, z) = z$ for $z \geq 0$ and $j \in J$, and assume that b is a regular detection function on J. Fix a cost $K > 0$ and let

$$\lambda = U^{-1}(K).$$

Then f_λ^* defined by

$$f_\lambda^*(j) = p_j^{-1}(\lambda) \quad \text{for} \quad j \in J$$

is optimal for cost K and $C[f_\lambda^*] = K$.

Proof. The proof follows in the same manner as Theorem 2.2.2 with the exception of the continuity of U and property (iii). The continuity of U is established as follows: As in Theorem 2.2.2,

$$U(\lambda) = \sum_{j \in J} p_j^{-1}(\lambda) \leq \sum_{j \in J} p(j)/\lambda \leq 1/\lambda \quad \text{for} \quad \lambda > 0.$$

Again one chooses a $\lambda_0 > 0$. For λ in the interval $[\tfrac{1}{2}\lambda_0, 2\lambda_0]$, one has $|p_x^{-1}(\lambda)| \leq 2p(j)/\lambda_0$. By a theorem of Weierstrass (Theorem A.3), the series defining U converges uniformly in the interval $[\tfrac{1}{2}\lambda_0, 2\lambda_0]$, and thus U, being a sum of continuous functions, is continuous in the interval by Theorem A.4. Since λ_0 is an arbitrary positive number, it follows that U is continuous on $(0, \infty)$. Property (iii) follows from the fact that each term in the series defining U approaches ∞ as $\lambda \downarrow 0$. This proves the theorem.

A closer look at Theorems 2.2.2 and 2.2.3 shows that we have really proved more than is stated. These theorems actually give methods of finding uniformly optimal plans.

Theorem 2.2.4. Assume that b is a regular detection function on X. For $x \in X$, let

$$c(x, z) = z \quad \text{and} \quad p_x(z) = p(x)b'(x, z) \quad \text{for} \quad z \geq 0,$$

$$p_x^{-1}(\lambda) = \begin{cases} \text{inverse of } p_x \text{ evaluated at } \lambda & \text{for} \quad 0 < \lambda \leq p_x(0), \\ 0 & \text{for} \quad \lambda > p_x(0), \end{cases}$$

$$U(\lambda) = \int_X p_x^{-1}(\lambda) \, dx \quad \text{for} \quad \lambda > 0.$$

If M is a cumulative effort function, then φ^* defined by

$$\varphi^*(x, t) = \rho_x^{-1}(U^{-1}(M(t))) \quad \text{for} \quad x \in X, \quad t \geq 0,$$

is uniformly optimal in $\Phi(M)$.

Proof. By taking $\lambda = U^{-1}(M(t))$ and $f_\lambda^* = \varphi^*(\cdot, t)$ in Theorem 2.2.2, we find

$$P[\varphi^*(\cdot, t)] = \max\{P[f] : f \in F(X) \text{ and } C[f] \leq M(t)\},$$

and

$$\int_X \varphi^*(x, t) \, dx = C[\varphi^*(\cdot, t)] = M(t) \quad \text{for} \quad t \geq 0.$$

Thus

$$P[\varphi^*(\cdot, t)] = \max\{P[\varphi(\cdot, t)] : \varphi \in \Phi(M)\} \quad \text{for} \quad t \geq 0.$$

It remains to show only that φ^* is a search plan. Clearly, $\varphi^*(\cdot, t) \in F(X)$ for $t \geq 0$. Recall that U is a strictly decreasing function on $(0, \lambda_u]$. As a result U^{-1} is strictly decreasing on $[0, \infty)$. Since M is increasing, $\gamma = U^{-1}(M)$ is a decreasing function. Since ρ_x^{-1} is decreasing, $\varphi^*(x, \cdot) = \rho_x^{-1}(\gamma)$ is increasing for $x \in X$. Thus φ^* is a search plan on X and the theorem is proved.

The analog of Theorem 2.2.4 for discrete search spaces is stated next without proof.

Theorem 2.2.5. *Assume that b is a regular detection function on J. For $j \in J$, let*

$$c(j, z) = z \quad \text{and} \quad \rho_j(z) = p(j)b'(j, z) \quad \text{for} \quad z \geq 0,$$

$$\rho_j^{-1}(\lambda) = \begin{cases} \text{inverse of } \rho_j \text{ evaluated at } \lambda & \text{for } 0 < \lambda \leq \rho_j(0), \\ 0 & \text{for } \lambda > \rho_j(0), \end{cases}$$

$$U(\lambda) = \sum_{j \in J} \rho_j^{-1}(\lambda) \quad \text{for} \quad \lambda > 0.$$

If M is a cumulative effort function, then φ^* defined by

$$\varphi^*(j, t) = \rho_j^{-1}(U^{-1}(M(t))) \quad \text{for} \quad j \in J, \quad t \geq 0,$$

is uniformly optimal in $\Phi(M)$.

Theorems 2.2.4 and 2.2.5 are key theorems for finding optimal search plans. The reason for this is that in most applications the detection functions are regular. Thus, for searches in which the target is stationary, the sensor has perfect discrimination, and effort is continuous, these theorems will in most cases (i.e., those in which the detection function is regular) provide a method

2.2 Regular Detection Functions

of finding optimal search plans. Often one cannot invert U analytically, and sometimes one cannot find a closed form for U itself. However, in these cases, the computation of U and U^{-1} can usually be performed by a computer. One can, for example, calculate U numerically, plot the graph of U, and then obtain U^{-1} from the graph.

We now show that uniformly optimal plans minimize the mean time to find the target. Let $\mu(\varphi)$ be the mean time to find the target when using search plan φ. In order to calculate $\mu(\varphi)$, we appeal to the following result, which is proved in Feller (1966, p. 148). Let **T** be a nonnegative random variable with distribution function D, i.e.,

$$D(t) = \Pr\{\mathbf{T} \leq t\} \quad \text{for} \quad t \geq 0.$$

Then

$$\mathbf{E}[\mathbf{T}] = \int_0^\infty [1 - D(t)]\, dt, \tag{2.2.15}$$

where **E** denotes expectation.

In the case where D has a derivative D' and $\lim_{t \to \infty} D(t) = 1$, one may show that (2.2.15) holds as follows. By the definition of expectation,

$$\mathbf{E}[\mathbf{T}] = \int_0^\infty t D'(t)\, dt. \tag{2.2.16}$$

Since $D(0) = 0$, one may use an integration by parts to obtain

$$\int_0^y t D'(t)\, dt = -y[1 - D(y)] + \int_0^y [1 - D(t)]\, dt \tag{2.2.17}$$

for finite $y > 0$. If $\mathbf{E}[\mathbf{T}] < \infty$, then observe that

$$0 = \lim_{y \to \infty} \int_y^\infty t D'(t)\, dt \geq \lim_{y \to \infty} y[1 - D(y)] \geq 0.$$

Thus, letting $y \to \infty$ in (2.2.17) one obtains (2.2.15). If $\mathbf{E}[\mathbf{T}] = \infty$,

$$\lim_{y \to \infty} \int_0^y [1 - D(t)]\, dt \geq \lim_{y \to \infty} \int_0^y t D'(t)\, dt = \infty,$$

and (2.2.15) holds.

Although the above derivation holds only for distribution functions D that have derivatives and for which $\lim_{t \to \infty} D(t) = 1$, the proof by Feller (1966) shows that (2.2.15) holds for arbitrary nonnegative random variables **T**, even those that have positive probability of being equal to infinity. As a result, one may write

$$\mu(\varphi) = \int_0^\infty (1 - P[\varphi(\cdot, t)])\, dt \tag{2.2.18}$$

for any search plan φ. That uniformly optimal plans minimize the mean time to find the target is an easy consequence of (2.2.18).

Theorem 2.2.6. *Let φ^* be uniformly optimal in $\Phi(M)$. Then*

$$\mu(\varphi^*) \leq \mu(\varphi) \quad \text{for} \quad \varphi \in \Phi(M).$$

Proof. Let $\varphi \in \Phi(M)$. Since φ^* is uniformly optimal within $\Phi(M)$,

$$P[\varphi^*(\cdot, t)] \geq P[\varphi(\cdot, t)] \quad \text{for} \quad t \geq 0,$$

and $\mu(\varphi^*) \leq \mu(\varphi)$ follows from (2.2.18). This proves the theorem.

Example 2.2.7. Let us examine some of the properties of the uniformly optimal search plan φ^* found in Example 2.2.1 for a circular normal distribution and exponential detection function. Recall that search effort is measured in swept area and that effort is applied at a constant rate Wv, where W is the sweep width and v the constant speed of the search sensor. In Example 2.2.1, we found that

$$\varphi^*((r, \theta), t) = \begin{cases} H\sqrt{t} - \dfrac{r^2}{2\sigma^2} & \text{for} \quad r \leq R(t), \\ 0 & \text{for} \quad r > R(t), \end{cases} \qquad (2.2.19)$$

where

$$R^2(t) = 2\sigma^2 H\sqrt{t} \quad \text{and} \quad H = (Wv/\pi\sigma^2)^{1/2}. \qquad (2.2.20)$$

Probability of Detection. Let us calculate $P[\varphi^*(\cdot, t)]$, the probability of detecting the target by time t using plan φ^*:

$$P[\varphi^*(\cdot, t)] = \int_0^{2\pi}\!\!\int_0^{\infty} \frac{1}{2\pi\sigma^2} \exp\!\left(-\frac{r^2}{2\sigma^2}\right)\{1 - \exp[-\varphi^*((r, \theta), t)]\} r \, dr \, d\theta$$

$$= 1 - \frac{1}{\sigma^2}\int_0^{R(t)} r \exp(-H\sqrt{t}) \, dr - \frac{1}{\sigma^2}\int_{R(t)}^{\infty} r \exp\!\left(-\frac{r^2}{2\sigma^2}\right) dr$$

$$= 1 - (1 + H\sqrt{t}) \exp(-H\sqrt{t}) \quad \text{for} \quad t \geq 0. \qquad (2.2.21)$$

From (2.2.21) we see that the probability of detecting the target approaches 1 as $t \to \infty$.

Expected Time to Detection. Let $\mu(\varphi^*)$ be the mean time to detect the target using φ^*. Then

$$\mu(\varphi^*) = \int_0^{\infty} (1 - P[\varphi^*(\cdot, t)]) \, dt = 6\pi\sigma^2/Wv. \qquad (2.2.22)$$

Note that $\mu(\varphi^*)$ is proportional to σ^2 and inversely proportional to Wv. The product Wv is often called the sweep rate of the sensor.

2.2 Regular Detection Functions

Posterior Target Distribution. Suppose the search has continued for a time T and the target has not been found. Let $\tilde{p}$ be the posterior probability density for the target location given that the target has not been found by time T. By Bayes' theorem

$$\tilde{p}(r, \theta) = \frac{p(r, \theta) \exp[-\varphi^*((r, \theta), T)]}{1 - P[\varphi^*(\cdot, T)]}.$$

Thus,

$$\tilde{p}(r, \theta) = \begin{cases} [2\pi\sigma^2(1 + H\sqrt{T})]^{-1} & \text{for } r \leq R(T), \\ \dfrac{\exp[-r^2/(2\sigma^2)] \exp(H\sqrt{T})}{2\pi\sigma^2(1 + H\sqrt{T})} & \text{for } r > R(T). \end{cases}$$

Observe that the posterior density of the target distribution is constant inside the circle of radius $R(T)$ centered at the origin. This is just the region to which search effort has been applied by time T. Thus the search is carried out in such a way that $p(r,\theta) \exp[-\varphi^*((r, \theta), T)]$ is constant inside the disc of radius $R(T)$ centered at the origin. As time increases and the target has not been detected, the disc over which the posterior density is constant expands and the value of the density in that disc decreases. The result is that the posterior target distribution is flattening and spreading as the target fails to be detected (see Fig. 2.2.2).

Expected Time Remaining to Search. Suppose search is extended for a time t beyond T by continuing to search according to the plan φ^*. Let $\tilde{P}_T(t)$ be the probability of finding the target by time $T + t$ given failure to find the target by time T. Then by the laws of conditional probability,

$$\tilde{P}_T(t) = \frac{P[\varphi^*(\cdot, T + t)] - P[\varphi^*(\cdot, T)]}{1 - P[\varphi^*(\cdot, T)]}$$

$$= 1 - \frac{1 - P[\varphi^*(\cdot, T + t)]}{1 - P[\varphi^*(\cdot, T)]}$$

$$= 1 - \frac{[1 + H(T + t)^{1/2}] \exp[-H(T + t)^{1/2}]}{[1 + H\sqrt{T}] \exp[-H\sqrt{T}]}.$$

It is interesting to compute $\mu_T(\varphi^*)$, the mean time beyond T required to detect the target using plan φ^* given that the target is not detected by time T. We have

$$\mu_T(\varphi^*) = \int_0^\infty [1 - \tilde{P}_T(t)] \, dt = \frac{6\pi\sigma^2}{Wv(1 + H\sqrt{T})} \left(1 + H\sqrt{T} + \frac{H^2T}{3}\right).$$

One may check that

$$\mu_T(\varphi^*) \geq 6\pi\sigma^2/Wv = \mu(\varphi^*)$$

and that $\mu_T(\varphi^*)$ increases as T increases.

This means that the longer one searches in this situation and fails to find the target, the worse one's prospects become, in the sense that the mean time remaining to detect is increasing. One would hope that the longer one searched, the shorter would be the remaining time to detect the target. In this example, the reverse happens. The explanation lies in the posterior distributions shown in Fig. 2.2.2. Notice that as T increases and the target fails to be detected, the

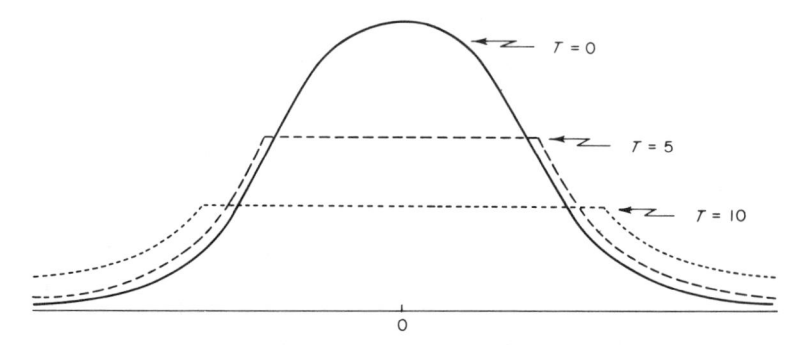

Fig. 2.2.2. Cross section of posterior target density.

posterior target distribution becomes flatter and more spread out. In some sense, the searcher is exhausting his information about the target location as he searches, unless he detects the target.

Additional Effort Density. Let us consider one final aspect of this search. Define

$$\varphi_T^*((r, \theta), t) = \varphi^*((r, \theta), T + t) - \varphi^*((r, \varphi), T) \quad \text{for} \quad x \in X, \; t \geq 0.$$

Then $\varphi_T^*((r, \theta), t)$ gives the additional effort density that accumulates at point (r, θ) in the interval $[T, T + t]$:

$$\varphi_T^*((r, \theta), t) = \begin{cases} H[(T + t)^{1/2} - \sqrt{T}] & \text{for} \quad 0 \leq r \leq R(T), \\ H(T + t)^{1/2} - \dfrac{r^2}{2\sigma^2} & \text{for} \quad R(T) < r \leq R(T + t), \\ 0 & \text{for} \quad R(T + t) < r < \infty. \end{cases}$$

Notice that the additional effort is placed uniformly over the region that has received search effort by time T, i.e., the disc of radius $R(T)$ centered at the origin. One may think of the plan φ^* as being obtained by placing thin uniform layers of search effort over a disc whose radius is increasing with time.

2.2 Regular Detection Functions

The result of this procedure is to cause effort density to accumulate near the origin and to decrease to 0 at the edge of the disc being searched.

Example 2.2.8. In this example an algorithm is described that can be used to find an allocation f^* that is optimal for cost K when the search space consists of a finite number of cells and the detection function is exponential.

Suppose there are k cells and that for $j = 1, \ldots, k$, $p(j)$ is the probability that the target is located in cell j. The detection function is given by

$$b(j, z) = 1 - e^{-\alpha_j z} \quad \text{for } z \geq 0, \ j = 1, \ldots, k,$$

where $\alpha_j > 0$. It is assumed that the cells are numbered so that

$$\alpha_1 p(1) \geq \alpha_2 p(2) \geq \cdots \geq \alpha_k p(k).$$

Let

$$y_1 = \ln[\alpha_1 p(1)] - \ln[\alpha_2 p(2)], \quad z_{11} = y_1/\alpha_1.$$

Observe that

$$p(1)b'(1, z_{11}) = \alpha_1 p(1) e^{-\alpha_j z_{11}} = \alpha_2 p(2) = p(2)b'(2, 0).$$

Let $\lambda = \alpha_2 p(2)$. Then f_λ^* defined by

$$f_\lambda^*(1) = z_{11}, \quad f_\lambda^*(j) = 0 \quad \text{for } j = 2, \ldots, k$$

satisfies (2.1.5) [i.e., (λ, f_λ^*) maximizes the pointwise Lagrangian], and by Theorem 2.1.2 f_λ^* is optimal for cost $C[f_\lambda^*] = z_{11}$. For $0 \leq w \leq z_{11}$, one may take $\lambda = p(1)b'(1, w)$ and use Theorem 2.1.2 to show the allocation

$$f_\lambda^*(1) = w, \quad f_\lambda^*(j) = 0 \quad \text{for } j = 2, \ldots, k$$

is optimal for cost w. Thus, all effort up to z_{11} should be placed entirely in cell 1. We shall see that the effort beyond z_{11} is split between cells 1 and 2.

Let

$$y_2 = \ln[\alpha_2 p(2)] - \ln[\alpha_3 p(3)],$$
$$z_{22} = y_2/\alpha_2, \quad z_{12} = y_2/\alpha_1.$$

Now let

$$\lambda = p(2)b'(2, z_{22}) = \alpha_3 p(3),$$
$$f_\lambda^*(1) = z_{11} + z_{12}, \quad f_\lambda^*(2) = z_{22}, \quad f_\lambda^*(j) = 0 \quad \text{for } j = 3, \ldots, k.$$

Then one may verify by the use of Theorem 2.1.2 that f_λ^* is optimal for cost $z_{11} + z_{12} + z_{22}$.

Let $0 \leq a \leq 1$, and consider

$$f_\lambda^*(1) = z_{11} + a z_{12}, \quad f_\lambda^*(2) = a z_{22}, \quad f_\lambda^*(j) = 0 \quad \text{for } j = 3, \ldots, k.$$

Then by taking

$$\lambda = \alpha_2 p(2) e^{-a z_{22}},$$

one may verify that f_λ^* is optimal for cost $z_{11} + a(z_{12} + z_{22})$. Thus, one can find an optimal allocation for any cost between 0 and $z_{11} + z_{12} + z_{22}$.

If one thinks about the search developing in time, then effort is applied only to cell 1 until z_{11} effort has been placed there. At this point, the additional effort is split between cells 1 and 2 so that the ratio of additional effort placed in cell 1 to that in cell 2 is α_2/α_1. If there are only two cells, then this gives a method of allocating any amount of effort.

Let

$$y_j = \ln[\alpha_j p(j)] - \ln[\alpha_{j+1} p(j+1)] \quad \text{for} \quad j = 1, \ldots, k-1.$$

For the general case of k cells, one may check that for $0 \leq a \leq 1$ and $1 \leq i \leq k-1$, f^* defined by

$$f^*(j) = \begin{cases} \dfrac{1}{\alpha_j}(y_j + y_{j+1} + \cdots + y_{i-1} + a y_i) & \text{for} \quad 1 \leq j \leq i, \\ 0 & \text{for} \quad i < j \leq k, \end{cases} \quad (2.2.23)$$

is optimal for its cost. Let

$$S(0) = 0, \quad S(i) = \sum_{j=1}^{i} \frac{1}{\alpha_j}(y_j + y_{j+1} + \cdots + y_i) \quad \text{for} \quad i = 1, 2, \ldots, k-1.$$

Then $S(i)$ is the amount of effort beyond which search spreads to cell $i+1$.

Thus to find an allocation that is optimal for cost K, one finds the value of i for which $S(i-1) \leq K \leq S(i)$. By setting

$$a = \frac{K - S(i-1)}{S(i) - S(i-1)}$$

in (2.2.23) the allocation f^* that is optimal for cost K is obtained. If $K > S(k-1)$, then taking

$$a = \frac{K - S(k-1)}{\sum_{j=1}^{k} 1/\alpha_j}$$

and letting

$$f^*(j) = \frac{1}{\alpha_j}(y_j + y_{j+1} + \cdots + y_{k-1} + a) \quad \text{for} \quad j = 1, \ldots, k$$

will yield an allocation that is optimal for cost K.

The above method is presented in the form of an algorithm written in the

2.2 Regular Detection Functions

style of a computer program (e.g., equal signs mean replacement just as in a programming language.)

Algorithm
1. Set $z_j = 0$ for $j = 1, \ldots, k$.
2. Set $S(0) = 0$.
3. Set $i = 1$.
4. Compute $y_i = \ln[\alpha_i p(i)] - \ln[\alpha_{i+1} p(i+1)]$.
5. Compute $S(i) = S(i-1) + y_i \sum_{j=1}^{i} 1/\alpha_j$.
6. If $S(i) \geq K$, go to step 10.
7. Set $z_j = z_j + y_i/\alpha_j$ for $1 \leq j \leq i$.
8. Set $i = i + 1$.
9. If $i = k$, go to step 12; otherwise, go to step 4.
10. Set $a = [K - S(i-1)]/[S(i) - S(i-1)]$.
11. Go to step 14.
12. Set $a = [K - S(k-1)]/\sum_{j=1}^{k} 1/\alpha_j$.
13. Set $y_k = 1$.
14. Set $f^*(j) = z_j + ay_i/\alpha_j$ for $j = 1, \ldots, i$.
15. Set $f^*(j) = 0$ for $i + 1 \leq j \leq k$.

An algorithm similar to this one is used by the US Coast Guard to allocate search effort for some searches at sea. Typically, the ocean region being searched is divided into k rectangles, and a probability $p(j)$ of the target or search object being in the jth rectangle is computed. Let A_j be the area of the jth rectangle and W_j the sweep width of the sensor when operating in the jth rectangle. Variations in sweep width from rectangle to rectangle can be caused by differing visibilities and meteorological conditions. The search in a rectangle is treated like a random search, so that

$$b(j, z) = 1 - e^{-zW_j/A_j} \quad \text{for} \quad z \geq 0, \quad j = 1, \ldots, k.$$

Here, effort z is measured in terms of track length and $\alpha_j = W_j/A_j$ for $j = 1, \ldots, k$. Each rectangle is treated like a cell in a discrete target distribution and effort is required to be spread uniformly within each rectangle. If the search vehicle can search for a time T at speed v, then the optimal allocation of $K = vT$ track length can be obtained by the algorithm.

The case where there is more than one type of search vehicle can be handled as follows. Suppose there are two vehicles. Vehicle 1 can search for a time T_1 at speed v_1 with a sensor having sweep width $\overline{W}_1$, while vehicle 2 can search for a time T_2 at speed v_2 with a sensor having sweep width $\overline{W}_2$. Then search effort is measured in swept area and $\alpha_j = 1/A_j$ for $j = 1, \ldots, k$. The algorithm can then be used to allocate optimally

$$K = \overline{W}_1 T_1 v_1 + \overline{W}_2 T_2 v_1$$

effort. The search planner still has to allocate the time of his two vehicles to obtained the desired effort (i.e., swept area) in each cell.

Extensions of the above to any number of vehicle types and number of vehicles within each type are easily accomplished. However, in order to use the algorithm with more than one type of vehicle, the search rate $\overline{W}_i v_i$ of the ith vehicle type must be the same in all search cells.

Let us consider a situation in which there is no uniformly optimal plan.

Example 2.2.9. Let $J = \{1, 2\}$, $p(1) = p(2) = \frac{1}{2}$, and

$$b(1, z) = \begin{cases} z & \text{for } 0 < z \le \frac{1}{2}, \\ \frac{1}{2} + \frac{1}{4}(z - \frac{1}{2}) & \text{for } \frac{1}{2} < z \le \frac{5}{2}, \\ 1 & \text{for } \frac{5}{2} < z \le \infty, \end{cases}$$

$$b(2, z) = \begin{cases} \frac{1}{2}z & \text{for } 0 \le z \le \frac{1}{2}, \\ \frac{1}{4} + \frac{3}{2}(z - \frac{1}{2}) & \text{for } \frac{1}{2} < z \le 1, \\ 1 & \text{for } 1 < z < \infty. \end{cases}$$

Let $M(t) = t$ for $t \ge 0$.

Suppose there is a uniformly optimal plan $\varphi^* \in \Phi(M)$. For $0 \le t \le \frac{1}{2}$, there is only one possibility, namely,

$$\varphi^*(1, t) = t, \qquad \varphi^*(2, t) = 0.$$

At time $t = \frac{1}{2}$, φ^* must stop searching in cell 1 and place all additional effort in cell 2 up to time $t = 1$. Recall that $\varphi^*(j, \cdot)$ must be increasing for $j = 1, 2$. Thus,

$$\begin{aligned}\varphi^*(1, t) &= \tfrac{1}{2} & \text{for } \tfrac{1}{2} \le t \le 1, \\ \varphi^*(2, t) &= t - \tfrac{1}{2} & \text{for } \tfrac{1}{2} \le t \le 1.\end{aligned}$$

Consider

$$P[\varphi^*(\cdot, 1)] = \tfrac{1}{2}b(1, \tfrac{1}{2}) + \tfrac{1}{2}b(2, \tfrac{1}{2}) = \tfrac{3}{8}.$$

Let φ be defined as follows:

$$\begin{aligned}\varphi(1, t) &= 0 & \text{for } t > 0, \\ \varphi(2, t) &= t & \text{for } t > 0.\end{aligned}$$

Then $\varphi \in \Phi(M)$ and

$$P[\varphi(\cdot, 1)] = \tfrac{1}{2} > P[\varphi*(\cdot, 1)] = \tfrac{3}{8}.$$

Thus there is no uniformly optimal plan φ^* for this search.

One reason that no uniformly optimal plan exists in this example is that $b(2, \cdot)$ is not concave and thus not regular. It is shown in Theorem 2.4.10 that if $b(j, \cdot)$ is concave for $j \in J$, then a uniformly optimal plan exists. In the case of a continuous target distribution, Theorem 2.4.6 shows that uniformly

2.3 Optimal Search with Uncertain Sweep Width

optimal plans exist under the rather weak condition that $b(x, \cdot)$ is increasing and right continuous for $x \in X$.

2.3. OPTIMAL SEARCH WITH UNCERTAIN SWEEP WIDTH

This section treats the problem of finding optimal search plans when the sweep width of the search sensor is not known with certainty but is assumed to be a random variable **W** with a known probability distribution.

There are many search situations in which the sweep width is uncertain. The uncertainty might arise from not knowing the condition of the target (e.g., whether it is intact or has disintegrated) or, in an ocean-bottom search, from not knowing the extent to which the target protrudes from the ocean floor. Moreover, sweep width estimates themselves are subject to testing errors, which provide an additional source of uncertainty.

The term *conditional detection function* will be used to refer to detection functions that are conditioned upon a specific value of sweep width. In this section uniformly optimal search plans are found in cases where the conditional detection functions corresponding to nonzero sweep widths are regular. Theorems 2.3.1 and 2.3.2, pertaining to search spaces X and J, respectively, show that in these cases the detection function (unconditional) for the search is also regular. Theorems 2.3.3 and 2.3.4 use this fact and the results of Section 2.2 to give uniformly optimal plans.

In Example 2.3.5, the uniformly optimal plan φ^* and mean time to detection $\mu(\varphi^*)$ are calculated for a search in which there is a circular normal target distribution, exponential conditional detection function, and gamma sweep width distribution. The results of this example are used to compare the mean time to detection for the uniformly optimal plan based on uncertain sweep width to the mean times resulting from plans based on a fixed sweep width. It is shown that for some situations there is a substantial penalty in mean time for using even the best fixed sweep width plan compared to the uniformly optimal plan.

Uniformly Optimal Search Plans When Sweep Width Is Uncertain

First we show that the problem of finding optimal search plans with uncertain sweep width may be reduced to the basic search problem given in Section 1.3. Then uniformly optimal search plans are found when the conditional detection functions corresponding to nonzero sweep widths are regular. The results hold for both search spaces X and J.

Assume that the sweep width of the search sensor is a random variable $\mathbf{W}(x)$ or $\mathbf{W}(j)$ whose distribution may depend on the location $x \in X$ or $j \in J$

of the target. Let $G(\cdot, x)$ or $G(\cdot, j)$ denote the conditional probability distribution function for sweep width given that the target is located at $x \in X$ or $j \in J$. Thus, for $x \in X$

$$G(w, x) = \Pr\{W(x) \leq w \mid \text{target located at } x\},$$

or for $j \in J$

$$G(w, j) = \Pr\{W(j) \leq w \mid \text{target located in cell } j\}.$$

Although **W** is allowed to depend on location, it is assumed not to vary over time.

For search space X, let B_w denote the conditional detection function that would hold if $W(x) = w$ for $x \in X$. For $w \geq 0$, B_w is assumed to be a detection function on X in the sense given in Section 1.2. For search space J, a similar definition of B_w holds.

The unconditional detection function b on X is the average of the functions B_w with respect to the conditional sweep width distributions, i.e.,

$$b(x, z) = \int_0^\infty B_w(x, z) G(dw, x) \qquad \text{for} \quad x \in X, \ z \geq 0. \tag{2.3.1}$$

The notation $G(dw, x)$ indicates integration with respect to the measure induced by the distribution function $G(\cdot, x)$. If $B_w(x, z)$ is a continuous function of w, then the integration in (2.3.1) becomes the standard Stieltjes integration.

For $f \in F(X)$, the probability of detection functional P is defined by

$$P[f] = \int_X \int_0^\infty p(x) B_w(x, f(x)) G(dw, x) \, dx,$$

and using (2.3.1), $P[f]$ may be written in the form

$$P[f] = \int_X p(x) b(x, f(x)) \, dx. \tag{2.3.2}$$

It is assumed that $c(x, z) = z$ for $z \geq 0$ and $x \in X$, so that the cost functional C is given by

$$C[f] = \int_X f(x) \, dx. \tag{2.3.3}$$

The corresponding quantities for the discrete search space J are

$$b(j, z) = \int_0^\infty B_w(j, z) G(dw, j) \qquad \text{for} \quad j \in J, \ z \geq 0, \tag{2.3.4}$$

$$P[f] = \sum_{j \in J} p(j) b(j, f(j)), \tag{2.3.5}$$

$$C[f] = \sum_{j \in J} f(j). \tag{2.3.6}$$

2.3 Optimal Search with Uncertain Sweep Width

The functionals P and C specified, respectively, by (2.3.2) and (2.3.3) or (2.3.5) and (2.3.6) are identical with the functionals specified by (1.3.2)–(1.3.5). Thus, having calculated b in (2.3.1) or (2.3.4), this problem is reduced to the basic search problem given in Section 1.3.

We will show that if B_w is regular for each w, then b defined by (2.3.1) or (2.3.4) is also regular. This being the case, Theorems 2.2.4 and 2.2.5 may be used to solve the problem of finding a uniformly optimal plan $\varphi^* \in \Phi(M)$ for a given cumulative effort function M.

We begin by considering the search space X. Let $B_w'(x, \cdot)$ denote the derivative of $B_w(x, \cdot)$, and recall that B_w is a regular detection function on X if $B_w(x, 0) = 0$ and $B_w'(x, \cdot)$ is continuous, positive, and strictly decreasing for $x \in X$. Note that $B_0 \equiv 0$ and hence is not regular. This will not be a problem as long as it is assumed that the event $W(x) = 0$ does not have probability 1 for some $x \in X$. For notational convenience, define B and B' by $B(w, x, z) = B_w(x, z)$ and $B'(w, x, z) = B_w'(x, z)$ for $w \geq 0$, $x \in X$, and $z \geq 0$.

Theorem 2.3.1. *Let G be a Borel function on $[0, \infty) \times X$ and let $G(\cdot, x)$ be a probability distribution function for each $x \in X$. Let B be a Borel function on $[0, \infty) \times X \times [0, \infty)$ and let $B'(\cdot, x, 0)$ be finitely integrable with respect to $G(\cdot, x)$ for each $x \in X$.*

If B_w is a regular detection function for all $w > 0$, and if, for $x \in X$, $G(\cdot, x)$ is not concentrated on zero, i.e., $\Pr\{W(x) > 0\} > 0$, then b is a regular detection function.

Proof. One must show that $b: X \times [0, \infty) \to [0, 1]$ is Borel, that $b(x, 0) = 0$, and that $b'(x, \cdot)$ is continuous, positive, and strictly decreasing for $x \in X$.

The fact that b is Borel follows from the hypothesis that both G and B are Borel on their respective domains. That $b(x, 0) = 0$ follows directly from (2.3.1) and the fact that $B_w(x, 0) = 0$ for $w \geq 0$.

Note that

$$\frac{1}{h}[B(w, x, z+h) - B(w, x, z)] = \frac{1}{h}\int_z^{z+h} B'(w, x, y)\, dy \leq B'(w, x, 0)$$

since $B'(w, x, \cdot)$ is assumed decreasing for all $w \geq 0$ and $x \in X$. Since $B'(\cdot, x, 0)$ is assumed finitely integrable with respect to $G(\cdot, x)$, the dominated convergence theorem permits one to interchange the order of differentiation and integration to obtain

$$b'(x, z) = \int_0^\infty B'(w, x, z) G(dw, x) \quad \text{for } z \geq 0, \quad x \in X.$$

Similarly, since $B'(w, x, \cdot)$ is continuous, $B'(\cdot, x, 0)$ is finitely integrable, and

$$B'(\cdot, x, 0) \geq B'(\cdot, x, z),$$

one may use the dominated convergence theorem to establish continuity of $b'(x, \cdot)$.

Positivity and strict monotonicity for $b'(x, \cdot)$ follow from the same properties of $B'(w, x, \cdot)$ for $w > 0$ and the fact that $\Pr\{W(x) > 0\} > 0$. This completes the proof.

The following theorem is the counterpart of Theorem 2.3.1 for search space J. Again for notational convenience we define B and B' by $B(w, j, z) = B_w(j, z)$ and $B'(w, j, z) = B_w'(j, z)$ for $w \geq 0, j \in J$, and $z \geq 0$.

Theorem 2.3.2. *Let $G(\cdot, j)$ be a probability distribution function for each $j \in J$. For $j \in J$ let $B'(\cdot, j, 0)$ be finitely integrable with respect to $G(\cdot, j)$ and let $B(\cdot, j, z)$ be a Borel function for $z \geq 0$.*

If B_w is a regular detection function for all $w > 0$, and if, for $j \in J$, $G(\cdot, j)$ is not concentrated on zero, i.e., $\Pr\{W(j) > 0\} > 0$, then b defined by (2.3.4) is a regular detection function.

Proof. The proof is basically the same as the proof of Theorem 2.3.1 with $j \in J$ substituted for $x \in X$. The only difference is that it is not necessary to show that b is Borel since this is not a part of the definition of a regular detection function on J. This completes the proof.

Theorems 2.3.3 and 2.3.4 give uniformly optimal search plans for the case of uncertain sweep width and regular conditional detection functions.

Theorem 2.3.3. *Suppose that the assumptions of Theorem 2.3.1 hold, and let*

$$b(x, z) = \int_0^\infty B_w(x, z) G(dw, x), \qquad c(x, z) = z, \qquad \text{for } x \in X, \quad z \geq 0.$$

Let M be a cumulative effort function. Then φ^* defined in Theorem 2.2.4 is uniformly optimal within $\Phi(M)$.

Proof. By Theorem 2.3.1, b is a regular detection function on X and the theorem follows from Theorem 2.2.4.

The analogous theorem for the discrete search space J is given next.

Theorem 2.3.4. *Suppose that the assumptions of Theorem 2.3.2 hold, and let*

$$b(j, z) = \int_0^\infty B_w(j, z) G(dw, j), \qquad c(j, z) = z, \qquad \text{for } j \in J, \quad z \geq 0.$$

Let M be a cumulative effort function. Then φ^* defined in Theorem 2.2.5 is uniformly optimal within $\Phi(M)$.

Proof. By Theorem 2.3.2, b is regular on J and the theorem follows from Theorem 2.2.5.

2.3 Optimal Search with Uncertain Sweep Width

Theorems 2.3.1–2.3.4 do not depend on any special properties of sweep width. Thus, the results of this section apply more generally to any case where the detection function depends on a stochastic parameter with a known probability distribution.

Additional discussion of optimal uncertain sweep width plans is provided by Richardson and Belkin (1972), Belkin (1975), and Stone et al. (1972, Section 5). The practical implementation of the plans is addressed in Chapter VII and in Richardson et al. (1971) where, among other things, it is shown how to approximate the optimal search plan when the sweep width distribution is discrete and concentrated on a finite number of sweep width values.

Applications

Here an example is presented in which Theorem 2.3.3 is applied to obtain the uniformly optimal search plan. The example gives calculations of probability of detection, expected time to detection, and the posterior marginal probability density for target location.

Example 2.3.5. Let the target density p be given by a circular normal distribution centered at the origin with standard deviation σ.

Thus, for $x = (x_1, x_2) \in X$,

$$p(x) = \frac{1}{2\pi\sigma^2} \exp\left(-\frac{x_1^2 + x_2^2}{2\sigma^2}\right). \tag{2.3.7}$$

Let sweep width be gamma distributed with convolution parameter $\nu > 0$ and scale parameter $\alpha > 0$, and let this distribution be independent of target location. Denoting the sweep width probability density function by $G'(w)$, we then have

$$G'(w) = w^{\nu-1}\alpha^\nu e^{-\alpha w}/\Gamma(\nu) \qquad \text{for} \quad w \geq 0, \tag{2.3.8}$$

where Γ is the standard mathematical gamma function. For integer ν, $\Gamma(\nu + 1) = \nu!$ Examples of gamma densities are plotted in Fig. 2.3.1.

If **W** denotes the sweep width random variable with density G' given by (2.3.8), then the expectation of **W** is $\mathbf{E}[\mathbf{W}] = \nu/\alpha$ and the variance of **W** is $\mathbf{Var}[\mathbf{W}] = \nu/\alpha^2$. Note that the ratio of the standard deviation to the mean is $(\mathbf{Var}[\mathbf{W}])^{1/2}/\mathbf{E}[\mathbf{W}] = 1/\sqrt{\nu}$, which approaches 0 as $\nu \to \infty$. In Fig. 2.3.1, this is reflected by the increasing tendency of the gamma density to peak about the mean as ν increases.

Assume that the conditional detection function depends on track length and is exponential and independent of target location (i.e., homogeneous). Then z denotes track length per unit area, and

$$B_w(x, z) = 1 - e^{-wz} \qquad \text{for} \quad w \geq 0 \quad x \in X, \quad z \geq 0.$$

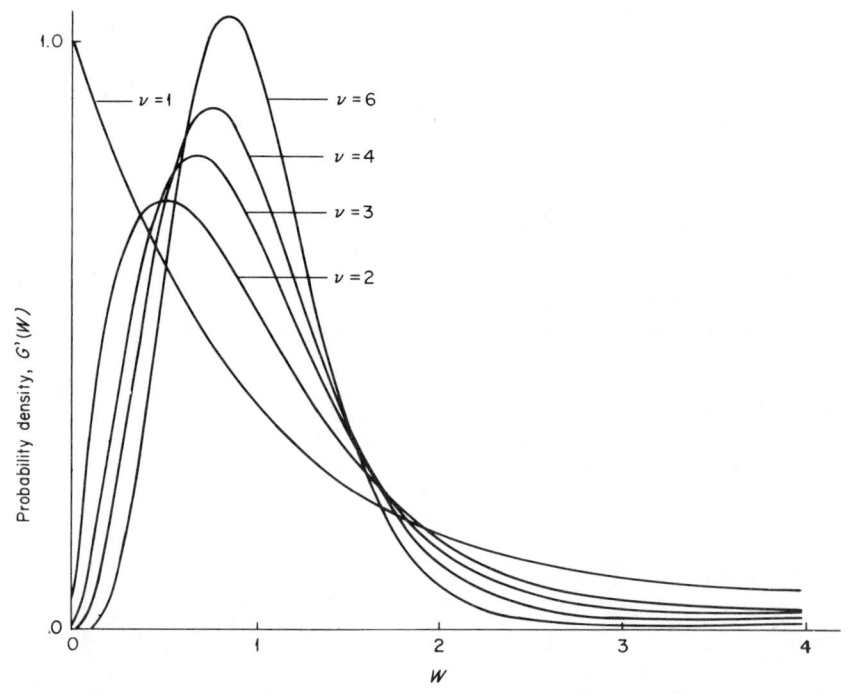

Fig. 2.3.1. Densities of gamma distributions. [*Note:* All distributions have mean 1 (i.e., $\nu/\alpha = 1$).]

An easy calculation shows that

$$b(x, z) = \int_0^\infty B_w(x, z)G'(w)\, dw = 1 - (1 + z/\alpha)^{-\nu}, \qquad (2.3.9)$$

and

$$b'(x, z) = \nu\alpha^\nu/(\alpha + z)^{\nu+1} \quad \text{for} \quad x \in X, \ z \geq 0. \qquad (2.3.10)$$

Uniformly Optimal Search Plan. Theorems 2.3.3 and 2.2.4 are now used to find the uniformly optimal search plan φ^* for the case of a circular normal target distribution and gamma sweep width distribution. From the form of B_w given above one may check that the conditions of Theorem 2.3.1 and hence Theorem 2.3.3 are satisfied, so that Theorem 2.2.4 may be used to find φ^*. Alternatively, one can check (2.3.9) and (2.3.10) to see directly that b is regular and that Theorem 2.2.4 is applicable. In either case in order to calculate φ^*, we first determine $\rho_x(z)$ and $\rho_x^{-1}(\lambda)$. Since

$$\rho_x(z) = p(x)b'(x, z) = p(x)\nu\alpha^\nu/(\alpha + z)^{\nu+1} \quad \text{for} \quad x \in X, \ z \geq 0,$$

2.3 Optimal Search with Uncertain Sweep Width

where $p(x)$ is given by (2.3.7), it follows that

$$\rho_x^{-1}(\lambda) = \begin{cases} \alpha[(\nu p(x)/\alpha\lambda)^{1/(\nu+1)} - 1] & \text{for } 0 < \lambda \leq \nu p(x)/\alpha, \\ 0 & \text{for } \lambda > \nu p(x)/\alpha. \end{cases} \quad (2.3.11)$$

Since $U(\lambda) = \int_X \rho_x^{-1}(\lambda)\, dx$ for $\lambda > 0$, we obtain

$$U(\lambda) = \nu(\nu + 1)\kappa\left[(\kappa\lambda)^{-1/(\nu+1)} + \frac{1}{\nu+1}\ln(\kappa\lambda) - 1\right]. \quad (2.3.12)$$

where

$$\kappa = 2\pi\sigma^2\alpha/\nu. \quad (2.3.13)$$

For the cumulative effort function $M(t) = \nu t$ for constant search speed $\nu > 0$, the uniformly optimal search plan φ^* is given in Theorem 2.2.4 as

$$\varphi^*(x, t) = \rho_x^{-1}(U^{-1}(\nu t)). \quad (2.3.14)$$

Unfortunately, U^{-1} is not simply expressible in terms of known and tabulated functions as was the case in Example 2.2.7.

From (2.3.11) and (2.3.14), one finds that the boundary of the region searched consists of all x for which $U^{-1}(\nu t) = \nu p(x)/\alpha$. Solving this equation for x, one finds that the boundary of the search region at time t is the circle consisting of points x for which $(x_1^2 + x_2^2)^{1/2} = R(t)$, where

$$R(t) = \sigma\{-2\ln[\kappa U^{-1}(\nu t)]\}^{1/2}. \quad (2.3.15)$$

Probability of Detection. Using (2.3.9) and converting to polar coordinates, one may calculate the probability of detection by time t in terms of $U^{-1}(\nu t)$ as

$$\begin{aligned} P[\varphi^*(\cdot, t)] &= \int_X p(x)b(x, \varphi^*(x, t))\, dx \\ &= 1 - \frac{1}{\sigma^2}\int_0^\infty \exp\left(-\frac{r^2}{2\sigma^2}\right)\left(1 + \frac{\varphi^*((r, \theta), t)}{\alpha}\right)^{-\nu} r\, dr \\ &= 1 - \frac{1}{\sigma^2}[\kappa U^{-1}(\nu t)]^{\nu/(\nu+1)}\int_0^{R(t)} \exp\left(-\frac{r^2}{2\sigma^2(\nu+1)}\right) r\, dr \\ &\quad - \frac{1}{\sigma^2}\int_{R(t)}^\infty \exp\left(-\frac{r^2}{2\sigma^2}\right) r\, dr \quad \text{for } t \geq 0. \end{aligned} \quad (2.3.16)$$

The definite integrals above are easily evaluated to obtain

$$P[\varphi^*(\cdot, t)] = 1 + \nu\kappa U^{-1}(\nu t) - (\nu + 1)[\kappa U^{-1}(\nu t)]^{\nu/(\nu+1)} \quad \text{for all } t > 0. \quad (2.3.17)$$

Since $U^{-1}(\nu t) \to 0$ as $t \to \infty$, $\lim_{t \to \infty} P[\varphi^*(\cdot, t)] = 1$.

Expected Time to Detection. The expected time to detection $\mu(\varphi^*)$ may be calculated using the equation

$$\mu(\varphi^*) = \int_0^\infty (1 - P[\varphi^*(\cdot, t)]) \, dt. \qquad (2.3.18)$$

Using $P[\varphi^*(\cdot, t)]$ given by (2.3.17), the integration is facilitated by a change of variables from t to $y = \kappa U^{-1}(vt)$. With this change of variables, (2.3.18) becomes

$$\mu(\varphi^*) = -\frac{1}{v\kappa} \int_0^1 [(v + 1)y^{v/(v+1)} - vy] U'(y/\kappa) \, dy, \qquad (2.3.19)$$

where U' denotes the derivative of U.

Evaluation of the integral in (2.3.19) using U given by (2.3.12) yields

$$\mu(\varphi^*) = \begin{cases} \dfrac{2\pi\sigma^2\alpha}{vv}\left(\dfrac{3v+1}{v-1}\right) & \text{for } v > 1, \\ \infty & \text{for } 0 < v \leq 1. \end{cases} \qquad (2.3.20)$$

If the sweep width were known to be equal to the mean v/α and one followed the optimal plan for this known sweep width, then the mean time to find the target would be obtained by setting $W = v/\alpha$ in (2.2.22), i.e., $6\pi\sigma^2\alpha/(vv)$. Note that

$$\frac{\mu(\varphi^*)}{6\pi\sigma^2\alpha/vv} = \frac{1}{3}\left(\frac{3v+1}{3v-1}\right) \geq 1,$$

which means that one pays a penalty in mean time because of the uncertainty in the sweep width.

However, it follows from (2.3.20) that for fixed $v/\alpha = W$,

$$\lim_{v \to \infty} \mu(\varphi^*) = 6\pi\sigma^2/Wv, \qquad (2.3.21)$$

which agrees with (2.2.22) in the known sweep width case. Intuitively, this is because as one lets $v \to \infty$ while holding $v/\alpha = W$, the sweep width density becomes increasingly concentrated about W and the known sweep width case is approached in the limit (see Fig. 2.3.1).

The result that $\mu(\varphi^*) = \infty$, when $0 < v \leq 1$, may be motivated by looking at the gamma densities G' given in (2.3.8). When $0 < v \leq 1$, the highest probability density occurs at $w = 0$. Thus, as the search progresses and one fails to find the target, the posterior sweep width distribution begins to peak about zero. However, when $v > 1$, $G'(0) = 0$, so that the density of the posterior sweep width distribution cannot peak about zero.

2.3 Optimal Search with Uncertain Sweep Width

Posterior Marginal Probability Density for Target Location. As in Section 2.2, let $\tilde{p}$ be the marginal posterior probability density for the target location given that the target has not been found by time T. Then

$$\tilde{p}(x) = \frac{p(x)^{1/\nu}(\alpha U^{-1}(vT)/\nu)^{\nu/(\nu+1)}}{1 - P[\varphi^*(\cdot, T)]} \quad \text{for} \quad x_1^2 + x_2^2 \leq R^2(T), \quad (2.3.22)$$

and

$$\tilde{p}(x) = \frac{p(x)}{1 - P[\varphi^*(\cdot, T)]} \quad \text{for} \quad x_1^2 + x_2^2 > R^2(T). \quad (2.3.23)$$

The term marginal density is used because, once search effort has been applied, the posterior sweep width and target location distributions are no longer independent.

Note that in contrast to the known sweep width case where the detection function is exponential, the posterior target density (2.3.22) within the circle searched is not constant. In fact, the posterior density still retains its central tendency with points nearer the origin having higher probability densities. This means that the optimal plan for this uncertain sweep width example does not proceed by always placing the next "small increment" of search where the posterior probability density is highest. In Section 3.2 it will be shown that always searching where the posterior target density is highest will produce optimal search plans if and only if the local detection function is homogeneous and exponential.

Comparison of Optimal Uncertain Sweep Width Plan with Fixed Sweep Width Plans

Recall that for a circular normal target distribution and known sweep width, the optimal plan has the simple analytic form given in (2.2.19). When the sweep width is gamma distributed, we are not able to specify the analytic form of the optimal plan φ^* because of our inability to invert U in (2.3.12) in an analytic fashion. The optimal plan for a gamma distributed sweep width must be calculated by inverting U numerically. This suggests the possibility of choosing a fixed sweep width w and using the plan φ_w that would be optimal if the sweep width were equal to w as a suboptimal alternative to the optimal plan φ^*. The advantage to using this suboptimal plan is its ease of calculation. The question is whether one can choose a value of w to yield a plan φ_w that is close in effectiveness to that of the optimal plan. Here we show that for some gamma distributions one can choose a fixed sweep width plan that is close in mean time to the optimal plan. However, there are situations in which even the best fixed sweep width plan exacts a large penalty in mean time to find the target.

Example 2.2.7 provides the optimal fixed sweep width plan in the case where effort is measured in terms of swept area. In the present section, effort is measured in terms of track length; the plan of Example 2.2.7 may be transformed into this kind of plan simply by dividing by sweep width. Thus, for sweep width w,

$$\varphi_w((r, \theta), t) = \begin{cases} \dfrac{1}{w}\left(H\sqrt{t} - \dfrac{r^2}{2\sigma^2}\right) & \text{for } r^2 \leq 2\sigma^2 H\sqrt{t}, \\ 0 & \text{for } r^2 > 2\sigma^2 H\sqrt{t}, \end{cases}$$

where

$$H = (wv/\pi\sigma^2)^{1/2}.$$

Let b be given by (2.3.9) and $P[\varphi_w(\cdot, t)] = \int_X p(x)b(x, \varphi_w(x, t)) \, dx$. The expected time to detection $\mu(\varphi_w)$ is then found by evaluating the expression

$$\mu(\varphi_w) = \int_0^\infty (1 - P[\varphi_w(\cdot, t)]) \, dt.$$

Actually the calculations required to compute $\mu(\varphi_w)$ are more easily performed if one first calculates $P_y[\varphi_w(\cdot, t)]$, the probability of detecting the target by time t, and $\mu_y(\varphi_w)$, the mean time to detect the target using plan φ_w given $\mathbf{W} = y$. Then

$$\mu(\varphi_w) = \int_0^\infty \mu_y(\varphi_w) G'(y) \, dy, \qquad (2.3.24)$$

where G' is given by (2.3.8). One can calculate that

$$P_y[\varphi_w(\cdot, t)] = \int_X p(x)[1 - e^{-y\varphi_w(x,t)}] \, dx$$

$$= 1 - \frac{1}{(y/w) - 1}\left[\frac{y}{w}\exp(-H\sqrt{t}) - \exp\left(-\frac{y}{w}H\sqrt{t}\right)\right],$$

and

$$\mu_y(\varphi_w) = \int_0^\infty (1 - P_y[\varphi_w(\cdot, t)]) \, dt$$

$$= 2\left(1 + \frac{w}{y} + \left(\frac{w}{y}\right)^2\right)\frac{\pi\sigma^2}{vw}.$$

Performing the integration in (2.3.24), one finds that for $0 < v \leq 2$, $\mu(\varphi_w) = \infty$ and for $v > 2$,

$$\mu(\varphi_w) = \frac{2\pi\sigma^2\alpha}{v}\left[\frac{1}{\alpha w} + \frac{1}{v - 1} + \frac{\alpha w}{(v - 1)(v - 2)}\right]. \qquad (2.3.25)$$

2.3 Optimal Search with Uncertain Sweep Width

By differentiating (2.3.25) with respect to w, one may find the value of w that minimizes the expected time to detection for a fixed sweep width plan. This value of w will be referred to as the *best fixed sweep width* and denoted w^*.

For $v > 2$, one finds that

$$w^* = \frac{1}{\alpha}[(v-1)(v-2)]^{1/2},$$

and

$$\mu(\varphi_w^*) = \frac{2\pi\sigma^2\alpha}{(v-1)v}\left\{1 + 2\left[\frac{(v-1)}{(v-2)}\right]^{1/2}\right\}.$$

Note that $w^* \leq v/\alpha = \mathbf{E}[W]$.

Thus, for $v > 2$,

$$\frac{\mu(\varphi_w^*)}{\mu(\varphi^*)} = \frac{v\{1 + 2[(v-1)/(v-2)]^{1/2}\}}{3v+1}.$$

This ratio is shown in Fig. 2.3.2 as a function of v.

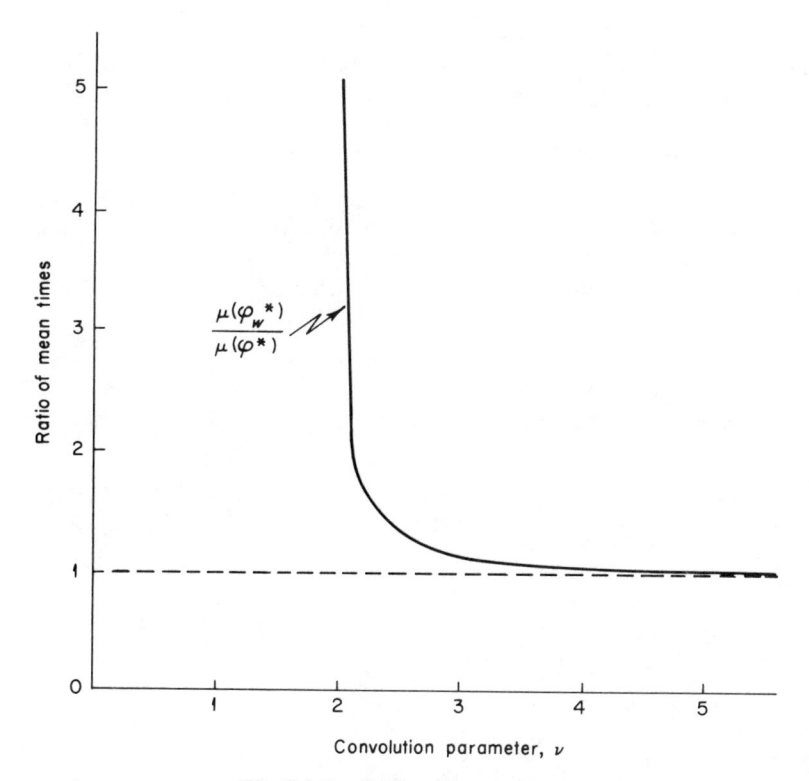

Fig. 2.3.2. Ratio of mean times.

Note that for small values of ν, use of the optimal plan φ^* provides substantial reductions in expected time to detection. The reductions become less significant as ν increases. This makes sense intuitively since, as one can see from Fig. 2.3.1, increasing ν corresponds to the increasing concentration of the gamma probability distribution about its mean value. In analytical terms this is apparent from the fact that the ratio of the standard deviation $\sqrt{\nu}/\alpha$ to the mean ν/α equals $1/\sqrt{\nu}$, which approaches 0 as $\nu \to \infty$.

*2.4. UNIFORMLY OPTIMAL SEARCH PLANS FOR THE GENERAL CASE

Theorems 2.4.3 and 2.4.4 extend Theorems 2.2.4 and 2.2.5, which give methods of computing uniformly optimal plans, to more general cost and payoff functions. Theorem 2.4.6 is concerned with the case where $c(x, z) = z$ for $z \geq 0$, $x \in X$, but b is any detection function such that $b(x, \cdot)$ is increasing and right continuous for $x \in X$. The theorem shows that a uniformly optimal search plan exists under these conditions. However, Example 2.2.9 shows that the conditions of Theorem 2.4.6 applied to a discrete search space no longer guarantee the existence of a uniformly optimal search plan. If, however, $b(j, \cdot)$ is concave and continuous for $j \in J$, then Theorem 2.4.10. shows that uniformly optimal search plans do exist for discrete search spaces.

Let g be a real-valued function defined on an interval I of real numbers. Then g is *absolutely continuous* if the derivative $g'(y)$ exists for a.e. $y \in I$ and

$$\int_a^b g'(y)\, dy = g(a) - g(b) \qquad \text{for} \quad a, b \in I.$$

For example, if g' is continuous on I, then g is absolutely continuous (see Rudin, 1953, Theorem 6.19).

We shall use e to denote either a Borel function $e: X \times [0, \infty) \to [0, \infty)$ or a function $e: J \times [0, \infty) \to [0, \infty)$. Define

$$E[f] = \int_X e(x, f(x))\, dx \qquad \text{for} \quad f \in F(X), \tag{2.4.1}$$

and

$$E[f] = \sum_{j \in J} e(j, f(j)) \qquad \text{for} \quad f \in F(J). \tag{2.4.2}$$

We also generalize the definition of ℓ so that

$$\ell(x, \lambda, z) = e(x, z) - \lambda c(x, z) \qquad \text{for} \quad x \in X,\ \lambda \geq 0,\ z \geq 0,$$
$$\ell(j, \lambda, z) = e(j, z) - \lambda c(j, z) \qquad \text{for} \quad j \in J,\ \lambda \geq 0,\ z \geq 0.$$

2.4 Optimal Plans for General Case

When $e(x, z) = p(x)b(x, z)$ for $x \in X$, $z \geq 0$, or $e(j, z) = p(j)b(j, z)$ for $j \in J$, $z \geq 0$, the above definition of ℓ reduces to the one given in Section 2.1.

Computation of Uniformly Optimal Plans

Analogs of Theorems 2.1.2 and 2.1.3 are now stated.

Theorem 2.4.1. *Suppose there is a finite $\lambda \geq 0$ and an $f_\lambda^* \in F(X)$ such that $C[f_\lambda^*] < \infty$ and*

$$\ell(x, \lambda, f_\lambda^*(x)) = \max\{\ell(x, \lambda, z) : z \geq 0\} \quad \text{for a.e.} \quad x \in X. \quad (2.4.3)$$

Then

$$E[f_\lambda^*] = \max\{E[f_\lambda^*] : C[f] \leq C[f_\lambda^*]\}. \quad (2.4.4)$$

Proof. If $E[f_\lambda^*] = \infty$, then (2.4.4) holds trivially. If $E[f_\lambda^*] < \infty$, the proof follows in the same manner as that of Theorem 2.1.3.

Theorem 2.4.2. *Suppose there is a finite $\lambda \geq 0$ and an $f_\lambda^* \in F(J)$ such that $C[f_\lambda^*] < \infty$ and*

$$\ell(j, \lambda, f_\lambda^*(j)) = \max\{\ell(j, \lambda, z) : z \geq 0\} \quad \text{for} \quad j \in J. \quad (2.4.5)$$

Then

$$E[f_\lambda^*] = \max\{E[f_\lambda^*] : C[f] \leq C[f_\lambda^*]\}. \quad (2.4.6)$$

Proof. After treating the case $E[f_\lambda^*] = \infty$ separately, the proof follows that of Theorem 2.1.2

Let us extend the definition of ρ that is given in Section 2.2. Let $e'(x, \cdot)$ denote the derivative of $e(x, \cdot)$ and define a rate of return function

$$\rho(x, z) = e'(x, z)/c'(x, z) \quad \text{for} \quad x \in X, \quad z \geq 0.$$

Suppose $\rho_x = \rho(x, \cdot)$ is continuous and strictly decreasing. Then letting $\bar{\rho}_x = \lim_{z \to \infty} \rho_x(z)$ for $x \in X$, we may define $\rho_x^{-1}(\lambda)$ for $\bar{\rho}_x < \lambda \leq \rho_x(0)$. For $\lambda > \rho_x(0)$ we take $\rho_x^{-1}(\lambda) = 0$, and for $0 < \lambda \leq \bar{\rho}_x$, we let $\rho_x^{-1}(\lambda) = \infty$. Define

$$\tilde{U}(\lambda) = \int_X c(x, \rho_x^{-1}(\lambda))\, dx \quad \text{for} \quad \lambda > 0.$$

For later use, define

$$\text{ess sup } \bar{\rho} = \inf\{k : k \geq \bar{\rho}_x \text{ for a.e. } x \in X\}.$$

If D is a Borel function $D: X \to [0, \infty)$ such that for $x \in X$, $e(x, z) \leq D(x)$ for $z \geq 0$, and $\int_X D(x)\, dx < \infty$, then we say that e is *integrably bounded by D*. Analogous definitions are understood to hold in the case where the search space is J, although the analog of integrably bounded will be called *summably bounded*.

The notion of a uniformly optimal search plan needs to be extended. A Borel function $\varphi: X \times [0, \infty) \to [0, \infty)$ is called an *allocation plan* if

(i) $\varphi(\cdot, t)$ is an allocation on X for $t \geq 0$, and
(ii) $\varphi(x, \cdot)$ is increasing for $x \in X$.

Let $M: [0, \infty) \to [0, \infty)$ be a cumulative effort function. Define $\tilde{\Phi}(M)$ to be the class of allocation plans such that

$$C[\varphi(\cdot, t)] = \int_X c(x, \varphi(x, t))\, dx = M(t) \quad \text{for } t \geq 0.$$

An allocation plan $\varphi^* \in \tilde{\Phi}(M)$ is said to be *uniformly optimal* if

$$E[\varphi^*(\cdot, t)] = \max\{E[\varphi(\cdot, t)]: \varphi \in \tilde{\Phi}(M)\} \quad \text{for } t \geq 0.$$

Theorem 2.4.3. *Let $e(x, \cdot)$ and $c(x, \cdot)$ be absolutely continuous. Suppose that e is integrably bounded by D and that $e'(x, \cdot)$ and $c'(x, \cdot)$ are positive functions. For $x \in X$, let*

$$\rho_x(z) = e'(x, z)/c'(x, z) \quad \text{for } z \geq 0,$$

$$\rho_x^{-1}(\lambda) = \begin{cases} \infty & \text{for } 0 < \lambda \leq \bar{\rho}_x \equiv \lim_{z \to \infty} \rho(x, z), \\ \text{inverse of } \rho_x \text{ evaluated at } \lambda & \text{for } \bar{\rho}_x < \lambda \leq \rho_x(0), \\ 0 & \text{for } \lambda > \rho_x(0), \end{cases}$$

and let

$$\tilde{U}(\lambda) = \int_X c(x, \rho_x^{-1}(\lambda))\, dx \quad \text{for } \lambda > 0.$$

Assume ρ_x is continuous and strictly decreasing for $x \in X$ and that ess sup $\bar{\rho} < \infty$. If M is a cumulative effort function, then φ^* given by

$$\varphi^*(x, t) = \rho_x^{-1}(\tilde{U}^{-1}(M(t))) \quad \text{for } x \in X, \ t \geq 0,$$

is uniformly optimal in $\tilde{\Phi}(M)$.

Proof. For ess sup $\bar{\rho} < \lambda < \infty$, f_λ^* defined by

$$f_\lambda^*(x) = \rho_x^{-1}(\lambda) \quad \text{for } x \in X$$

satisfies (2.4.3). Thus by Theorem 2.4.1

$$E[f^*] = \max\{E[f]: f \in F(X) \text{ and } C[f] \leq C[f_\lambda^*]\}.$$

The main step in finishing the proof is to show that for λ in the interval (ess sup $\bar{\rho}, \infty$), $\tilde{U}$ takes all values between 0 and ∞ and that it has an inverse function defined on $[0, \infty)$.

2.4 Optimal Plans for General Case

Since ρ_x is decreasing,

$$\lambda = \rho_x(\rho_x^{-1}(\lambda)) \le e'(x, z)/c'(x, z) \quad \text{for} \quad 0 \le z \le \rho_x^{-1}(\lambda),$$

and we have

$$D(x) \ge e(x, \rho_x^{-1}(\lambda)) = \int_0^{\rho_x^{-1}(\lambda)} e'(x, z)\, dz \ge \lambda \int_0^{\rho_x^{-1}(\lambda)} c'(x, z)\, dz$$
$$= \lambda c(x, \rho_x^{-1}(\lambda)).$$

Thus

$$c(x, \rho_x^{-1}(\lambda)) \le D(x)/\lambda \quad \text{for} \quad x \in X,$$

and

$$\tilde{U}(\lambda) = \int_X c(x, \rho_x^{-1}(\lambda))\, dx \le \frac{1}{\lambda} \int_X D(x)\, dx.$$

Since $\int_X D(x)\, dx < \infty$, one may apply the dominated convergence theorem and the continuity of ρ_x^{-1} to show that $\tilde{U}$ is continuous.

Observe that ρ_x^{-1} is strictly decreasing for $\bar{\rho}_x < \lambda \le \rho_x(0)$ and that $\lim_{\lambda \downarrow \bar{\rho}_x} \rho_x^{-1}(\lambda) = \infty$ and $\lim_{\lambda \uparrow \rho_x(0)} \rho_x^{-1}(\lambda) = 0$. Thus U is strictly decreasing for $\lambda_\ell \le \lambda \le \lambda_u$, where

$$\lambda_\ell = \sup\{\lambda : U(\lambda) = \infty\}, \quad \lambda_u = \inf\{\lambda : U(\lambda) = 0\}.$$

Furthermore $\lim_{\lambda \to \lambda_\ell} U(\lambda) = \infty$ and $\lim_{\lambda \to \lambda_u} U(\lambda) = 0$. Thus one may define $\tilde{U}^{-1}$ on $[0, \infty)$, and $C[\varphi^*(\cdot, t)] = M(t)$.

Since $\rho_x^{-1} \ge 0$ and $\gamma = U^{-1}(M)$ is a decreasing function, $\varphi^*(x, \cdot)$ is increasing for $x \in X$, and $\varphi^*(\cdot, t)$ is an allocation on X for $t \ge 0$. Thus $\varphi^* \in \tilde{\Phi}(M)$ and the theorem is proved.

Theorem 2.4.4 gives the analog of Theorem 2.4.3 for discrete search space.

Theorem 2.4.4. *Let $e(j, \cdot)$ and $c(j, \cdot)$ be absolutely continuous. Suppose that e is summably bounded by D and that $e'(x, \cdot)$ and $c'(x, \cdot)$ are positive functions. For $j \in J$, let*

$$p_j(z) = e'(j, z)/c'(j, z) \quad \text{for} \quad z \ge 0,$$

$$p_j^{-1}(\lambda) = \begin{cases} \infty & \text{for} \quad 0 < \lambda \le \bar{p}_j \equiv \lim_{z \to \infty} p_j(z), \\ \text{inverse of } p_j \text{ evaluated at } \lambda & \text{for} \quad \bar{p}_j < \lambda \le p_j(0), \\ 0 & \text{for} \quad \lambda > p_j(0). \end{cases}$$

and let

$$\tilde{U}(\lambda) = \sum_{j \in J} c(j, p_j^{-1}(\lambda)) \quad \text{for} \quad \lambda > 0.$$

Assume that ρ_j is continuous and strictly decreasing for $j \in J$ and that $\sup_{j \in J} \bar{\rho}_j < \infty$. If M is a cumulative effort function, then φ^* given by

$$\varphi^*(j, t) = \rho_j^{-1}(\tilde{U}^{-1}(M(t))) \quad \text{for } j \in J, \ t \geq 0.$$

is uniformly optimal in $\tilde{\Phi}(M)$.

The proof of Theorem 2.4.4 is similar to that of Theorem 2.4.3 and is not given.

Existence of Uniformly Optimal Plans

We now consider the case where $c(x, z) = z$ for $z \geq 0$ and $x \in X$ and show that a uniformly optimal search plan exists whenever $b(x, \cdot)$ is increasing and right continuous for $x \in X$. In order to do this we extend the definition of $b(x, \cdot)$ so that

$$b(x, \infty) = \lim_{z \to \infty} b(x, z) \quad \text{for } x \in X.$$

Define

$$\zeta(x, \lambda) = \max\{\ell(x, \lambda, z) : z \geq 0\} \quad \text{for } x \in X, \ \lambda \geq 0.$$

Since $b(x, \cdot)$ is increasing and right continuous, $\ell(x, \lambda, \cdot)$ is upper semicontinuous on the interval $[0, \infty]$. The interval $[0, \infty]$ is compact in the topology of the extended real numbers so that the upper semicontinuity of $\ell(x, \lambda, \cdot)$ guarantees that the supremum of this function is achieved on $[0, \infty]$. If $\lambda > 0$, then since $b(x, z) \leq 1$ for $z \geq 0$, the maximum does not occur at $z = \infty$.

Let

$$g(x, \lambda) = \sup\{z : z \geq 0 \text{ and } \ell(x, \lambda, z) = \zeta(x, \lambda)\} \quad \text{for } x \in X, \ \lambda > 0.$$

Note that $g(x, \lambda) < \infty$ for $x \in X$, $\lambda > 0$. Let

$$g_\ell(x, \lambda) = \lim_{\lambda' \downarrow \lambda} g(x, \lambda') \quad \text{for } x \in X, \ \lambda > 0,$$

and define

$$I(\lambda) = \int_X g(x, \lambda) \, dx \quad \text{and} \quad I_\ell(\lambda) = \int_X g_\ell(x, \lambda) \, dx \quad \text{for } \lambda > 0.$$

In order to prove the existence of uniformly optimal plans the following lemma is used.

Lemma 2.4.5. *Suppose that b is a detection function on X such that $b(x, \cdot)$ is increasing and right continuous for $x \in X$. Then the following hold:*

(a) $g(\cdot, \lambda) \in F(X)$ and $I(\lambda) < 1/\lambda$ for $\lambda > 0$;
(b) $g(x, \cdot)$ is decreasing for $x \in X$ and I is decreasing;

2.4 Optimal Plans for General Case

(c) $g(x, \cdot)$ is left continuous for $x \in X$ and I is left continuous;
(d) For $\lambda > 0$, there exists $h: X \times [I_\ell(\lambda), I(\lambda)] \to [0, \infty)$ such that
   (1) $h(x, \cdot)$ is increasing for $x \in X$,
   (2) for $t \in [I_\ell(\lambda), I(\lambda)]$, $C[h(\cdot, t)] = t$, $h(x, t) \leq g(x, \lambda)$ for $x \in X$, and
   (3) $\ell(x, h(x, t), \lambda) = \zeta(x, \lambda)$ for $x \in X$.

Proof. Since $b(x, \cdot)$ is increasing,

$$\ell(x, \lambda, g(x, \lambda)) = \zeta(x, \lambda) \quad \text{for} \quad x \in X, \quad \lambda \geq 0. \tag{2.4.7}$$

Define $\Pi(x, z) = x$ for $x \in X$ and $z \geq 0$. Take $\lambda > 0$, and let a be a real number. By 2.2.13 of Federer (1969),

$$\{x : \zeta(x, \lambda) > a\} = \Pi\{(x, z) : \ell(x, \lambda, z) > a\}$$

is Lebesgue measurable, and by 2.3.6 of Federer (1969) $\zeta(\cdot, \lambda)$ is a.e. equal to a Borel function $\tilde{\zeta}(\cdot, \lambda)$. Similarly,

$$\{x : g(x, \lambda) > a\} = \Pi\{(x, z) : \ell(x, \lambda, z) = \tilde{\zeta}(x, \lambda) \text{ and } z > a\}$$

is Lebesgue measurable so that $g(\cdot, \lambda)$ is a.e. equal to a Borel function. Thus we may take $g(\cdot, \lambda)$ to be Borel, and it follows that $g(\cdot, \lambda) \in F(X)$.

Since $b(x, \cdot)$ is increasing,

$$p(x)b(x, \infty) - p(x)b(x, 0) \geq p(x)b(x, g(x, \lambda)) - p(x)b(x, 0)$$
$$\geq \lambda g(x, \lambda) \quad \text{for} \quad x \in X,$$

where the last inequality follows from (2.4.7). Thus

$$g(x, \lambda) \leq \frac{p(x)}{\lambda}[b(x, \infty) - b(x, 0)] \leq \frac{p(x)}{\lambda} \quad \text{for} \quad x \in X,$$

and

$$I(\lambda) = \int_X g(x, \lambda)\,dx \leq \frac{1}{\lambda} \quad \text{for} \quad \lambda > 0.$$

This proves (a).

Suppose $0 < \lambda_1 < \lambda_2 < \infty$. Fix $x \in X$ and let $y_1 = g(x_1, \lambda_1)$ and $y_2 = g(x_2, \lambda_2)$. Then

$$\ell(x, y_1, \lambda_1) \geq \ell(x, y_2, \lambda_1) \quad \text{and} \quad \ell(x, y_2, \lambda_2) \geq \ell(x, y_1, \lambda_2),$$

which implies

$$\lambda_1(y_1 - y_2) \leq p(x)[b(x, y_1) - b(x, y_2)] \leq \lambda_2(y_1 - y_2). \tag{2.4.8}$$

Since $\lambda_1 < \lambda_2$, we must have $y_1 \geq y_2$, or (2.4.8) would yield a contradiction. Thus $g(x, \cdot)$ is a decreasing function for $x \in X$, and I is decreasing also. This proves (b).

To prove (c), we first show that $\zeta(x, \cdot)$ is continuous for $x \in X$. Choose $0 < \lambda_1 < \lambda_2 < \infty$ and fix $x \in X$. Then $g(x, \lambda_1) \geq g(x, \lambda_2)$. Suppose $\zeta(x, \lambda_1) \geq \zeta(x, \lambda_2)$. Since

$$\ell(x, \lambda_2, g(x, \lambda_1)) \leq \ell(x, \lambda_2, g(x, \lambda_2)) = \zeta(x, \lambda_2),$$

it follows that

$$|\zeta(x, \lambda_1) - \zeta(x, \lambda_2)| \leq |\ell(x, \lambda_1, g(x, \lambda_1)) - \ell(x, \lambda_2, g(x, \lambda_1))|$$
$$\leq (\lambda_2 - \lambda_1)g(x, \lambda_1).$$

Since $g(x, \lambda_1) \geq g(x, \lambda_2)$ a similar argument shows that this same inequality holds for the case of $\zeta(x, \lambda_1) < \zeta(x, \lambda_2)$. In fact, for $0 < \lambda_1 \leq \lambda_1' \leq \lambda_2'$,

$$|\zeta(x, \lambda_1') - \zeta(x, \lambda_2')| \leq |\lambda_1' - \lambda_2'|g(x, \lambda_1). \tag{2.4.9}$$

Since $|g(x, \lambda)| < \infty$ for $0 < \lambda < \infty$, and $x \in X$ is arbitrary, (2.4.9) implies that $\zeta(x, \cdot)$ is continuous for $x \in X$.

To show that $g(x, \cdot)$ is left continuous, let $\lambda_i \uparrow \lambda_0$ and define $y_i = g(x, \lambda_i)$ for $i = 0, 1, \ldots$. Then $\{y_i\}_{i=1}^{\infty}$ is a decreasing sequence with a limit $z \geq y_0$. By the continuity of $\zeta(x, \cdot)$,

$$\lim_{i \to \infty} \ell(x, \lambda_i, y_i) = \lim_{i \to \infty} \zeta(x, \lambda_i) = \zeta(x, \lambda_0) = \ell(x, \lambda_0, y_0).$$

However, $b(x, \cdot)$ is right continuous, so

$$\ell(x, \lambda_0, y_0) = \lim_{i \to \infty} \ell(x, \lambda_i, y_i) = \ell(x, \lambda_0, z).$$

Hence, by the definition of $y_0 = g(x, \lambda_0)$, $z \leq y_0$. Since we noted above that $z \geq y_0$, it follows that $z = y_0$ and that $g(x, \cdot)$ is left continuous for $x \in X$. The monotone convergence theorem may be used to show that I is left continuous. This proves (c).

We claim that

$$\ell(x, \lambda, g_\ell(x, \lambda)) = \zeta(x, \lambda) \quad \text{for} \quad x \in X, \quad \lambda > 0. \tag{2.4.10}$$

Recall that $g_\ell(x, \lambda) = \lim_{\lambda' \downarrow \lambda} g(x, \lambda')$. Since $b(x, \cdot)$ is increasing and $g(x, \cdot)$ is decreasing

$$\ell(x, \lambda, g_\ell(x, \lambda)) \geq \lim_{\lambda' \downarrow \lambda} \ell(x, \lambda', g(x, \lambda')) = \lim_{\lambda' \downarrow \lambda} \zeta(x, \lambda') = \zeta(x, \lambda).$$

Since $\ell(x, \lambda, g_\ell(x, \lambda)) \leq \zeta(x, \lambda)$ by definition, (2.4.10) holds.

To prove (d) we fix $\lambda > 0$ and define

$$u_r(x) = \begin{cases} g(x, \lambda) & \text{for} \quad \|x\| \leq r \\ g_\ell(x, \lambda) & \text{for} \quad \|x\| > r \end{cases} \quad \text{for} \quad x \in X, \quad r \geq 0,$$

2.4 Optimal Plans for General Case

where $\|x\|$ denotes the Euclidean norm of x in n-space. Since $g_\ell(x, \lambda) \le g(x, \lambda)$ for $x \in X$, one may use the monotone convergence theorem to show that $\hat{I}$ defined by

$$\hat{I}(r) = \int_X u_r(x)\, dx \quad \text{for} \quad r \ge 0$$

is continuous. It is easy to see that $\hat{I}$ increases from $I_\ell(\lambda)$ to $I(\lambda)$ and that the existence of the function h in (d) is assured. This proves the lemma.

Theorem 2.4.6. *Suppose that b is a detection function on X such that $b(x, \cdot)$ is increasing and right continuous for $x \in X$. Let $c(x, z) = z$ for $x \in X$, $z \ge 0$, and let M be a cumulative effort function. Then there exists a search plan φ^* that is uniformly optimal within $\Phi(M)$.*

Proof. We prove the theorem for the case $M(t) = t$ for $t \ge 0$. The extension to arbitrary increasing M is obvious.

From (a) of Lemma 2.4.5, we have $\lim_{\lambda \to \infty} I(\lambda) = 0$. Let $\bar{I} = \lim_{\lambda \to 0} I(\lambda)$. We consider first the case where $\bar{I} = \infty$. Since I is monotone it has only a countable number of discontinuities. Let N be a countable index set such that $\{\lambda_n : n \in N\}$ is the set of discontinuity points of I. Let $L_n = [I_\ell(\lambda_n), I(\lambda_n)]$ for $n \in N$. Then the intervals L_n are disjoint and are the jump intervals at the discontinuity points of I. For $t \in (0, \infty) - \bigcup_{n \in N} L_n$, let

$$\hat{\gamma}(t) = \sup\{\lambda : I(\lambda) = t\}.$$

By the left continuity of I, $I(\hat{\gamma}(t)) = t$. For $t \in L_n$, let $\hat{\gamma}(t) = \lambda_n$. By (d) of Lemma 2.4.5, we may find a function h_n defined on $X \times L_n$ such that $h_n(x, \cdot)$ is increasing for $x \in X$, and for $t \in L_n$

$$C[h_n(\cdot, t)] = t,$$

$h_n(x, t) \le g(x, \lambda_n)$ and $\ell(x, \lambda_n, h_n(x, t)) = \zeta(x, \lambda_n)$ for $x \in X$.

Now define $\varphi^*(x, 0) = 0$ for $x \in X$ and

$$\varphi^*(x, t) = \begin{cases} g(x, \hat{\gamma}(t)) & \text{if } t \in (0, \infty) - \bigcup_{n \in N} L_n, \\ h_n(x, t) & \text{if } t \in L_n \text{ for some } n \in N. \end{cases}$$

Then $C[\varphi^*(\cdot, t)] = t$ for $t \ge 0$, and for $t > 0$, $\lambda = \hat{\gamma}(t)$ and $f_\lambda^* = \varphi^*(\cdot, t)$ satisfy (2.1.9), so by Theorem 2.1.3

$$P[\varphi^*(\cdot, t)] = \max\{P[\varphi(\cdot, t)] : \varphi \in \Phi(M)\} \quad \text{for} \quad t \ge 0.$$

Thus to show that φ^* is uniformly optimal within $\Phi(M)$, we need to show only that $\varphi^* \in \Phi(M)$. Clearly $\varphi^*(\cdot, t) \in F(X)$, and we have noted that $C[\varphi^*(\cdot, t)] = t = M(t)$ for $t \ge 0$. Thus it remains only to show that $\varphi(x, \cdot)$ is increasing for a.e. $x \in X$.

Since I is decreasing, $\hat{\gamma}$ is decreasing.

Let $t > s > 0$. If $\hat{\gamma}(t) < \hat{\gamma}(s)$, then for $x \in X$, $\varphi^*(x, t) \geq g_\ell(x, \hat{\gamma}(t)) \geq g(x, \hat{\gamma}(s)) \geq \varphi^*(x, s)$. If $\hat{\gamma}(t) = \hat{\gamma}(s)$, then t and s are both in the same L_n for some $n \in N$, and $\varphi(x, t) \geq \varphi(x, s)$ for $x \in X$ by construction. This proves the theorem for the case where $\bar{I} = \infty$.

If $\bar{I} < \infty$, then for a.e. $x \in X$, $\lim_{\lambda \to 0} g(x, \lambda) < \infty$. Let

$$g(x, 0) = \lim_{\lambda \to 0} g(x, \lambda) \quad \text{for} \quad x \in X.$$

Then $b(x, g(x, 0)) = \max\{b(x, z) : z \geq 0\}$. Thus for $0 \leq t < \bar{I}$, one may define φ^* as above. For $t \geq \bar{I}$, consider the case where X is 2-space, and let

$$\varphi^*(x, t) = \begin{cases} g(x, 0) + 1 & \text{for} \quad \|x\| \leq \left(\dfrac{t - \bar{I}}{\pi}\right)^{1/2}, \\ g(x, 0) & \text{for} \quad \|x\| > \left(\dfrac{t - \bar{I}}{\pi}\right)^{1/2} \end{cases}$$

Then $C[\varphi^*(\cdot, t)] = t$ for $t \geq \bar{I}$. Since $b(x, g(x, 0) + 1) = b(x, g(x, 0))$ by the increasing nature of $b(x, \cdot)$ for $x \in X$, it follows that φ^* is uniformly optimal in $\Phi(M)$. A similar device may be employed for general n-space, and the theorem is proved.

Remark 2.4.7. Example 2.2.9 shows that Theorem 2.4.6 is not true when the search space X is replaced by the discrete space J. However, one can show that uniformly optimal plans exist for searches involving discrete search spaces when $b(j, \cdot)$ is concave and continuous for $j \in J$. In Stone (1972a) a generalization of Theorem 2.4.6 is given for more general search spaces and for cost functions $c(x, \cdot)$ that are continuous and increasing for $x \in X$.

However, the assumption in Theorem 2.4.6 that $b(x, \cdot)$ is right continuous for $x \in X$ cannot be removed, as the following example shows.

Example 2.4.8. In this example we show that if $b(x, \cdot)$ is not right continuous, there may be no allocation that is optimal for a cost K. Suppose X is the real line. For $x \in X$, let

$$p(x) = \begin{cases} 1 & \text{for } 0 \leq x \leq 1, \\ 0 & \text{otherwise,} \end{cases}$$

$$c(x, z) = z \quad \text{for} \quad z \geq 0,$$

$$b(x, z) = \begin{cases} \tfrac{1}{2}z & \text{for } 0 \leq z \leq 1, \\ 1 & \text{for } z > 1. \end{cases}$$

2.4 Optimal Plans for General Case

For $\varepsilon > 0$, let

$$f_\varepsilon(x) = \begin{cases} 1 + \varepsilon & \text{for } 0 \le x \le \dfrac{1}{1+\varepsilon}, \\ 0 & \text{otherwise.} \end{cases}$$

We see that

$$C[f_\varepsilon] = 1, \qquad P[f_\varepsilon] = \frac{1}{1+\varepsilon} \qquad \text{for } \varepsilon > 0.$$

Thus, one may find an allocation with cost 1 and probability of detection arbitrarily close to 1. However, there is no f^* such that $C[f^*] = 1$ and $P[f^*] = 1$. Thus, there is no allocation that is optimal for cost $K = 1$.

We finish this section by showing that uniformly optimal search plans exist for discrete spaces when the detection function is concave and continuous.

Let h be a real-valued concave function defined on an interval L of real numbers. It is shown in Hardy *et al.* (1964) that if h is continuous, then the derivative $h'(z)$ exists for a.e. $z \in L$ and is decreasing.

Let $b(j, \cdot)$ be concave, continuous, and increasing for $j \in J$. Define

$$\begin{aligned} g(j, \lambda) &= \sup\{z : z = 0 \text{ or } b'(j, z) \ge \lambda\} & \text{for } j \in J,\ \lambda > 0, \\ g_\ell(j, \lambda) &= \inf\{z : b'(j, z) \le \lambda\} & \text{for } j \in J,\ \lambda > 0. \end{aligned} \qquad (2.4.11)$$

Let

$$\zeta(j, \lambda) = \max\{\ell(j, \lambda, z) : z \ge 0\} \qquad \text{for } j \in J,\ \lambda > 0.$$

Then one may verify that

$$\begin{aligned} g(j, \lambda) &= \sup\{z : \ell(j, \lambda, z) = \zeta(j, \lambda)\}, \\ g_\ell(j, \lambda) &= \lim_{\lambda' \downarrow \lambda} g(j, \lambda') \qquad \text{for } j \in J, \qquad \lambda > 0. \end{aligned}$$

Let

$$I(\lambda) = \sum_{j \in J} g(j, \lambda) \qquad \text{for } \lambda > 0, \qquad (2.4.12)$$

and

$$I_\ell(\lambda) = \sum_{J \in j} g_\ell(j, \lambda) \qquad \text{for } \lambda > 0.$$

The analogs of properties (a)–(c) of Lemma 2.4.5 may be shown to hold for g and I as defined in (2.4.11) and (2.4.12) by the same methods as given in the proof of Lemma 2.4.5. However, property (d) requires a somewhat different treatment.

Lemma 2.4.9. *Suppose b is a detection function on J such that $b(j, \cdot)$ is concave, continuous, and increasing for $j \in J$. Then for $\lambda > 0$, there exists an $h: J \times [I_\ell(\lambda), I(\lambda)] \to [0, \infty)$ such that*

(1) $h(j, \cdot)$ *is continuous and increasing for $j \in J$;*
(2) *for $t \in [I_\ell(\lambda), I(\lambda)]$, $C[h(\cdot, t)] = t$, $h(j, t) \leq g(x, \lambda)$ for $x \in X$; and*
(3) $\ell(j, \lambda, h(j, t)) = \zeta(j, \lambda)$ *for $j \in J$.*

Proof. Fix $\lambda > 0$. Let $j_0 \in J$. For any y such that $g_\ell(j_0, \lambda) \leq y \leq g(j_0, \lambda)$, the decreasing nature of $b'(j_0, \cdot)$ yields

$$\ell'(j_0, \lambda, z) \geq \lambda \quad \text{for} \quad 0 < z < y,$$
$$\leq \lambda \quad \text{for} \quad y < z < \infty.$$

Thus $\ell(j_0, \lambda, y) = \zeta(j_0, \lambda)$.

For $0 \leq \theta \leq 1$, let

$$u_\theta(j) = (1 - \theta)g_\ell(j, \lambda) + \theta g(j, \lambda) \quad \text{for} \quad j \in J.$$

Let $\hat{I}$ be defined by

$$\hat{I}(\theta) = \sum_{j \in J} u_\theta(j) \quad \text{for} \quad 0 \leq \theta \leq 1.$$

By the monotone convergence theorem $\hat{I}$ is continuous. In addition, $\hat{I}(0) = I_\ell(\lambda)$ and $\hat{I}(1) = I(\lambda)$. The existence of the function h is now assured and the lemma is proved.

Theorem 2.4.10. *Suppose b is a detection function J such that $b(j, \cdot)$ is continuous, concave, and increasing for $j \in J$. Let $c(j, z) = z$ for $j \in J$, $z \geq 0$, and let M be a cumulative effort function. Then there is a search plan φ^* that is uniformly optimal within $\Phi(M)$.*

Proof. The proof follows that of Theorem 2.4.6.

Observe that in order to obtain the counterpart of property (d) of Lemma 2.4.5 for a discrete search space it is necessary to assume concavity of $b(j, \cdot)$ for $j \in J$, which guarantees that any $y \in [g_\ell(j, \hat{\lambda}), g(j, \lambda)]$ maximizes $\ell(j, \lambda, \cdot)$. This in turn enables us to obtain the function h of Lemma 2.4.9. In the case of the continuous search space, concavity of $b(x, \cdot)$ for $x \in X$ is not required. This is because property (d) of Lemma 2.4.5 can be obtained by the use of a collection of sets $\{S_\alpha : \alpha \in [0, 1]\}$ and a measure ν on X with the following properties:

(i) $S_0 = \varphi$, $S_1 = X$, and $\alpha < \beta$ implies $S_\alpha \subset S_\beta$,
(ii) $\nu(S_\alpha) < \infty$ for $0 \leq \alpha \leq 1$,
(iii) $\lim_{\alpha \to \beta} \nu(S_\alpha) = \nu(S_\beta)$ for $\beta \in [0, 1]$, and
(iv) $\lim_{\alpha \uparrow \beta} S_\alpha = S_\beta$ for $\beta \in [0, 1]$.

The existence of such a collection of sets is equivalent to assuming that ν is a nonatomic and σ-finite measure on X. [See Stone (1972a) and Halkin (1964).] In the case of Lemma 2.4.5, ν is Lebesgue measure on n-dimensional Euclidean space, and the sets S_α are given by

$$S_\alpha = \left\{ x : \|x\| < \frac{\alpha}{1 - \alpha} \right\} \quad \text{for} \quad 0 \leq \alpha \leq 1.$$

NOTES

The use of Lagrange multipliers to solve constrained extremal problems without differentiability assumptions has a long history. In statistics, there is the Neyman–Pearson lemma (Neyman and Pearson, 1933) and its generalizations as given, for example, by Lehmann (1959), Dantzig and Wald (1951), and Wagner (1969). In nonlinear programming, the earliest such use appears to be by Slater (1950). The articles by Everett (1963) and Zahl (1963) were the first in the operations research literature to point out this use of Lagrange multipliers. The approach used in Section 2.1 resembles that of Everett. The necessary conditions in Theorems 2.1.4 and 2.1.5 are taken from Wagner and Stone (1974), whose proof is essentially an application of a proof of Aumann and Perles (1965) to the more general situation considered by Wagner and Stone (1974). Necessity results corresponding to the Neyman–Pearson lemma are given by Dantzig and Wald (1951).

The terminology "uniformly optimal" was introduced by Arkin (1964a, b). The optimal allocation of effort for an exponential detection function and circular normal target distribution given in Example 2.2.1 first appeared in the work of Koopman (1946). In a later work Koopman (1957) gave a method for finding optimal allocations for more general target distributions and exponential detection functions. DeGuenin (1961) essentially identified the class of regular detection functions and for these detection functions found a necessary condition for an optimal allocation involving the function ρ of Section 2.2. He also suggested a method for finding optimal allocations that is similar to that in Theorem 2.2.4. Versions of Theorem 2.2.4 were obtained by Karlin (1959) and by Wagner (1969, Example 3) for special one-dimensional target distributions. The present version of Theorem 2.2.4 first appeared as a special case of Theorem 2 of Stone and Stanshine (1971). The algorithm in Example 2.2.8 is an adaptation of the one given by Charnes and Cooper (1958).

The problem of computing detection probabilities when sweep width or sensor capability is uncertain is discussed by Koopman (1956b). The analysis in Section 2.3 is based on the work of Richardson and Belkin (1972). This analysis was motivated by problems encountered during the search for the submarine Scorpion because of the large uncertainties concerning the detection capabilities of the sensors employed [see Richardson and Stone (1971)]. The optimum allocation and mean time to detection $\mu(m^*)$ in Example 2.3.5 were first obtained in an unpublished manuscript (Stone, 1969), and the best fixed sweep width plan and its mean time $\mu(\varphi_w^*)$ were obtained by Stone and Rosenberg (1968).

The proofs of Lemma 2.4.5 and Theorem 2.4.6 are taken from Stone (1972a). The result in Theorem 2.4.6 was first proved by Stone (1973a).

There are many interesting variations on the basic search problem that are not considered in this book. One might investigate the problem of optimal search when there is a cost for switching search effort from one cell to another as Gilbert (1959), Kisi (1966), and Onaga (1971) have done. Searches where one has the option of switching from one sensor to another but where both cannot be used simultaneously are considered by Persinger (1973).

When the target is located along the line and the searcher is constrained to move along this line, one is faced with the linear search problem. This problem has been investigated by Beck (1964, 1965), Beck and Newman (1970), Beck and Warren (1973), Franck (1965), and Fristedt and Heath (1974) when the target is stationary.

Search for a particle that is random walking on the integers is considered by McCabe (1974). In particular, the particle's position is described by a simple Bernoulli random walk $\{X_n, n = 1, 2, \ldots\}$ with the particle's position at time 1 being given by a specified distribution. A search plan ξ consists of a sequence of integers $(\xi_1, \xi_2, \ldots)$ such that

$$|\xi_{n+1} - \xi_n| \leq 1 \quad \text{for} \quad n = 1, 2, \ldots, \qquad (*)$$
$$\xi_1 = 0,$$

i.e., the searcher can move a distance of at most one in each time unit and he must start at the origin. Let τ_ξ be the first time that the searcher and the particle either meet or cross paths, i.e.,

$$\tau_\xi = \begin{cases} 1 & \text{if } \xi_1 = X_1, \\ \min_{n \geq 2}\{n : (X_n - \xi_n)(X_{n-1} - \xi_{n-1}) \leq 0\}, \\ +\infty & \text{otherwise.} \end{cases}$$

McCabe shows that there exists a plan ξ such that $E[\tau_\xi] < \infty$ provided that the particle's position at time 1 is within a finite interval with probability 1. An argument communicated by B. Fristedt and D. Heath shows that an optimal search plan exists, i.e., a plan that minimizes $E[\tau_\xi]$ among plans satisfying relation $(*)$. An outline of the argument proceeds as follows: Let Ξ^* be the class of plans satisfying $(*)$, and give Ξ^* the product topology induced

Notes

by pointwise convergence, i.e., $\xi^k \to \xi$ iff for each n, $\xi_n^k = \xi_n$ for all k sufficiently large (k may depend on n). Then Ξ^* is compact in this topology.

Let $I(\xi, n)$ be the indicator function of the set $\{\tau_\xi \geq n\}$. By the Fatou–Lebesgue theorem,

$$E[\tau_\xi] = E\left[\sum_{n=1}^\infty I(\xi, n)\right] = E\left[\sum_{n=1}^\infty \liminf_{k \to \infty} I(\xi^k, n)\right]$$
$$\leq \liminf_{k \to \infty} E[\tau_{\xi^k}]$$

for any sequence $\xi^k \to \zeta$ in Ξ^*. Thus, the mapping $\xi \to E[\tau_\xi]$ is lower semicontinuous on Ξ^* and assumes its minimum on this space.

References considering additional variations on the basic search problem may be found in the two published bibliographies on search by Dobbie (1968) and Enslow (1966).

Chapter III

Properties of Optimal Search Plans

In this chapter, the necessary conditions for optimality given in Section 2.1 are used to explore the properties of optimal allocations and uniformly optimal plans. In Section 3.1 an interpretation of Lagrange multipliers is given. Section 3.2 considers plans that at each instant in time place effort in the cell or cells having the highest posterior probability of containing the target. It is shown that such plans are uniformly optimal if and only if the detection function is homogeneous and exponential. Section 3.3 investigates plans generated in an incremental fashion by allocating a number of increments of effort so that each increment yields the maximum increase in probability of detection. For a continuous search space it is shown that these incremental plans produce an optimal allocation of the total effort contained in the increments. Section 3.4 considers extensions of the results on incremental plans to situations with multiple constraints.

3.1. AN INTERPRETATION OF LAGRANGE MULTIPLIERS

For discrete search space this section gives an interpretation of Lagrange multipliers in terms of marginal rates of return. This interpretation is then used to construct the optimal plans of Section 2.2 in an alternate fashion.

3.1 An Interpretation of Lagrange Multipliers

Assume that the search space J is discrete and that b is a detection function on J such that $b'(j, \cdot)$ is continuous and decreasing for $j \in J$. Let $c'(j, \cdot)$ be positive, continuous, and increasing for $j \in J$, and let f^* be optimal for cost $C[f^*]$ in the interior of the range of C. By Corollary 2.1.6, there is $\lambda \geq 0$ such that

$$\begin{aligned} p(j)b'(j,f^*(j)) &= \lambda c'(j,f^*(j)) & \text{for } f^*(j) > 0, \\ p(j)b'(j,f^*(j)) &\leq \lambda c'(j,f^*(j)) & \text{for } f^*(j) = 0. \end{aligned} \quad (3.1.1)$$

Suppose that we add an increment h to $f^*(j)$. The increase in detection probability is

$$p(j)[b(j,f^*(j) + h) - b(j,f^*(j))] \simeq hp(j)b'(j,f^*(j)).$$

The resulting increase in cost is

$$c(j,f^*(j) + h) - c(j,f^*(j)) \simeq hc'(j,f^*(j)).$$

In the limit as $h \to 0$, the incremental or marginal increase in probability of detection divided by the increase in cost becomes

$$p(j)b'(j,f^*(j))/c'(j,f^*(j)).$$

Generalize the definition of ρ given in Section 2.2 so that

$$\rho(j, z) = p(j)b'(j, z)/c'(j, z) \quad \text{for } z \geq 0, \ j \in J. \quad (3.1.2)$$

When $c(j, z) = z$ for $z \geq 0, j \in J$, the above definition reduces to the one given in Section 2.2. Then $\rho(j, z)$ is the ratio of the marginal increase in probability to the increase in cost when z effort has been placed in cell j and a small amount of effort is then added. The ratio $\rho(j, z)$ is called the *marginal rate of return* when z effort has been placed in cell j. Since $c'(j, z) > 0$ for $z \geq 0, j \in J$, (3.1.1) may be written as

$$\begin{aligned} \rho(j,f^*(j)) &= \lambda & \text{for } f^*(j) > 0, \\ \rho(j,f^*(j)) &\leq \lambda & \text{for } f^*(j) = 0. \end{aligned} \quad (3.1.3)$$

Thus, for the allocation f^*, the Lagrange multiplier gives the marginal rate of return in those cells in which effort has been placed.

Suppose that $\rho(j, \cdot)$ is continuous and strictly decreasing. Then λ uniquely determines f^* to satisfy (3.1.3). Thus, one could think of the allocation f^* as being applied as follows: In the jth cell effort is applied until the ratio $\rho(j, z)$ reaches λ. If $\rho(j, 0) \leq \lambda$, then no effort is placed in that cell.

Now consider a situation in which $\rho(j, \cdot)$ is continuous and decreasing and $c'(j, \cdot) > 0$ for $j \in J$. Let $\varphi^\#$ be a search plan that develops by adding the next "small increment" of effort to the cell or cells having the highest value of $\rho(j, \varphi^\#(j, t))$ at each time t, that is, the search is maximizing the increase in probability of detection resulting from each small increment of effort. Is such a plan shortsighted, or does it produce optimal allocations?

To answer this question, we make the above definition of $\varphi^\#$ more precise. Let $\varphi^\#$ be a search plan such that at each time t

$$\rho(j, \varphi^\#(j, t)) = \rho(i, \varphi^\#(i, t)) \quad \text{if } \varphi^\#(j, t) > 0 \text{ and } \varphi^\#(i, t) > 0,$$
$$\rho(j, \varphi^\#(j, t)) \leq \rho(i, \varphi^\#(i, t)) \quad \text{if } \varphi^\#(j, t) = 0 \text{ and } \varphi^\#(i, t) > 0.$$

Then $\varphi^\#$ is called *locally optimal*. This plan would be the result of following the intuitive description of $\varphi^\#$ and letting the increment of effort approach zero.

Fix $t > 0$. Let $j_0 \in J$ be such that $\varphi^\#(j_0, t) > 0$, and set

$$\lambda(t) = \rho(j_0, \varphi^\#(j_0, t)).$$

Since $\rho(j, \cdot)$ is decreasing and $c'(j, \cdot) > 0$, it follows that for $j \in J$

$$p(j)b'(j, z) - \lambda(t)c'(j, z) \geq 0 \quad \text{for} \quad 0 \leq z < \varphi^\#(j, t),$$
$$p(j)b'(j, z) - \lambda(t)c'(j, z) \leq 0 \quad \text{for} \quad \varphi^\#(j, t) < z < \infty.$$

Thus $(\lambda(t), \varphi^\#(\cdot, t))$ maximizes the pointwise Lagrangian and $\varphi^\#(\cdot, t)$ is optimal for cost $C[\varphi^\#(\cdot, t)]$ by Theorem 2.1.2. Hence at each time t, $\varphi^\#$ produces an allocation that is optimal for cost $C[\varphi^\#(\cdot, t)]$.

The above result is stated in the form of a theorem.

Theorem 3.1.1. *Let ρ_j be continuous and decreasing and $c'(j, \cdot) > 0$ for $j \in J$. If $\varphi^\#$ is a locally optimal plan, then for each $t > 0$, $\varphi^\#(\cdot, t)$ is optimal for cost $C[\varphi^\#(\cdot, t)]$.*

If b is a regular function on J and $c(j, z) = z$ for $z \geq 0$, $j \in J$, then $\varphi^\#$ becomes the plan φ^* obtained in Theorem 2.2.5. Recall that φ^* is uniformly optimal in $\Phi(M)$, where $M(t) = C[\varphi^\#(\cdot, t)]$ for $t \geq 0$.

Example 3.1.2. For a two-cell target distribution and exponential detection function, the plan $\varphi^\#$ can be illustrated as follows: Let $J = \{1, 2\}$, $p(1) > p(2) > 0$, and

$$c(j, z) = z, \quad b(j, z) = 1 - e^{-z} \quad \text{for } z \geq 0, \ j \in J.$$

Then

$$\rho(j, z) = p(j)e^{-z} \quad \text{for } z \geq 0, \ j \in J.$$

Since $p(1) > p(2)$, search begins in cell 1 and is concentrated there until z_1 effort has been placed in cell 1, where z_1 satisfies

$$\rho(1, z_1) = \rho(2, 0),$$

i.e.,

$$z_1 = \ln(p(1)/p(2)).$$

At this point, additional effort is split between cells 1 and 2 in a manner that keeps the ratios in the two cells equal. Let d_1 and d_2 be increments of effort added to cells 1 and 2, respectively, after z_1 effort has been placed in cell 1. Then

$$\rho(1, z_1 + d_1) = \rho(2, d_2),$$

which implies

$$d_1 = d_2.$$

Let $M(t) = t$ for $t \geq 0$. The plan $\varphi^\#$ may be described as follows: Up to time $t = z_1$ all effort is placed in cell 1. Beyond $t = z_1$, additional effort is split evenly and simultaneously between cell 1 and cell 2. Thus,

$$\varphi^\#(1, t) = \begin{cases} t & 0 \leq t \leq z_1, \\ \dfrac{t + z_1}{2} & z_1 < t, \end{cases} \qquad \varphi^\#(2, t) = \begin{cases} 0 & 0 \leq t \leq z_1, \\ \dfrac{t - z_1}{2} & z_1 < t. \end{cases}$$

By Theorem 3.1.1, $\varphi^\#$ is uniformly optimal. In addition, $\varphi^\#$ is equal to φ^* obtained from Theorem 2.2.5.

3.2. MAXIMUM PROBABILITY SEARCH

Suppose a search plan progresses so that it is always placing effort in the cell or cells having the highest posterior probability of containing the target. This type of search plan as well as its analog for continuous search spaces, which will be precisely defined below, is called a *maximum probability search plan*. Does a maximum probability search plan produce a uniformly optimal search plan? For a circular normal target distribution and a homogeneous exponential detection function, the uniformly optimal plan proceeds exactly as the maximum probability search plan would (see Example 2.2.7). The uniformly optimal plan in Example 3.1.2 also proceeds in this manner. In this section, it is shown that uniform optimality of maximum probability search plans is characteristic of a search with a homogeneous exponential detection function.

Assume that b is a detection function on J such that $b'(j, \cdot)$ is continuous and decreasing for $j \in J$. Let $c(j, z) = z$ for $z \geq 0, j \in J$, and suppose that φ is a search plan that always searches in the cell or cells having the highest posterior probability of containing the target. The result of such a search is that the posterior probabilities are all equal in the cells in which search effort has been placed, that is,

$$\frac{p(j)[1 - b(j, \varphi(j, t))]}{1 - P[\varphi(\cdot, t)]} = \frac{p(i)[1 - b(i, \varphi(i, t))]}{1 - P[\varphi(\cdot, t)]}$$

$$\text{if} \quad \varphi(j, t) > 0 \quad \text{and} \quad \varphi(i, t) > 0,$$

$$\leq \frac{p(i)[1 - b(i, \varphi(i, t))]}{1 - P[\varphi(\cdot, t)]}$$

$$\text{if} \quad \varphi(j, t) = 0 \quad \text{and} \quad \varphi(i, t) > 0.$$

Note that the denominators in the two inequalities above are all the same. As a result these relations could be written excusively in terms of the numerators. Thus a search plan φ is called a *maximum probability search plan* on J if for each $t \geq 0$, there exists a $\hat{\lambda}(t) \geq 0$ such that

$$\begin{align} p(j)[1 - b(j, \varphi(j, t))] &= \hat{\lambda}(t) & \text{if} \quad \varphi(j, t) > 0, \\ &\leq \hat{\lambda}(t) & \text{if} \quad \varphi(j, t) = 0. \end{align} \quad (3.2.1)$$

Suppose $f^* = \varphi(\cdot, t)$ is optimal for cost $C[f^*]$ and $0 < C[f^*] < \infty$. Since $b'(j, \cdot)$ is continuous and decreasing and $c(j, z) = z$ for $z \geq 0, j \in J$, it follows from Corollary 2.1.6 that there exists a $\lambda \geq 0$ such that λ and f^* satisfy

$$\begin{align} p(j)b'(j, f^*(j)) &= \lambda & \text{for} \quad f^*(j) > 0, \\ &\leq \lambda & \text{for} \quad f^*(j) = 0. \end{align} \quad (3.2.2)$$

Let φ be a maximum probability search plan such that $0 < C[\varphi(\cdot, t)] < \infty$ and $\varphi(\cdot, t)$ is optimal for cost $C[\varphi(\cdot, t)]$ for $t > 0$. From (3.2.1) and (3.2.2) it follows that there exists a function $k: [0, \infty) \to [0, \infty)$ such that for $t > 0$

$$b'(j, \varphi(j, t)) = k(t)[1 - b(j, \varphi(j, t))] \quad \text{for} \quad \varphi(j, t) > 0. \quad (3.2.3)$$

We are now in a position to prove the following theorem.

Theorem 3.2.1. *Suppose that J has at least two cells and that b is a detection function on J such that $b'(j, \cdot)$ is continuous and decreasing and $b(j, 0) = 0$ for $j \in J$. Let $c(j, z) = z$ for $z \geq 0$, $j \in J$, and let M be a cumulative effort function that takes on all finite values between 0 and ∞. Each maximum probability search plan for each target distribution on J is uniformly optimal within $\Phi(M)$ if and only if b is a homogeneous exponential detection function.*

Proof. Suppose that the maximum probability search plan for each target distribution is uniformly optimal in $\Phi(M)$. Define

$$\hat{k}_j(z) = b'(j, z)/[1 - b(j, z)] \quad \text{for} \quad j \in J, \quad z \geq 0.$$

3.2 Maximum Probability Search

Then $\hat{k}_j$ is continuous, and

$$b(j, z) = 1 - \exp\left[-\int_0^z \hat{k}_j(y)\, dy\right] \quad \text{for } j \in J. \tag{3.2.4}$$

Let z_1 and z_2 be positive. Let the target distribution be such that $p(j) = 0$ for $j \neq 1, 2$ and

$$p(1)[1 - b(1, z_1)] = p(2)[1 - b(2, z_2)].$$

Let t be such that $M(t) = z_1 + z_2$. Then there is a maximum probability search plan φ such that $\varphi(1, t) = z_1$ and $\varphi(2, t) = z_2$. By (3.2.3),

$$b'(1, z_1) = k(t)[1 - b(1, z_1)], \quad b'(2, z_2) = k(t)[1 - b(1, z_2)].$$

Differentiating (3.2.4), one finds that $\hat{k}_1(z_1) = \hat{k}_2(z_2) = k(t)$. Since z_1 and z_2 are arbitrary positive numbers, it follows that there exists $\alpha \geq 0$ such that $\alpha = \hat{k}_1(z) = \hat{k}_2(z)$ for $z \geq 0$. Extending this argument to all $j \in J$, one has

$$b(j, z) = 1 - e^{-\alpha z} \quad \text{for } z \geq 0.$$

This proves the only if part of the theorem.

Suppose b is a homogeneous exponential detection function. Then there is an $\alpha > 0$ such that $b(j, z) = 1 - e^{-\alpha z}$ for $z \geq 0$, $j \in J$. Let φ be a maximum probability search. Since $\alpha[1 - b(j, z)] = b'(j, z)$ for $z \geq 0$, (3.2.1) yields for each $t \geq 0$ a λ, namely $\alpha\hat{\lambda}(t)$, such that $\lambda \geq 0$ and

$$\begin{aligned} p(j)b'(j, \varphi(j, t)) &= \lambda & \text{for } \varphi(j, t) > 0, \\ &\leq \lambda & \text{for } \varphi(j, t) = 0. \end{aligned}$$

Letting $f = \varphi(\cdot, t)$, one finds that (λ, f) maximizes the pointwise Lagrangian and that by Theorem 2.1.2, f is optimal for cost $C[\varphi(\cdot, t)] = M(t)$. Thus φ is uniformly optimal within $\Phi(M)$ and the theorem is proved.

The counterpart of the maximum probability search in the case of a continuous search space is a search that is always placing effort where the posterior target density is the highest. More precisely, we say that a search plan m is a *maximum probability search plan* on X if for each $t \geq 0$, there exists a $\hat{\lambda}(t) > 0$ such that

$$\begin{aligned} p(x)[1 - b(x, \varphi(x, t))] &= \hat{\lambda}(t) & \text{if } \varphi(x, t) > 0, \\ &\leq \hat{\lambda}(t) & \text{if } \varphi(x, t) = 0. \end{aligned} \tag{3.2.5}$$

Suppose that $b'(x, \cdot)$ is continuous for $x \in X$ and that $c(x, z) = z$ for $z \geq 0$, $x \in X$. If f^* is optimal for cost $C[f^*]$ in the interior of the range of C, then by Corollary 2.1.7, there is a $\lambda \geq 0$ such that λ and f^* satisfy

$$\begin{aligned} p(x)b'(x, f^*(x)) &= \lambda & \text{for a.e. } x \in X \text{ such that } f^*(x) > 0, \\ &\leq \lambda & \text{for a.e. } x \in X \text{ such that } f^*(x) = 0. \end{aligned} \tag{3.2.6}$$

If φ is a maximum probability search plan such that $f^* = \varphi(\cdot, t)$ is optimal for cost $0 < C[f^*] < \infty$ for $t > 0$, then (3.2.5) and (3.2.6) yield a function $k: [0, \infty) \to [0, \infty)$ such that for $t > 0$

$$b'(x, \varphi(x, t)) = k(t)[1 - b(x, \varphi(x, t))] \qquad \text{for a.e. } x \text{ such that } \varphi(x, t) > 0. \tag{3.2.7}$$

Theorem 3.2.2 gives the counterpart of Theorem 3.2.1 for a continuous search space, However, for Theorem 3.2.2 we must broaden slightly the definitions of homogeneous and exponential detection function over those given in Section 1.2. Let b be a detection function on X. Then b is *homogeneous* if there is an $x_0 \in X$ such that for a.e. $x \in X$, $b(x_0, z) = b(x, z)$ for $z \geq 0$, and b is *exponential* if there exists for a.e. $x \in X$ an $\alpha(x) > 0$ such that $b(x, z) = 1 - e^{-\alpha(x)z}$ for $z \geq 0$. Note that changing the definition of $b(x, \cdot)$ for x in a set of measure zero leaves the functional P unchanged.

Theorem 3.2.2. *Suppose b is a detection function on X such that $b(x, 0) = 0$ and $b'(x, \cdot)$ is continuous for $x \in X$. Let $c(x, z) = z$ for $z \geq 0$ and $x \in X$, and let M be a cumulative effort function that takes on all finite values between 0 and ∞. Each maximum probability search plan for each target distribution on X is uniformly optimal within $\Phi(M)$ if and only if b is a homogeneous exponential detection function.*

Proof. Using (3.2.7) the proof proceeds in the same manner as that of Theorem 3.2.1 except in the only if part, where one obtains an $\alpha \geq 0$ such that

$$b(x, z) = 1 - e^{-\alpha z} \qquad \text{for } z \geq 0 \text{ and a.e. } x \in X. \tag{3.2.8}$$

Note that $b'(x, \cdot)$ is not required to be decreasing in Theorem 3.2.2, while this requirement is made on $b(j, \cdot)$ in Theorem 3.2.1. The reason lies in Corollaries 2.1.6 and 2.1.7 and ultimately in the differences between Theorems 2.1.4 and 2.1.5, which are discussed in Section 2.1. Recall that these corollaries are used to obtain (3.2.2) and (3.2.6), which are used in the proofs of Theorems 3.2.1 and 3.2.2, respectively.

From Remark 2.1.8 it follows that Theorem 3.2.1 could be proved without assuming $b'(j, \cdot)$ is decreasing for $j \in J$ by restricting J to have a finite number of members and using the Kuhn–Tucker necessary conditions to obtain (3.2.2).

Remark 3.2.3. If we consider a discrete effort analog to Theorem 3.2.1, we can use a dynamic programming argument to show that a maximum probability search will produce a uniformly optimal plan, provided the detection function is the discrete counterpart of the exponential detection function.

3.3 Incremental Optimization

Suppose that α is the conditional probability of detecting the target on a single look in cell j given that the target is in cell j and that each look is independent of previous looks. The distribution of the number of looks in cell j required to find the target given that it is in cell j is geometric, i.e.,

Pr{detecting on or before the nth look in cell j | target in cell j} = $1 - (1 - \alpha)^n$.

As with the exponential distribution, this distribution has "no memory," that is, the distribution of the number of looks remaining to detect given that the target is in cell j and has not been detected by the first n looks in that cell is still geometric with single-look detection probability α.

If the search consists of one look, it is clear that this look should be made in the cell with the highest probability of containing the target. Fix a positive integer n and suppose that for a search consisting of n looks it is uniformly optimal to follow a maximum probability search. Consider an optimal search with $n + 1$ looks. Let j be the cell with maximum probability and suppose this search places its first look in cell $i \neq j$. If this looks fails to find the target, then by the principle of dynamic programming, the search plan for the remaining n looks must be optimal for the posterior distributions, given failure to find the target on the first look. Because of the geometric distribution's lack of memory, the detection function remains unchanged. Only the target distribution changes. Thus, by our assumption, a maximum probability search is optimal for the remaining n looks. Since cell j was not searched on the first look, it is the highest probability cell in the posterior target distribution, and we may assume that the second look is in cell j. Let $r(n - 1)$ be the probability of finding the target in the remaining $(n - 1)$ looks given failure on the first two. The probability of detection for this plan is given by

$$\alpha p(i) + \alpha p(j) + [1 - \alpha p(j) - \alpha p(i)] r(n - 1).$$

Observe that if we reverse the order of the first two looks, we obtain the same probability of detection. Thus, there must be an optimal plan that looks first in cell j. By our discussion above concerning the principle of dynamic programming, we may assume that it continues by looking in the cell with the highest posterior probability. Thus, a maximum probability plan is optimal (and, in fact, uniformly optimal) for $n + 1$ looks. By the principle of mathematical induction, this result holds for any positive integer n.

3.3. INCREMENTAL OPTIMIZATION

This section considers search plans that are obtained sequentially by allocating a number of increments of effort so that each increment yields the maximum increase in detection probability considering the previous increments. The object of this section is to find the relationship between incremental plans and plans that are designed to allocate the total effort contained in the increments in an optimal fashion. For continuous target distributions, it is shown that any plan that allocates each increment optimally produces an optimal allocation of the total effort. In the case of a discrete target distribution, this result is proved under the additional assumption that $b(j, \cdot)$ is concave and $c(j, \cdot)$ convex for $j \in J$.

Suppose a search planner is given authorization to expend K_1 amount of effort on a search. He chooses an allocation h_1 of this effort, which is optimal for cost K_1, but after applying effort according to h_1 fails to find the target. He is then given authorization to continue the search and to expend K_2 additional effort. He allocates the additional effort according to an incremental allocation h_2. An *incremental allocation* is one that is applied on top of a previous allocation. Suppose that h_2 is optimal for cost K_2 given the failure of the allocation h_1 to find the target. The two incremental allocations h_1 and h_2 combine to produce an allocation $f = h_1 + h_2$ of the total effort $K_1 + K_2$ and a resulting probability $P[f]$ of detecting the target. At this point the search planner may ask if he could have performed the search more efficiently in order to produce a higher probability of detection if he knew in advance that the total $K_1 + K_2$ amount of effort was going to be available. In this section we answer that question.

Let the target distribution be continuous and $c(x, z) = z$ for $z \geq 0$, $x \in X$. Suppose that the allocation h_1 has failed to detect the target. Then the posterior target distribution is given by the density

$$\tilde{p}(x) = \frac{p(x)[1 - b(x, h_1(x))]}{1 - P[h_1]} \quad \text{for} \quad x \in X.$$

Let

$$\hat{b}(x, z) = \frac{b(x, z + h_1(x)) - b(x, h_1(x))}{1 - b(x, h_1(x))} \quad \text{for} \quad z \geq 0, \quad x \in X.$$

Then $\hat{b}(x, z)$ is the conditional probability of detecting the target with additional effort density z given that the target is located at x and has failed to be detected by effort density $h_1(x)$. Let

$$\tilde{P}[h] = \int_X \tilde{p}(x)\hat{b}(x, h(x)) \, dx \quad \text{for} \quad h \in F(x).$$

Then $\tilde{P}[h]$ gives the conditional probability of detecting the target with the incremental allocation h given that the allocation h_1 has been applied and the target has not been detected. The incremental allocation h_2 is *conditionally optimal* for K_2 given h_1 if

$$C[h_2] \leq K_2 \quad \text{and} \quad \tilde{P}[h_2] = \max\{\tilde{P}[h] : h \in F(X) \text{ and } C[h] \leq K_2\}. \quad (3.3.1)$$

3.3 Incremental Optimization

If h_1 is optimal for $C[h_1]$ and h_2 is conditionally optimal for $C[h_2]$ given h_1, then (h_1, h_2) is called a *conditionally optimal pair*.

Let (h_1, h_2) be a conditionally optimal pair. Consider the total allocation $f = h_1 + h_2$. It produces a probability $P[f]$ of detecting the target with cost $C[f] = C[h_1] + C[h_2]$. If f is optimal for cost $C[f]$, then (h_1, h_2) is called a *totally optimal pair*. Analogous definitions of conditionally optimal and totally optimal pairs are assumed to hold for discrete target distributions.

The question asked by the search planner may now be phrased as follows: When is a conditionally optimal pair of allocations a totally optimal pair?

Before proving Theorem 3.3.1, it is convenient to note that the maximization of $\tilde{P}$ in (3.3.1) may be expressed in an alternate form. Observe that

$$\tilde{P}[h] = \frac{P[h + h_1] - P[h_1]}{1 - P[h_1]},$$

so that

$$\tilde{P}[h_2] = \max\{\tilde{P}[h] : h \in F(X) \text{ and } C[h] \leq K_2\} \qquad (3.3.2)$$

is equivalent to

$$P[h_2 + h_1] = \max\{P[h + h_1] : h \in F(X) \text{ and } C[h] \leq K_2\}. \qquad (3.3.3)$$

Let $Z(x) = [h_1(x), \infty)$ for $x \in X$. Then

$$\hat{F}(X) = \begin{bmatrix} \text{set of Borel functions } f: X \to [0, \infty) \\ \text{such that } h_1(x) \leq f(x) < \infty \text{ for } x \in X \end{bmatrix}. \qquad (3.3.4)$$

By letting $f^* = h_1 + h_2$, one can see that finding $h_2 \in F(X)$ to satisfy (3.3.3) is equivalent to finding $f^* \in \hat{F}(X)$ to satisfy

$$P[f^*] = \max\{P[f] : f \in \hat{F}(X) \text{ and } C[f] \leq K_1 + K_2\}. \qquad (3.3.5)$$

Theorem 3.3.1. *Let the search space be X and $c(x, z) = z$ for $z \geq 0$, $x \in X$. If (h_1, h_2) is a conditionally optimal pair such that $0 < C[h_1] < \infty$ and $0 < C[h_2] < \infty$, then (h_1, h_2) is totally optimal.*

Proof. Let (h_1, h_2) be a conditionally optimal pair. In order to show that (h_1, h_2) is totally optimal, we must prove that $f^* = h_1 + h_2$ is optimal for cost $C[f^*]$. This is accomplished by finding a finite $\lambda \geq 0$ such that (λ, f^*) maximizes the pointwise Lagrangian with respect to $F(X)$ and invoking Theorem 2.1.3. Since h_1 is optimal within $F(X)$ for cost $C[h_1]$, there exists by Theorem 2.1.5 a finite $\lambda_1 \geq 0$ such that (λ_1, h_1) maximizes the pointwise Lagrangian, that is, for a.e. $x \in X$,

$$p(x)b(x, h_1(x)) - \lambda_1 h_1(x) \geq p(x)b(x, z) - \lambda_1 z \qquad \text{for } z \geq 0. \qquad (3.3.6)$$

Since h_1 and h_2 are Borel, $f^* = h_1 + h_2$ is Borel and the set $\{(x, z) : z \in Z(x)$

for $x \in X\}$ is Borel. Since (h_1, h_2) is a conditionally optimal pair and (3.3.3) is equivalent to (3.3.5), it follows that for $\hat{F}(X)$ given in (3.3.4), f^* is optimal within $\hat{F}(X)$ for cost $C[f^*]$. By Theorem 2.1.5, there exists a finite $\lambda_2 \geq 0$ such that

$$p(x)b(x, f^*(x)) - \lambda_2 f^*(x) \geq p(x)b(x, z) - \lambda_2 z \quad \text{for} \quad z \geq h_1(x). \quad (3.3.7)$$

Evaluating (3.3.7) at $z = h_1(x)$, one obtains

$$p(x)[b(x, f^*(x)) - b(x, h_1(x))] \geq \lambda_2[f^*(x) - h_1(x)] = \lambda_2 h_2(x)$$
$$\text{for a.e.} \quad x \in X. \quad (3.3.8)$$

Evaluating (3.3.6) at $z = f^*(x)$, one finds

$$\lambda_1 h_2(x) \geq p(x)[b(x, f^*(x)) - b(x, h_1(x))] \quad \text{for a.e.} \quad x \in X. \quad (3.3.9)$$

Since $C[h_2] > 0$, $h_2(x) > 0$ on a set of positive measure, and (3.3.8) combined with (3.3.9) implies $\lambda_1 \geq \lambda_2$.

Since $\lambda_1 \geq \lambda_2$, it follows from (3.3.6) that for a.e. $x \in X$,

$$0 \leq p(x)[b(x, h_1(x)) - b(x, z)] - \lambda_1[h_1(x) - z]$$
$$\leq p(x)[b(x, h_1(x)) - b(x, z)] - \lambda_2[h_1(x) - z] \quad \text{for} \quad 0 \leq z \leq h_1(x).$$

Thus for a.e. $x \in X$, the above inequality and that in (3.3.8) yield

$$\ell(x, \lambda_2, z) \leq \ell(x, \lambda_2, h_1(x)) \leq \ell(x, \lambda_2, f^*(x)) \quad \text{for} \quad 0 \leq z \leq h_1(x). \quad (3.3.10)$$

Combining (3.3.7) and (3.3.10), one obtains for a.e. $x \in X$,

$$\ell(x, \lambda_2, f^*(x)) \geq \ell(x, \lambda_2, z) \quad \text{for} \quad z \geq 0.$$

Thus, (λ_2, f^*) maximizes the pointwise Lagrangian with respect to $F(X)$ and, by Theorem 2.1.3, f^* is optimal for cost $C[f^*]$. This proves the theorem.

Observe that in Theorem 3.3.1, the only hypothesis on the detection function b or the target density function p is that they be Borel functions, this very weak hypothesis being understood as part of the search model. This stands in contrast to the following theorem for discrete target distributions.

Theorem 3.3.2. *Let $b(j, \cdot)$ be concave and $c(j, z) = z$ for $j \in J$. If (h_1, h_2) is a conditionally optimal pair such that $0 < C[h_1] < \infty$ and $0 < C[h_2] < \infty$, then (h_1, h_2) is totally optimal.*

Proof. The proof proceeds in the same manner as that of Theorem 3.3.1. However, one must invoke Theorem 2.1.4 to obtain λ_1 and λ_2. This is the reason that one must assume that $b(j, \cdot)$ is concave for $j \in J$.

Corollary 3.3.3. *Let the target distribution be continuous and $c(x, z) = z$ for $z \geq 0$, $x \in X$. If $(h_1, h_2, \ldots)$ is a sequence of incremental allocations such*

3.3 Incremental Optimization

that $0 < C[h_i] < \infty$ for $i = 1, 2, \ldots$, and h_1 is optimal for cost $C[h_1]$ with h_{i+1} being conditionally optimal for cost $C[h_{i+1}]$ given $f_i = \sum_{k=1}^{i} h_k$ for $i = 1, 2, \ldots$, then f_i is optimal for cost $C[f_i]$ for $i = 1, 2, \ldots$.

Proof. The proof follows by induction on i. For $i = 1$, f_i is optimal for cost $C[f_1]$ by assumption, since $f_1 = h_1$. Suppose that f_i is optimal for cost $C[f_i]$. Then (f_i, h_{i+1}) is a conditionally optimal pair and, by Theorem 3.3.1, $f_{i+1} = f_i + h_{i+1}$ is optimal for $C[f_{i+1}]$. By the principle of mathematical induction, the theorem is proved.

Corollary 3.3.4. *Let $b(j, \cdot)$ be concave and $c(j, z) = z$ for $j \in J$. If $(h_1, h_2, \ldots)$ is a sequence of incremental allocations such that $0 < C[h_i] < \infty$ for $i = 1, 2, \ldots$, and h_1 is optimal for cost $C[h_1]$ with h_{i+1} being conditionally optimal for cost $C[h_{i+1}]$ given $f_i = \sum_{k=1}^{i} h_k$, then f_i is optimal for cost $C[f_i]$ for $i = 1, 2, \ldots$.*

Proof. The proof follows in the same manner as Corollary 3.3.3 with the use of Theorem 3.3.2.

In Stone (1972a) it is shown that Theorem 3.3.1 may be extended to allow very general separable payoff and cost functionals provided that $c(x, \cdot)$ is increasing for $x \in X$. In contrast, the following example shows that the conclusion of Theorem 3.3.2 is no longer true if one drops the assumption that $b(j, \cdot)$ is concave for $j \in J$.

Example 3.3.5. Let $J = \{1, 2\}$, $p(1) = p(2) = \frac{1}{2}$. Define

$$b(1, z) = \begin{cases} \frac{1}{2}z & \text{for } 0 \le z \le 2, \\ 1 & \text{for } z \ge 2, \end{cases}$$

$$b(2, z) = \begin{cases} \frac{1}{2}z & \text{for } 0 \le z \le 1, \\ \frac{1}{4}z^2 + \frac{1}{4} & \text{for } 1 < z \le \sqrt{3}, \\ 1 & \text{for } z > \sqrt{3}, \end{cases}$$

and

$$c(1, z) = c(2, z) = z \quad \text{for } z \ge 0.$$

The allocation h_1 defined by

$$h_1(1) = 1, \quad h_1(2) = 0,$$

is clearly optimal for cost 1, Let h_2 be the incremental allocation defined by

$$h_2(1) = 0, \quad h_2(2) = 1.$$

Then h_2 is conditionally optimal for cost 1 given h_1. The easiest way to see this is to observe that the equivalence of (3.3.2) and (3.3.3) holds for discrete target distributions also.

However, $f = h_1 + h_2$ is not optimal for cost 2. To see this, let f^* be defined by

$$f^*(1) = 2 - \sqrt{3}, \quad f^*(2) = \sqrt{3},$$

Then $C[f^*] = 2$ and

$$P[f^*] = 1 - \tfrac{1}{4}\sqrt{3} > \tfrac{1}{2} = P[f].$$

In a sense, the reason that (h_1, h_2) fails to be totally optimal is that although h_1 is optimal for cost 1 there is no finite $\lambda_1 \geq 0$ such that (λ_1, h_1) maximizes the pointwise Lagrangian. To see this, we suppose such a λ_1 exists. Then since $h_1(1) > 0$,

$$0 = \ell(1, \lambda_1, h_1(1)) = \tfrac{1}{2} b'(1, 1) - \lambda_1 = \tfrac{1}{4} - \lambda_1.$$

Thus $\lambda_1 = \tfrac{1}{4}$. However,

$$\ell(2, \tfrac{1}{4}, 0) = 0 < \ell(2, \tfrac{1}{4}, \sqrt{3}) = \tfrac{1}{2} - \tfrac{1}{4}\sqrt{3}.$$

Thus, (λ_1, h_1) does not maximize the pointwise Lagrangian, contrary to the assumption.

3.4. MULTIPLE CONSTRAINTS

In this section the notion of constrained optimization is extended from a single constraint to multiple constraints. The concept of incremental optimality is introduced, which generalizes conditional optimality. It is shown here that Theorem 3.3.1 cannot be extended to multiple constraints even if one assumes concavity of b and convexity of c.

Let E be the generalized payoff functional introduced in Section 2.4, that is,

$$E[f] = \int_X e(x, f(x))\, dx \quad \text{for } f \in F(X),$$

where $e: X \times [0, \infty) \to [0, \infty)$ is a fixed Borel function.

The cost functional C is vector valued. There is a fixed function $c = (c^1, c^2, \ldots, c^k)$, where $c^i: X \times [0, \infty) \to [0, \infty)$ is Borel measurable for $i = 1, \ldots, k$, such that

$$C[f] = \int_X c(x, f(x))\, dx \quad \text{for } f \in F(x).$$

3.4 Multiple Constraints

The above equality is understood to mean

$$C^i[f] = \int_X c^i(x, f(x))\, dx \qquad \text{for} \quad i = 1, \ldots, k,$$

where C^i denotes the ith component of C.

Similarly,

$$E[f] = \sum_{j \in J} e(j, f(j)) \qquad \text{for} \quad f \in F(J),$$

$$C[f] = \sum_{j \in J} c(j, f(j)) \qquad \text{for} \quad f \in F(J).$$

The notion of a pointwise Lagrangian also generalizes to multiple constraints. Let $\mathscr{E}_k$ denote Euclidean k-space. If α and β are k-vectors of real numbers, then α^i and β^i denote the ith components of α and β, respectively, and $\alpha \leq \beta$ means $\alpha^i \leq \beta^i$ for $i = 1, \ldots, k$. The dot product $\alpha \cdot \beta = \sum_{i=1}^{k} \alpha^i \beta^i$. We shall use 0 to designate the origin in k-space as well as the number 0. Define

$$\ell(x, \lambda, z) = e(x, z) - \lambda \cdot c(x, z) \qquad \text{for} \quad x \in X, \quad \lambda \in \mathscr{E}_k, \quad z \geq 0,$$

and

$$\ell(j, \lambda, z) = e(j, z) - \lambda \cdot c(x, z) \qquad \text{for} \quad j \in J, \quad \lambda \geq \mathscr{E}_k, \quad z \geq 0.$$

If $f \in \hat{F}(X)$, then (λ, f) is said to *maximize the pointwise Lagrangian* with respect to $\hat{F}(X)$ if

$$\ell(x, \lambda, f(x)) = \max\{\ell(x, \lambda, z) : z \in Z(x)\} \qquad \text{for a.e.} \quad x \in X.$$

Similarly, if $f \in \hat{F}(J)$, then (λ, f) *maximizes the pointwise Lagrangian* with respect to $\hat{Q}(J)$ if

$$\ell(j, \lambda, f(j)) = \max\{\ell(j, \lambda, z) : z \in Z(x)\} \qquad \text{for} \quad j \in J.$$

The allocation $f^* \in F$ is called *optimal within $\hat{F}$ for cost K*, where $K = (K^1, K^2, \ldots, K^k)$ if $C[f^*] \leq K$ and

$$E[f^*] = \max\{E[f] : f \in \hat{F} \text{ and } C[f] \leq K\}.$$

The following theorems generalize the results of Section 2.1 to multiple constraints. These theorems are stated without proof, but they follow from Corollary 5.2 and Theorem 5.3 of Wagner and Stone (1974).

Theorem 3.4.1. *Let $f^* \in \hat{F}(X)$. Then the following hold:*

(i) *If there is a finite $\lambda \geq 0$ such that (λ, f^*) maximizes the pointwise Lagrangian with respect to $\hat{F}(X)$, then f^* is optimal within $\hat{F}(X)$ for cost $C[f^*]$.*

(ii) *Suppose $\{(x, z) : x \in X, z \in Z(x)\}$ is Borel measurable. If $E[f^*] < \infty$, $C[f^*]$ is in the interior of the image of C, and f^* is optimal for cost $C[f^*]$, then there exists a finite $\lambda \geq 0$ such that (λ, f^*) maximizes the pointwise Lagrangian with respect to $\hat{F}(X)$.*

Theorem 3.4.2. *Let $f^* \in \hat{F}(J)$. Then the following hold:*

(i) *If there is a finite $\lambda \geq 0$ such that (λ, f^*) maximizes the pointwise Lagrangian with respect to $\hat{F}(J)$, then f^* is optimal within $\hat{F}(J)$ for cost $C[f^*]$.*

(ii) *Suppose $e(j, \cdot)$ is concave, $c^i(j, \cdot)$ is convex for $i = 1, \ldots, k$, and $Z(j)$ is an interval for $j \in J$. If $E[f^*] < \infty$, $C[f^*]$ is in the interior of the image of C, and f^* is optimal within $\hat{F}(J)$ for cost $C[f^*]$, then there exists a $\lambda \geq 0$ such that (λ, f^*) maximizes the pointwise Lagrangian.*

A generalization of the notion of conditional optimality is now considered. For $i = 1, 2, \ldots$, let $f_i \in F(X)$ be such that $f_1 \leq f_2 \leq \ldots$. For notational convenience let $f_0 = 0$ and $C[f_0] = 0$. If for $i = 1, 2, \ldots$,

$$E[f_i] = \max\{E[f] : f \geq f_{i-1}, f \in F(X) \text{ and } C[f_{i-1}] \leq C[f] \leq C[f_i]\},$$

then $(f_1, f_2, \ldots)$ is called an *incrementally optimal* sequence. If for each $i = 1, 2, \ldots, f_i$ is optimal for cost $C[f_i]$, then the sequence $(f_1, f_2, \ldots)$ is *totally optimal*. Analogous definitions of incrementally and totally optimal are understood to hold for sequences $(f_1, f_2, \ldots)$ such that $f_i \in F(J)$ for $i = 1, 2, \ldots$. Conceptually, an incrementally optimal sequence $(f_1, f_2, \ldots)$ is one such that for $i = 1, 2, \ldots$, the incremental allocation $h_i = f_i - f_{i-1}$ obtains the maximum increase in the payoff function within the total cost $C[f_i]$ given the previous allocation f_{i-1}. The mathematical advantage of the notion of incremental optimality is that it does not rely on probabilities and thus may be applied to more general payoff functions than the notion of conditional optimality. In addition, more general cost functionals may be used.

In the case where $e(x\ z) = p(x)b(x, z)$, $c(x, z) = z$ for $z \geq 0$, $x \in X$, and $k = 1$, then $E[f] = P[f]$, and the concepts of conditional and incremental optimality are related as follows. Let $(f_1, f_2, \ldots)$ be an incrementally optimal sequence. Define $h_i = f_i - f_{i-1}$ for $i = 1, 2, \ldots$. By the equivalence of (3.3.2) and (3.3.3) it follows that h_i is conditionally optimal for cost $C[h_i]$ given f_{i-1} for $i = 1, 2, \ldots$.

The following theorem, the proof of which is not given, is a consequence of Stone (1972a, Theorem 3.1). The proof in Stone (1972a) is very similar to the one given in Theorem 3.3.1.

Theorem 3.4.3. *Let the search space be X, $k = 1$, and $c(x, \cdot) = c^1(x, \cdot)$ be increasing for $x \in X$. If $(f_1, f_2, \ldots)$ is an incrementally optimal sequence such that for $i = 1, 2, \ldots, |E(f_i)| < \infty$ and $C[f_i]$ is in the interior of the range of C, then $(f_1, f_2, \ldots)$ is a totally optimal sequence.*

3.4 Multiple Constraints

From Theorem 3.4.3, one sees that when X is the search space, incrementally optimal sequences are totally optimal under very weak conditions provided that there is a single constraint, i.e., the cost functional is a 1-vector. The following example shows that even if we require $e(x, \cdot)$ to be concave and $c^i(x, \cdot)$ to be convex for $i = 1, \ldots, k$ and $x \in X$, Theorem 3.4.3 cannot be extended to multiple constraints.

Example 3.4.4. Let X be the real line. For $z \geq 0$ let

$$
\begin{array}{llll}
e(x, z) = z, & c^1(x, z) = 4z, & c^2(x, z) = 2z, & \text{for } 0 \leq x \leq 1, \\
e(x, z) = \tfrac{3}{4}z, & c^1(x, z) = z, & c^2(x, z) = 2z, & \text{for } 1 < x \leq 2, \\
e(x, z) = \tfrac{1}{2}z, & c^1(x, z) = z, & c^2(x, z) = \tfrac{9}{8}z, & \text{for } 2 < x \leq 3, \\
e(x, z) = 0, & c^1(x, z) = z, & c^2(x, z) = 4z, & \text{for } 3 < x \leq 4, \\
e(x, z) = 0, & c^1(x, z) = 0, & c^2(x, z) = 0, & \text{for } x \notin [0, 4].
\end{array}
$$

Let (y_1, y_2) represent a point in 2-space. One can verify that the range of C is the convex region in the nonnegative quadrant of 2-space that lies between the lines $y_2 = \tfrac{1}{2}y_1$ and $y_2 = 4y_1$.
Let

$$ f_1(x) = \begin{cases} 1 & \text{for } 1 < x \leq 2, \\ 0 & \text{otherwise,} \end{cases} \qquad f_2(x) = \begin{cases} 1 & \text{for } 0 \leq x \leq 2, \\ 0 & \text{otherwise.} \end{cases} $$

Note that $C[f_1] = (1, 2)$ and $C[f_2] = (5, 4)$ are in the interior of the range of C. By choosing $\lambda_1 = (\tfrac{1}{4}, \tfrac{1}{4})$ and $\lambda_2 = (\tfrac{1}{40}, \tfrac{9}{20})$, one may check that for $x \in X$

$$ e(x, f_1(x)) - \lambda_1 \cdot c(x, f_1(x)) \geq e(x, z) - \lambda_1 \cdot c(x, z) \qquad \text{for } z \geq 0, $$

and

$$ e(x, f_2(x)) - \lambda_2 \cdot c(x, f_2(x)) \geq e(x, z) - \lambda_2 \cdot c(x, z) \qquad \text{for } z \geq f_1(x). $$

Thus, it follows from Theorem 3.4.1 that (f_1, f_2) is an incrementally optimal sequence. However, f_2 is not optimal for cost $(5, 4)$. To see this, define

$$ f^*(x) = \begin{cases} 1 & \text{for } \tfrac{7}{48} \leq x \leq 1, \text{ and for } 1 + \tfrac{5}{12} \leq x \leq 3, \\ 0 & \text{otherwise.} \end{cases} $$

Then one may check that $C(f^*) = (5, 4)$ but $E[f^*] = 1 + \tfrac{19}{24} > 1 + \tfrac{3}{4} = E[f_2]$.

A reader familiar with mathematical programming problems might suspect from the beginning that Theorem 3.4.3 could not be extended to multiple constraints without substantial reduction in generality. For if it could, then one could solve separable mathematical programming problems by maximizing with respect to one constraint at a time. However, it still remains an interesting problem to see if one can find conditions under which Theorem 3.4.3 can be extended to multiple constraints.

NOTES

The interpretation of Lagrange multipliers as the marginal rate of return at the optimal solution is a standard one in economics and nonlinear programming. The results on incremental optimization in Section 3.3 developed in the following way. Koopman (1957, p. 617), observed that Theorem 3.3.1 holds for the case of a homogeneous exponential detection function. The term conditionally optimal is due to Dobbie (1963), who suggested that the result of Theorem 3.3.1 should hold for detection functions b such that $b(x, \cdot)$ is concave and differentiable for $x \in X$. The results, given in Theorems 3.3.1 and 3.3.2, and Example 3.3.5 were first proved in Stone (1973a). Theorem 3.4.3 and Example 3.4.4 are taken from Stone (1972a). Stone (1972a, Example 3.2) shows that the conclusion of Theorem 3.4.3 is no longer true if one drops the assumptions that $c(x, \cdot)$ is increasing for $x \in X$. Further historical discussion is given by Stone (1973a).

Chapter IV

Search with Discrete Effort

In contrast to Chapters I–III, where effort can be applied continuously, this chapter considers searches in which effort is applied in discrete units. For this situation, uniformly optimal search plans are found, as well as plans that minimize the mean cost to find the target.

The class of locally optimal discrete search plans is introduced. These plans proceed by always looking in the cell that yields the highest ratio of the increment in probability of detection to the increment in cost. If in each cell this ratio decreases with the number of looks in the cell, then a locally optimal plan minimizes the mean cost to find the target. If in addition cost is measured in number of looks, then a locally optimal plan maximizes the probability of detecting the target in n looks for $n = 1, 2, \ldots$, that is, a locally optimal plan is uniformly optimal.

This chapter also considers whereabouts searches, whose object is to maximize the probability of correctly specifying the cell containing the target within a given constraint on cost. A whereabouts search succeeds by either detecting the target or correctly guessing which cell contains the target after failing to detect the target within the constraint on cost. In Section 4.4 it is shown that an optimal whereabouts search may be obtained by specifying a cell to be guessed and performing an optimal detection search in the remaining cells.

4.1. MODEL FOR SEARCH WITH DISCRETE EFFORT

In this section the basic model for discrete search is given. The model is illustrated with two examples.

The target distribution is assumed to be discrete as defined in Section 1.1. A search proceeds by a sequence of looks or searches in the cells. A *discrete search plan* is a sequence $\xi = (\xi_1, \xi_2, \xi_3, \ldots)$ that is to be interpreted as meaning: Look first in cell ξ_1. If the target is not found on the first look, then continue by looking next in cell ξ_2. Similarly, if the target is not found in the first $k - 1$ looks, then the search is continued by looking next in cell ξ_k. It is not assumed that a look in the cell containing the target will always result in detection. Instead, there is a function $b: J \times \{0, 1, 2, \ldots\} \to [0, 1]$ such that

$b(j, k) =$ probability of detecting the target on or before the kth look in cell j given that the target is in cell j.

Of course, $b(j, 0) = 0$ for $j \in J$. In addition, there is a function $c: J \times \{0, 1, 2, \ldots\} \to [0, \infty)$ such that

$c(j, k) =$ cost of performing k searches in cell j.

Again $c(j, 0) = 0$ for $j \in J$.

The functions b and c given here are just the detection and cost functions of Chapters I–III restricted, in the second variable, to the nonnegative integers. Thus, one is not allowed to make only "half a search" in some cell. The search effort must be allocated in discrete units in each cell so that one has the choice of the number n of searches in a given cell, where $n = 0, 1, 2, \ldots$.

Let

$$\beta(j, k) = b(j, k) - b(j, k - 1)$$
$$\gamma(j, k) = c(j, k) - c(j, k - 1)$$
for $j \in J, \quad k = 1, 2, \ldots$.

Then $\beta(j, k)$ is the probability of failing to detect the target on the first $k - 1$ looks in cell j and succeeding on the kth, given that the target is in cell j. The cost of the kth search in cell j is given by $\gamma(j, k)$. It is assumed that $\gamma(j, k) \geq 0$ for $j \in J, k = 1, 2, \ldots$.

To illustrate the model of discrete search consider the following examples.

Example 4.1.1. An object is lost in shallow waters, Suppose that the target distribution is approximated by a discrete one with a finite number of cells in the manner described in Section 1.1, and the method of search is to send a team of divers into a selected cell. The team operates by trying to swim through the cell in a line abreast keeping a constant distance between them. The cell sizes are chosen so that a team can make one pass through the cell

4.1 Model for Search with Discrete Effort

before it must surface. Suppose that for operational reasons one would not perform a partial pass through a cell. Thus one effectively has a discrete search.

Let α_j be the probability of detecting the target on a single pass or look in cell j given that the target is in cell j. Each look has an independent probability of finding the target, so that

$$\beta(j, k) = \alpha_j(1 - \alpha_j)^{k-1} \quad \text{for} \quad j \in J, \quad k = 1, 2, \ldots.$$

If one is interested in maximizing the probability of detecting the target in a fixed number K of looks, then for the cost of a single look one would take

$$\gamma(j, k) = 1 \quad \text{for} \quad j \in J, \quad k = 1, 2, \ldots.$$

Suppose that one decided to perform $f(j)$ looks in cell j for $j \in J$, where

$$\sum_{j \in J} f(j) = K.$$

Then the probability of detecting the target with the allocation f is

$$\sum_{j \in J} p(j) b(j, f(j)) = \sum_{j \in J} p(j)[1 - (1 - \alpha_j)^{f(j)}],$$

where $p(j)$ is the probability that the target is in cell j.

The following example is taken from Kadane (1968).

Example 4.1.2. Suppose that one is looking for a gold coin located in one of a finite number of boxes. The jth box contains $n_j > 0$ coins, and there are $N = \sum_{j \in J} n_j$ coins in total. Of these coins, $N - 1$ are copper, and the remaining one is gold. Let p_j be the probability that the gold coin is located in the jth box. How should one choose K coins (without replacement) to maximize the probability of finding the gold coin?

If the box containing the gold coin has a total of n coins in it, the probability that a coin drawn from this box will be the gold coin is $1/n$. If the jth box contains the gold coin, then the probability of obtaining the gold coin on (and not before) the kth draw from the box is computed as follows:

The probability of failing on the rth draw given failure on the first $r - 1$ draws is

$$\frac{n_j - r}{n_j - r + 1} \quad \text{for} \quad 1 \leq r \leq n_j.$$

The probability of succeeding on the rth draw given failure on the first $r - 1$ draws is

$$1/(n_j - r + 1).$$

Thus, the probability of failing on the first $k - 1$ draws and succeeding on the kth is

$$\beta(j, k) = \frac{n_j - 1}{n_j} \frac{n_j - 2}{n_j - 1} \cdots \frac{n_j - k + 1}{n_j - k + 2} \frac{1}{n_j - k + 1},$$

$$= \frac{1}{n_j} \quad \text{for} \quad 1 \leq k \leq n_j.$$

As a result,

$$\beta(j, k) = \begin{cases} \dfrac{1}{n_j} & \text{for} \quad 1 \leq k \leq n_j, \\ 0 & \text{for} \quad k > n_j. \end{cases}$$

In addition,

$$\gamma(j, k) = 1 \quad \text{for} \quad k = 1, 2, \ldots.$$

If $f(j) \leq n_j$ gives the number of draws to be made from box j, then

$$\sum_{j \in J} p(j) f(j) / n_j$$

gives the probability of drawing the gold coin in $\sum_{j \in J} f(j)$ draws.

4.2. UNIFORMLY OPTIMAL SEARCH PLANS

This section presents conditions under which one may find a search plan that maximizes the probability of detection at each number of looks. To do this, we make use of the Lagrange multiplier theorems of Section 2.1.

For discrete search problems, the class of allocations $\hat{F}(J)$ is the set of all functions $f: J \to \{0, 1, 2, \ldots\}$, that is, $Z(j) = \{0, 1, 2, \ldots\}$ for $j \in J$, so that allocations in $\hat{F}(J)$ are restricted to take values in the nonnegative integers. This is in contrast to searches in which continuous effort is allowed and allocations may take values in the nonnegative reals. For $f \in \hat{F}(J)$, one has as before

$$P[f] = \sum_{j \in J} p(j) b(j, f(j)), \qquad C[f] = \sum_{j \in J} c(j, f(j)).$$

An allocation $f^* \in \hat{F}(J)$ is said to be *optimal for cost K* if $C[f^*] \leq K$ and

$$P[f^*] = \max\{P[f] : f \in \hat{F}(J) \text{ and } C[f] \leq K\}.$$

The following theorem gives a sufficient condition for an allocation to be optimal. This theorem is essentially a form of the Neyman–Pearson lemma adapted to search problems.

4.2 Uniformly Optimal Search Plans

Theorem 4.2.1. *Suppose there is a finite $\lambda \geq 0$ and an $f^* \in \hat{F}(J)$ such that $C[f^*] < \infty$ and for $j \in J$*

$$\frac{p(j)\beta(j,k)}{\gamma(j,k)} \geq \lambda \quad \text{for} \quad 1 \leq k \leq f^*(j), \tag{4.2.1}$$

$$\leq \lambda \quad \text{for} \quad f^*(j) < k < \infty.$$

Then f^* satisfies

$$P[f^*] - \lambda C[f^*] \geq P[f] - \lambda C[f] \quad \text{for} \quad f \in \hat{F}(J) \text{ such that } C[f] < \infty, \tag{4.2.2}$$

$$P[f^*] = \max\{P[f] : f \in \hat{F}(J) \text{ and } C[f] \leq C[f^*]\}. \tag{4.2.3}$$

Proof. Let $f \in \hat{F}(J)$. Observe that

$$p(j)b(j,f(j)) = \sum_{k=1}^{f(j)} p(j)\beta(j,k),$$

$$c(j,f(j)) = \sum_{k=1}^{f(j)} \gamma(j,k).$$

Suppose $f^*(j) > f(j)$. Then by (4.2.1)

$$p(j)[b(j,f^*(j)) - b(j,f(j))] = \sum_{k=f(j)+1}^{f^*(j)} p(j)\beta(j,k)$$

$$\geq \lambda \sum_{k=f(j)+1}^{f^*(j)} \gamma(j,k)$$

$$= \lambda[c(j,f^*(j)) - c(j,f(j))].$$

If $f^*(j) \leq f(j)$, the same result is obtained so that

$$P[f] - P[f] \geq \lambda(C[f^*] - C[f]).$$

Thus f^* satisfies (4.2.2), and by Theorem 2.1.1, f^* is optimal for cost $C[f^*]$. This proves the theorem.

Remark 4.2.2. The ratio $p(j)\beta(j,k)/\gamma(j,k)$ is the discrete search counterpart of the rate of marginal rate of return $\rho(j,z)$ defined in (3.1.2), that is, $\beta(j,k)$ corresponds to $b'(j,z)$ and $\gamma(j,k)$ corresponds to $c'(j,z)$. Recall Corollary 2.1.6, which gives necessary conditions for an allocation to be optimal within $F(J)$ when $b'(j,\cdot)$ is decreasing and $c'(j,\cdot)$ increasing for $j \in J$. One might expect that a similar result would hold for discrete search; that is, one might suppose that if $\beta(j,\cdot)$ is decreasing and $\gamma(j,\cdot)$ increasing for $j \in J$, then for each allocation $f^* \in \hat{F}(J)$ that is optimal for cost $C[f^*]$, there is a $\lambda \geq 0$ such that for $j \in J$

$$p(j)\beta(j,k) - \lambda\gamma(j,k) \geq 0 \quad \text{for} \quad 1 \leq k \leq f^*(j),$$
$$\leq 0 \quad \text{for} \quad f^*(j) < k < \infty. \tag{4.2.4}$$

If such a λ exists, then by the argument given in Theorem 4.2.1, f^* and λ satisfy (4.2.2). However, Wagner (1969, Example 1) shows that there exist allocations f^* that are optimal for cost $C[f^*]$ but for which there is no λ such that f^* and λ satisfy (4.2.2) [and hence (4.2.4)] even under the assumption that $\beta(j, \cdot)$ is decreasing and $\gamma(j, \cdot)$ increasing for $j \in J$. Thus the hoped for discrete version of Corollary 2.1.6 does not hold. When using Theorem 4.2.1 to find allocations that are optimal for a given cost, this means that there may be allocations $f^* \in \hat{F}(J)$ that are optimal for cost $C[f^*]$ that cannot be obtained by using Lagrange multiplier techniques.

There are methods of dealing with the "gaps" left by the Lagrange multiplier methods. Such methods are discussed by Loane (1971), Everett (1963), and Kettelle (1962). Alternatively, optimal allocations may be found by branch and bound (see Kadane, 1968) or other integer programming methods.

If the cost of a search in a cell is independent of the cell and the number of times the cell has been searched, then a counterpart to Corollary 2.1.6 does exist.

Theorem 4.2.3. *Let γ_0 be a fixed positive number such that $\gamma(j, k) = \gamma_0$ for $j \in J$ and $k = 1, 2, \ldots$. Suppose that $\beta(j, \cdot)$ is decreasing for $j \in J$ and that $f^* \in \hat{F}(J)$ is such that $0 < C[f^*] < \infty$. Then a necessary and sufficient condition for f^* to be optimal for cost $C[f^*]$ is that there exist a $\lambda \geq 0$ such that for $j \in J$*

$$\frac{1}{\gamma_0} p(j)\beta(j, k) \geq \lambda \quad \text{for} \quad 1 \leq k \leq f^*(j), \quad (4.2.5)$$

$$\leq \lambda \quad \text{for} \quad f^*(j) < k < \infty. \quad (4.2.6)$$

Proof. The sufficiency of (4.2.5) and (4.2.6) follows from Theorem 4.2.1. For the necessity, let $f^* \in \hat{F}(J)$ be optimal for cost $C[f^*]$. Let $J^+ = \{j : j \in J \text{ and } f^*(j) > 0\}$ and

$$\lambda = \inf_{j \in J^+} p(j)\beta(j, f^*(j))/\gamma_0.$$

Since $\beta(j, k) \geq 0$ for $j \in J$ and $k = 1, 2, \ldots$, it follows that $\lambda \geq 0$. By the definition of λ and the decreasing nature of $\beta(j, \cdot)$, (4.2.5) is satisfied.

To complete the proof, we reason by contradiction and suppose (4.2.6) does not hold. Let $j_1 \in J$ be such that $p(j_1)\beta(j_1, k)/\gamma_0 > \lambda$ for some $k > f^*(j_1)$. By the decreasing nature of $\beta(j_1, \cdot)$,

$$\frac{1}{\gamma_0} p(j_1)\beta(j_1, f^*(j_1) + 1) > \lambda. \quad (4.2.7)$$

4.2 Uniformly Optimal Search Plans

Then by the definition of λ, there is a $j_2 \in J^+$ such that

$$\frac{1}{\gamma_0} p(j_2)\beta(j_2, f(j_2)) < \frac{1}{\gamma_0} p(j_1)\beta(j_1, f^*(j_1) + 1). \tag{4.2.8}$$

Let f be defined by

$$f(j) = \begin{cases} f^*(j_1) + 1 & \text{for } j = j_1, \\ f^*(j_2) - 1 & \text{for } j = j_2, \\ f^*(j) & \text{otherwise.} \end{cases}$$

Then $f \in \hat{F}(J)$, $C[f] = C[f^*]$, and by (4.2.8) $P[f] > P[f^*]$, contrary to the assumption that f^* is optimal for cost $C[f^*]$. Thus (4.2.6) must hold and the theorem is proved.

Remark 4.2.4. As we noted before, Theorem 4.2.1. is similar to the Neyman–Pearson lemma of statistics (see Lehmann, 1959). In fact, one might view the problem of finding an allocation that is optimal for a cost K as that of choosing a set S of pairs (j, k) to maximize

$$\sum_{(j,k) \in S} p(j)\beta(j, k)$$

subject to

$$\sum_{(j,k) \in S} \gamma(j, k) \leq K.$$

This is, in fact, done by Kadane (1968). When taking this point of view, one has to be concerned about the possibility of including in S the pair $(j, k+1)$ without including (j, k). The approach used in Theorem 4.2.1 is similar to that of Wagner (1969).

Recall that a discrete search plan $\xi = (\xi_1, \xi_2, \ldots)$ is a sequence such that $\xi_k \in J$ for $k = 1, 2, \ldots$, and one that specifies that the kth look should be in cell ξ_k if the previous $k-1$ looks have failed to find the target. Let Ξ be the set of discrete search plans. Define $\hat{P}[n, \xi]$ to be the probability of detecting the target on or before the nth look when using plan ξ, and $\hat{C}[n, \xi]$ the cost of the first n looks when using plan ξ.

Let $\gamma(j, k) = 1$ for $j \in J$, $k = 1, 2, \ldots$, so that one is interested in maximizing the probability of detecting the target in a given number K of looks. A plan $\xi^* \in \Xi$ such that

$$\hat{P}[n, \xi^*] = \max\{\hat{P}[n, \xi] : \xi \in \Xi\} \quad \text{for } n = 1, 2, \ldots,$$

is called *uniformly optimal*. Thus, a uniformly optimal discrete search plan proceeds in a manner that maximizes the probability of detection for each number of looks taken.

Note that uniformly optimal plans are defined only when $\gamma(j, k) = 1$ for

$j \in J$ and $k = 1, 2, \ldots$. However, if $\gamma(j, k) = \gamma_0 > 0$ for $j \in J$, $k = 1, 2$, one may assume that $\gamma_0 = 1$ by a simple change of units. The following corollary gives a method of finding uniformly optimal plans.

Let $r(j, n, \xi)$ be the number of looks out of the first n that are placed in cell j by plan ξ. Construct a discrete search plan ξ as follows. Let ξ_1 be such that

$$\frac{p(\xi_1)\beta(\xi_1, 1)}{\gamma(\xi_1, 1)} = \max_{j \in J} \frac{p(j)\beta(j, 1)}{\gamma(j, 1)}.$$

Having determined the first $n - 1$ looks (i.e., $\xi_1, \xi_2, \ldots, \xi_{n-1}$), let i be such that

$$\frac{p(i)\beta(i, r(i, n-1, \xi) + 1)}{\gamma(i, r(i, n-1, \xi) + 1)} = \max_{j \in J} \frac{p(j)\beta(j, r(j, n-1, \xi) + 1)}{\gamma(j, r(j, n-1, \xi) + 1)}.$$

Let $\xi_n = i$, so that the search plan ξ always takes the next look in a cell that yields the highest ratio on that look of probability of detection to cost. A discrete search plan constructed in the above manner is called *locally optimal*.

For a search plan $\xi = (\xi_1, \xi_2, \ldots)$, let $k_n = r(\xi_n, n, \xi)$. Then the nth look of plan ξ is made in cell ξ_n and constitutes the k_nth time this cell has received a look. Let

$$\hat{\lambda}_n(\xi) = \frac{p(\xi_n)\beta(\xi_n, k_n)}{\gamma(\xi_n, k_n)} \quad \text{for} \quad n = 1, 2, \ldots, \quad \xi \in \Xi.$$

Theorem 4.2.5. *Let $p(j)\beta(j, \cdot)/\gamma(j, \cdot)$ be decreasing for $j \in J$. If $\xi^* \in \Xi$ is locally optimal, then*

$$\hat{P}[n, \xi^*] - \hat{\lambda}_n(\xi^*)\hat{C}[n, \xi^*] \geq P[f] - \hat{\lambda}_n(\xi^*)C[f] \quad \text{for} \quad f \in \hat{F}(J), \tag{4.2.9}$$

$$\hat{P}[n, \xi^*] = \max\{P[f] : f \in \hat{F}(J) \text{ and } C[f] \leq \hat{C}[n, \xi^*]\} \quad \text{for} \quad n = 1, 2, \ldots. \tag{4.2.10}$$

If, in addition, $\gamma(j, k) = 1$ for $j \in J$, $k = 1, 2, \ldots$, then ξ^* is uniformly optimal.

Proof. Fix $n > 0$. Let $f^*(j) = r(j, n, \xi^*)$ for $j \in J$, and $\lambda = \hat{\lambda}_n(\xi^*)$. Since ξ^* is a locally optimal plan and $p(j)\beta(j, \cdot)/\gamma(j, \cdot)$ is decreasing for $j \in J$, λ and f^* satisfy (4.2.1). Since $\hat{P}[n, \xi^*] = P[f^*]$, $\hat{C}[n, \xi^*] = C[f^*]$, and n is arbitrary, Theorem 4.2.1 yields the result that (4.2.9) and (4.2.10) hold. If $\gamma(j, k) = 1$ for $j \in J$, $k = 1, 2, \ldots$, then (4.2.10) implies that ξ^* is uniformly optimal, and the theorem is proved.

When $\gamma(j, k) = 1$ for $j \in J$, $k = 1, 2, \ldots$, the assumption that $p(j)\beta(j, \cdot)/\gamma(j, \cdot)$ is decreasing for $j \in J$ is equivalent to assuming that $\beta(j, \cdot)$ is decreasing for $j \in J$. A discrete detection function satisfying the latter property may be

4.2 Uniformly Optimal Search Plans

thought of as a discrete analog of a regular detection function and Theorem 4.2.5 as a discrete analog of Theorem 3.1.1.

From Remark 4.2.2, we know that under the conditions of Theorem 4.2.5, a locally optimal plan may miss some allocations that are optimal for their own cost, that is, there may exist an $f^* \in \hat{F}(J)$ that is optimal for cost $C[f^*]$ but for which there is no integer n such that $\hat{C}[n, \xi^*] = C[f^*]$. This possibility is illustrated in Fig. 4.2.1. The dots indicate the pairs $(\hat{C}[n, \xi^*], \hat{P}[n, \xi^*])$ for $n = 1, 2, \ldots$. Note that the slope of the dashed line connecting the $n - 1$st point with the nth point is just $\hat{\lambda}_n(\xi^*)$. The ✗ in the figure gives the coordinates of a pair $(C[f^*], P[f^*])$ such that f^* is optimal for cost $C[f^*]$, but one that is missed by ξ^*. Note that the point lies below the dashed line. This means there is no λ such that λ and f^* satisfy (4.2.2).

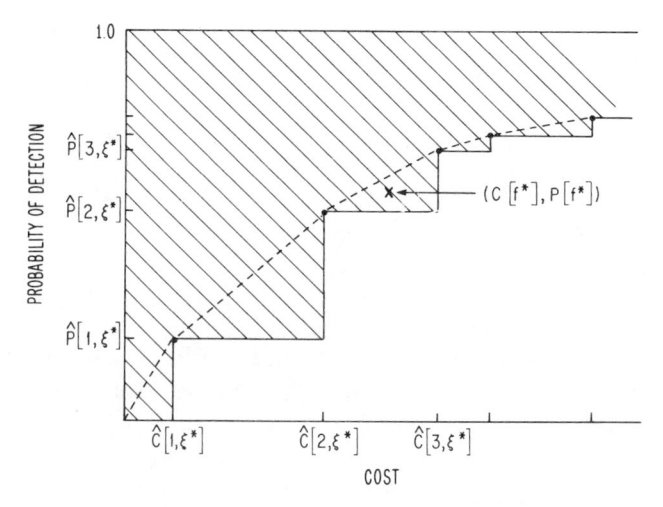

Fig. 4.2.1. Graph of cost and probability pairs. [*Notes:* ξ^* is a locally optimal search plan; f^* is an optimal allocation missed by ξ^*.]

If $\gamma(j, k) = 1$ for $j \in J$, $k = 1, 2, \ldots$, then problems of missing pairs like $(C[f^*], P[f^*])$ cannot occur. This is the reason that the definition of uniform optimality is restricted to this situation.

Example 4.2.6. In this example the uniformly optimal plan is found for the discrete search described in Example 4.1.1. Recall that

$$\beta(j, k) = \alpha_j(1 - \alpha_j)^{k-1}, \qquad \gamma(j, k) = 1 \qquad \text{for} \quad j \in J, \quad k = 1, 2, \ldots.$$

Thus $p(j)\beta(j, \cdot)/\gamma(j, \cdot)$ is decreasing for $j \in J$, and Theorem 4.2.5 may be applied to find a uniformly optimal plan ξ^* as follows: Make the nth look in a cell i such that

$$p(i)\alpha_i(1 - \alpha_i)^{r(i,n-1,\xi^*)} = \max_{j \in J} p(j)\alpha_j(1 - \alpha_j)^{r(j,n-1,\xi^*)}.$$

Since the number of cells is finite, such an i is guaranteed to exist.

Example 4.2.7. Here the uniformly optimal plan is found for the search described in Example 4.1.2. Recall that for $j \in J$

$$\beta(j, k) = \begin{cases} \dfrac{1}{n_j} & \text{for } 1 \leq k \leq n_j, \\ 0 & \text{for } k > n_j, \end{cases}$$

$$\gamma(j, k) = 1 \quad \text{for} \quad k = 1, 2, \ldots.$$

Since $p(j)\beta(j, \cdot)/\gamma(j, \cdot)$ is decreasing for $j \in J$, Theorem 4.2.5 again applies. In this case, a uniformly optimal plan is obtained by first ordering the cells so that

$$\frac{p_{j_1}}{n_{j_1}} \geq \frac{p_{j_2}}{n_{j_2}} \geq \frac{p_{j_3}}{n_{j_3}} \geq \cdots.$$

The first n_{j_1} looks are made in box j_1. If the gold coin is not found, then the subsequent n_{j_2} looks are made in box j_2, and so on.

4.3. MINIMIZING EXPECTED COST TO FIND THE TARGET

In this section it is shown that when $p(j)\beta(j, \cdot)/\gamma(j, \cdot)$ is decreasing for $j \in J$, a locally optimal search plan minimizes the mean cost to find the target.

Let $\mu(\xi)$ be the expected cost to find the target when using plan ξ. Again, let $\xi = (\xi_1, \xi_2, \ldots)$ and $k_n = r(\xi_n, n, \xi)$, so that the nth look of plan ξ is in cell ξ_n and constitutes the k_nth time this cell has received a look. If $\lim_{n \to \infty} \tilde{P}[n, \xi] = 1$, then

$$\mu(\xi) = \sum_{n=1}^{\infty} \hat{C}[n, \xi](\hat{P}[n, \xi] - \hat{P}[n - 1, \xi])$$

$$= \sum_{n=1}^{\infty} \sum_{r=1}^{n} \gamma(\xi_r, k_r) p(\xi_n) \beta(\xi_n, k_n)$$

$$= \sum_{r=1}^{\infty} \sum_{n=r}^{\infty} \gamma(\xi_r, k_r) p(\xi_n) \beta(\xi_n, k_n)$$

$$= \sum_{r=1}^{\infty} \gamma(\xi_r, k_r)(1 - \hat{P}[r - 1, \xi]),$$

where $\hat{P}[0, \xi] = 0$.

4.3 Minimizing Expected Cost to Find the Target

If $\lim_{n \to \infty} \hat{P}[n, \xi] < 1$, then we shall take $\mu(\xi) = \infty$ by convention. Thus,

$$\mu(\xi) = \sum_{n=1}^{\infty} \gamma(\xi_n, k_n)(1 - \hat{P}[n-1, \xi]) \qquad (4.3.1)$$

holds whenever $\lim_{n \to \infty} \hat{P}[n, \xi] = 1$ or γ is bounded away from 0. The latter is true because if γ is bounded away from 0 and $\lim_{t \to \infty} \hat{P}[n, \xi] < 1$, then the summation in (4.3.1) will be infinite.

If $\gamma(j, k) = 1$ for $j \in J$, $k = 1, 2, \ldots$, then (4.3.1) becomes

$$\mu(\xi) = \sum_{n=0}^{\infty} (1 - \hat{P}[n, \xi]). \qquad (4.3.2)$$

In this case it is easy to show that a uniformly optimal plan ξ^* minimizes the expected number of looks to find the target. Since ξ^* is uniformly optimal, $\hat{P}[n, \xi^*] \geq \hat{P}[n, \xi]$ for $\xi \in \Xi$, $n = 0, 1, 2, \ldots$. From (4.3.2) it is clear that $\mu(\xi^*) \leq \mu(\xi)$ for $\xi \in \Xi$. If $p(j)\beta(j, \cdot)$ is decreasing for $j \in J$, then by Theorem 4.2.5 any locally optimal plan is uniformly optimal and minimizes the expected number of looks to find the target. Similarly, a uniformly optimal plan minimizes the expected value of any increasing function that depends only on the number of looks taken in the search.

For general cost functions, the situation is considered where γ is bounded away from 0 [so that (4.3.1) holds] and $p(j)\beta(j, \cdot)/\gamma(j, \cdot)$ is decreasing for $j \in J$. In this situation a locally optimal plan minimizes the expected cost to find the target. However, the simple argument given when (4.3.2) holds is not applicable here because $\gamma(\xi_r, k_r)$ may depend on ξ when general cost functions are allowed.

The following two theorems give necessary and sufficient conditions for a plan to minimize expected cost to find the target. First, it is established that when $p(j)\beta(j, \cdot)/\gamma(j, \cdot)$ is decreasing for $j \in J$, one must look among the locally optimal plans to find one that minimizes expected cost.

Theorem 4.3.1. *Suppose γ is bounded away from 0 and $p(j)\beta(j, \cdot)/\gamma(j, \cdot)$ is decreasing for $j \in J$. If $\xi^* \in \Xi$ is such that $\mu(\xi^*) < \infty$ and*

$$\mu(\xi^*) \leq \mu(\xi) \qquad \text{for} \quad \xi \in \Xi, \qquad (4.3.3)$$

then ξ^* is locally optimal.

Proof. Let $\xi^* = (\xi_1^*, \xi_2^*, \ldots)$. Then

$$\mu(\xi^*) = \sum_{n=1}^{\infty} \gamma(\xi_n^*, k_n) \left[1 - \sum_{i=1}^{n-1} p(\xi_i^*)\beta(\xi_i^*, k_i) \right].$$

Suppose that ξ^* is not locally optimal. Then at some stage the next look is not chosen to be in the cell with the highest ratio. Suppose the omitted look corresponds to the kth look in cell j. Since γ is bounded away from 0 and $\mu(\xi^*) < \infty$, ξ^* must eventually include this look. For if this look were permanently omitted, the fact that $p(j)\beta(j, k) > 0$ and γ is bounded away from 0 would imply that $\mu(\xi^*) = \infty$. Thus, since $p(j)\beta(j, \cdot)/\gamma(j, \cdot)$ is decreasing for $j \in J$, there must be an $n \geq 0$ such that $\hat{\lambda}_n(\xi^*) < \hat{\lambda}_{n+1}(\xi^*)$.

Let ξ be the plan that follows ξ^* except for placing the nth look in ξ^*_{n+1} and $n + 1$st look in ξ_n^*. Then

$$\mu(\xi^*) - \mu(\xi) = \gamma(\xi_n^*, k_n)\left[1 - \sum_{i=1}^{n-1} p(\xi_i^*)\beta(\xi_i^*, k_i)\right]$$
$$+ \gamma(\xi^*_{n+1}, k_{n+1})\left[1 - \sum_{i=1}^{n} p(\xi_i^*)\beta(\xi_i^*, k_i)\right]$$
$$- \gamma(\xi^*_{n+1}, k_{n+1})\left[1 - \sum_{i=1}^{n-1} p(\xi_i^*)\beta(\xi_i^*, k_i)\right]$$
$$- \gamma(\xi_n^*, k_n)\left[1 - \sum_{i=1}^{n-1} p(\xi_i^*)\beta(\xi_i^*, k_i) - p(\xi^*_{n+1})\beta(\xi^*_{n+1}, k_{n+1})\right]$$
$$= \gamma(\xi_n^*, k_n)p(\xi^*_{n+1})\beta(\xi^*_{n+1}, k_{n+1})$$
$$- \gamma(\xi^*_{n+1}, k_{n+1})p(\xi_n^*)\beta(\xi_n^*, k_n) > 0,$$

where the inequality follows from $\hat{\lambda}_n(\xi^*) < \hat{\lambda}_{n+1}(\xi^*)$. However, $\mu(\xi^*) > \mu(\xi)$ contradicts (4.3.3). Thus it must be true that ξ^* is locally optimal and the theorem is proved.

The following theorem contains the main result on minimizing expected cost to find the target. The theorem gives conditions under which any locally optimal plan minimizes the mean cost to find the target.

Theorem 4.3.2. *Suppose γ is bounded away from 0 and $p(j)\beta(j, \cdot)/\gamma(j, \cdot)$ is decreasing for $j \in J$. If $\xi^* \in \Xi$ is locally optimal, then*

$$\mu(\xi^*) \leq \mu(\xi) \quad \text{for} \quad \xi \in \Xi. \tag{4.3.4}$$

Proof. Let $\xi^* = (\xi_1^*, \xi_2^*, \ldots)$. Then $\hat{C}[n, \xi^*] - \hat{C}[n - 1, \xi^*] = \gamma(\xi_n^*, k_n^*)$, $\hat{P}[n, \xi^*] - \hat{P}[n - 1, \xi^*] = p(\xi_n^*)\beta(\xi_n^*, k_n^*)$ for $n = 1, 2, \ldots$, and

$$\mu(\xi^*) = \sum_{n=1}^{\infty} \gamma(\xi_n^*, k_n^*)(1 - \hat{P}[n - 1, \xi^*]). \tag{4.3.5}$$

From (4.3.5) one can readily see that $\mu(\xi^*)$ is the area of the shaded portion of Fig. 4.2.1. Let h^* be the function that gives a linear interpolation between the points $(\hat{C}[n, \xi^*], \hat{P}[n, \xi^*])$ and $(\hat{C}[n + 1, \xi^*], \hat{P}[n + 1, \xi^*])$ for $n = 1, 2, \ldots$. If $C^* = \lim_{n \to \infty} C[n, \xi^*]$ is finite, then define $h(y) = 1$ for $y \geq C^*$. The graph of h^* is shown by the dashed line in Fig. 4.2.1. Then $\int_0^{\infty}[1 - h^*(y)]dy$

4.3 Minimizing Expected Cost to Find the Target

gives the area above the dashed line and $\frac{1}{2}\sum_{n=1}^{\infty} \gamma(\xi_n^*, k_n^*)p(\xi_n^*)\beta(\xi_n^*, k_n^*)$ gives the area of the triangles remaining in the shaded area. Thus,

$$\mu(\xi^*) = \int_0^\infty [1 - h^*(y)]\, dy + \tfrac{1}{2}\sum_{n=1}^{\infty} \gamma(\xi_n^*, k_n^*)p(\xi_n^*)\beta(\xi_n^*, k_n^*). \quad (4.3.6)$$

Let $\xi = (\xi_1, \xi_2, \ldots) \in \Xi$ and let h be the function that is obtained by linear interpolation between the cost and probability of detection pairs for ξ. Then

$$\mu(\xi) = \int_0^\infty [1 - h(y)]\, dy + \tfrac{1}{2}\sum_{n=1}^{\infty} \gamma(\xi_n, k_n)p(\xi_n)\beta(\xi_n, k_n). \quad (4.3.7)$$

If $\lim_{n \to \infty} \hat{P}[n, \xi] < 1$, then since γ is bounded away from 0, $\mu(\xi) = \infty$ and $\mu(\xi^*) \le \mu(\xi)$ holds trivially. If $\lim_{n \to \infty} \hat{P}[n, \xi] = 1$, then for any pair (j, k) such that $p(j)\beta(j, k) > 0$, the plan ξ must include a kth look in cell j. As a result, the summation on the right-hand side of (4.3.7) contains all possible nonzero terms and must be at least as large as the summation in (4.3.6). Consequently, to prove (4.3.4) holds, it is sufficient to show that $h^*(y) \ge h(y)$ for $y \ge 0$.

Observe that the conditions of Theorem 4.2.5 are satisfied, and suppose $h(y) > h^*(y)$ for some $y \ge 0$. Then there must be a cost point $\hat{C}[n, \xi]$ such that

$$\hat{P}[n, \xi] > h^*(\hat{C}[n, \xi]). \quad (4.3.8)$$

If $\hat{C}[n, \xi] = \hat{C}[n^*, \xi^*]$ for some n^*, then

$$\hat{P}[n^*, \xi^*] = h^*(\hat{C}[n, \xi]),$$

and (4.3.8) violates conclusion (4.2.10) of Theorem 4.2.5. Thus there is an n^* such that

$$\hat{C}[n^*, \xi^*] < \hat{C}[n, \xi] < \hat{C}[n^* + 1, \xi^*],$$

and the point $(\hat{C}[n, \xi], \hat{P}[n, \xi])$ lies above the chord joining $(\hat{C}[n^*, \xi^*], \hat{P}[n^*, \xi^*])$ and $(\hat{C}[n^* + 1, \xi^*], \hat{P}[n^* + 1, \xi^*])$. This chord has slope $\hat{\lambda}_{n+1}(\xi^*)$, and it follows that

$$\frac{\hat{P}[n^* + 1, \xi^*] - \hat{P}[n, \xi]}{\hat{C}[n^* + 1, \xi^*] - \hat{C}[n, \xi]} < \hat{\lambda}_{n+1}(\xi^*). \quad (4.3.9)$$

Define $f(j) = r(j, n, \xi)$. Then (4.3.9) implies

$$\hat{P}[n^* + 1, \xi^*] - \hat{\lambda}_{n+1}(\xi^*)\hat{C}[n^* + 1, \xi^*] < P[f] - \hat{\lambda}_{n+1}(\xi^*)C[f],$$

which contradicts conclusion (4.2.9) of Theorem 4.2.5. Thus it must be true that $h^*(y) \ge h(y)$ for $y \ge 0$ and that $\mu(\xi^*) \le \mu(\xi)$. This proves the theorem.

Observe that if the hypotheses of Theorem 4.3.2 are satisfied and if a locally optimal plan ξ^* does not satisfy $\lim_{n \to \infty} \hat{P}[n, \xi^*] = 1$, then there is no plan ξ such that $\mu(\xi) < \infty$. This is clear, since $\mu(\xi^*) = \infty$ and $\mu(\xi^*) \le \mu(\xi)$ for $\xi \in \Xi$.

By Theorem 4.3.2 or the discussion at the beginning of this section, the uniformly optimal plans found in Examples 4.2.6 and 4.2.7 minimize the expected cost (i.e., expected number of looks) to detect the target.

4.4. OPTIMAL WHEREABOUTS SEARCH

In a *whereabouts search* the objective is to maximize the probability of correctly stating which cell contains the target subject to a constraint K on cost. There are two ways in which a searcher may correctly state the target's position. First, he may detect the target; or second, failing to detect the target by the end of the search, he may guess which cell contains the target. This is in contrast to *detection searches*, where the objective is maximize the probability of detecting the target subject to a constraint on cost.

A *whereabouts search strategy* consists of a pair (f, i), where $f \in \hat{F}(J)$ is an allocation and i the cell to be guessed if the target is not detected by the allocation f. Recall that $f(j)$ gives the number of looks to be made in cell j so that the whereabouts search strategy gives the number of looks to be made in each cell and a cell to guess if the target is not detected.

The probability of detecting the target with allocation f is

$$\sum_{j \in J} p(j)b(j, f(j)).$$

The probability that the target is in cell i and not detected is $p(i)[1 - b(i, f(i))]$. Thus, the probability of success using strategy (f, i) is

$$p(i) - p(i)b(i, f(i)) + \sum_{j \in J} p(j)b(j, f(j)) = p(i) + \sum_{j \neq i} p(j)b(j, f(j)). \quad (4.4.1)$$

If there is a constraint K on cost, then f must satisfy $C[f] \leq K$. A whereabouts search strategy that maximizes (4.4.1) subject to $C[f] \leq K$ is called an *optimal whereabouts search strategy for cost K*.

In this section it is shown that an optimal whereabouts search strategy (f^*, i^*) never looks in cell i^*, and furthermore the allocation f^* maximizes the probability of detecting the target in the cells other than i subject to the constraint K. Conditions are found under which one may specify the cell to be guessed. Having done this, the whereabouts search problem is reduced to an optimal detection search problem.

Theorem 4.4.1. *If (f^*, i^*) is an optimal whereabouts search strategy for cost K, then f^* may be chosen so that $f^*(i^*) = 0$. In addition, f^* maximizes*

$$\sum_{j \neq i^*} p(j)b(j, f(j)) \qquad (4.4.2)$$

subject to $C[f] \leq K$ and $f \in \hat{F}(J)$.

4.4 Optimal Whereabouts Search

Proof. The probability of success when using (f^*, i^*) is

$$p(i^*) + \sum_{j \neq i^*} p(j)b(j, f^*(j)). \tag{4.4.3}$$

From (4.4.3) it is clear that looking in cell i^* does not increase the probability of success. Thus one may take $f^*(i^*) = 0$ without affecting the success probability or optimality of (f^*, i^*). It is also clear from (4.4.3) that f^* must maximize (4.4.2) over all allocations $f \in \hat{F}(J)$ such that $C[f] \leq K$ in order for (f^*, i^*) to be optimal for cost K. This proves the theorem.

If there is a finite number of cells and for each cell i^* one can find an allocation $f^* \in \hat{F}(J)$ that maximizes (4.4.2) subject to $C[f] \leq K$, then one can find an optimal whereabouts search strategy by computing the success probability associated with each choice of i^* and choosing a pair (f^*, i^*) that produces the highest probability of success.

In the remainder of this section conditions are found under which the cell i^* to be chosen at the end of an unsuccessful detection search can be easily specified.

Lemma 4.4.2. *Let γ be bounded away from 0, and suppose i^* satisfies*

$$c(i^*, k) \geq c(j, k) \tag{4.4.4}$$
$$p(i^*)[1 - b(i^*, k)] \geq p(j)[1 - b(j, k)] \qquad \text{for } j \in J, \; k = 1, 2, \ldots. \tag{4.4.5}$$

If there is an optimal whereabouts search strategy for cost K, then there is one that is optimal for cost K and chooses i^* if the detection search fails.

Proof. Let (f, i) be optimal for cost K, where $i \neq i^*$. By Theorem 4.4.1, one may take $f(i) = 0$. Let $g(i)$ be the largest integer such that $c(i, g(i)) \leq c(i^*, f(i^*))$. Define

$$g(i^*) = 0 \qquad \text{and} \qquad g(j) = f(j) \qquad \text{for } j \neq i \text{ or } i^*.$$

Note that $C[g] \leq C[f] \leq K$, and by (4.4.4), $f(i^*) \leq g(i)$.

The difference in the success probabilities of (g, i^*) and (f, i) is

$$\begin{aligned}
& p(i^*) + P[g] - [p(i) + P[f]] \\
&= p(i^*) - p(i^*)b(i^*, f(i^*)) - [p(i) - p(i)b(i, g(i))] \\
&\geq p(i^*)[1 - b(i^*, f(i^*))] - p(i)[1 - b(i, f(i^*))] \\
&\geq 0,
\end{aligned}$$

where the next to last inequality follows from $f(i^*) \leq g(i)$ and the last from (4.4.5). Thus, (g, i^*) is an optimal whereabouts search strategy for cost K, and the lemma is proved.

Theorem 4.4.3. *Suppose for all* $i, j \in J$, $c(i, k) = c(j, k)$ *and* $b(i, k) = b(j, k)$ *for* $k = 1, 2, \ldots$. *Let* i^* *be such that*

$$p(i^*) = \max_{j \in J} p(j).$$

If there exists an optimal whereabouts search strategy for cost K, then there is one that is optimal for cost K and guesses i^* *if the detection search fails.*

Proof. The hypotheses of Lemma 4.4.2 are satisfied and the theorem follows.

Suppose that in Example 4.1.1, $\alpha_j = \alpha_i$ for $i, j \in J$. Then Theorem 4.4.3 may be applied to find an optimal whereabouts search strategy for K looks. To do this, one chooses a cell i^* with the highest probability of containing the target. One then performs K looks of a locally optimal search plan in the remaining cells. By Theorem 4.2.5, this will result in an allocation that maximizes the probability of finding the target in the cells other than i^* in K looks. If this search fails to detect the target, then one guesses cell i^* for the location of the target.

Remark 4.4.4. Consider the following variation on the whereabouts search problem. In addition to the costs of searching in cells, there is a penalty cost for failing to identify the target's position correctly. In this case the problem is how to search and when to stop (i.e., guess) in order to minimize expected cost. In this form the whereabouts search problem resembles the optimal search and stop problem considered in the next chapter. This problem appears not to have been investigated in the search literature.

Whereabouts searches in a false target environment have been considered by Richardson (1973).

NOTES

The uniformly optimal search plan found in Theorem 4.2.5 was obtained by Chew (1967) for the case of $\beta(j, k) = \alpha_j(1 - \alpha_j)^{k-1}$ and $\gamma(j, k) = 1$ for $j \in J$, $k = 1, 2, \ldots$. The result for arbitrary β such that $\beta(j, \cdot)$ is decreasing was proved by Kadane (1968). Theorem 4.3.2 was first proved by Blackwell for the case where $\beta(j, k) = \alpha_j(1 - \alpha_j)^{k-1}$ and $\gamma(j, k) = \gamma_j$ for $j \in J$, $k = 1, 2, \ldots$, by using dynamic programming methods (see Matula, 1964). The method of proving Theorem 4.3.2 used here is a generalization of the argument given by Black (1965) to prove Blackwell's result without using dynamic programming. Theorem 4.3.1, which finds a necessary condition for a search plan to minimize expected cost to find the target, is adapted from DeGroot (1970, pp. 425–426). Section 4.4 on optimal whereabouts search is based on Kadane (1971).

Notes

In Matula (1964) the question of ultimately periodic searches is considered. Sweat (1970) considers a search problem similar to the one in Example 4.1.1 except that the searcher has a probability σ_j of being able to continue the search after looking in cell j for $j \in J$. This probability is independent of previous looks. Sweat shows that an optimal policy for maximizing the probability of finding the target (with no time limit) is to perform a locally optimal search resulting from replacing the target location probability $p(j)$ with $p(j)/(1 - \sigma_j)$ for $j \in J$. Sweat's problem is a special case of what Kadane (1969) calls a "quiz show" problem. Kadane shows that a generalization of Sweat's optimal policy is optimal for quiz show problems.

Optimal search plans have been found for situations where the detection probabilities $\beta(j, k)$ are random variables whose values become known to the searcher after looking in a cell. When the $\beta(j, k)$, $j \in J$, $k = 1, 2, \ldots$, are independent identically distributed random variables and $c(j, k) = 1$, for $j \in J$, $k = 1, 2, \ldots$, Kan (1972) has shown that a uniformly optimal plan is obtained by always making the next look in the cell that yields the highest expected probability of detecting the target given the past values of the detection probabilities. Hall (1973) also obtained this result but without requiring the detection probabilities to be identically distributed.

Chapter V

Optimal Search and Stop

In this chapter searches are considered in which the target has a value and there is a cost associated with searching. The search planner wishes to find an optimal search and stop plan, i.e., a plan that tells how to search and when to stop in order to maximize the expected return to the searcher.

Section 5.1 considers searches with discrete effort. It is shown that optimal search and stop plans exist and satisfy the principle of optimality of dynamic programming. Conditions are found that guarantee that it is optimal to search until the target is found. Conditions are also given under which it is optimal not to search at all. The main result, Theorem 5.1.11, shows that if the value of the target is independent of time and location, the search space J is finite, and $p(j)\beta(j, \cdot)/\gamma(j, \cdot)$ is decreasing for $j \in J$, then an optimal search and stop plan follows a locally optimal plan as long as it continues to search.

Section 5.2 considers optimal search and stop when search effort is continuous. In contrast to Section 5.1, the cost of search is assumed to be a function of time rather than the search plan. However, plans are required to be in the class $\Phi(M)$. It is shown that for a fixed stopping time, it is optimal to follow a plan φ^* that is uniformly optimal within $\Phi(M)$ up to that stopping time. This reduces the optimal search and stop problem to an optimal stopping problem. Theorem 5.2.5 provides sufficient conditions for a stopping time to be optimal.

5.1. OPTIMAL SEARCH AND STOP WITH DISCRETE EFFORT

This section considers the problem of optimal search and stop when search effort is discrete. There is a reward for finding the target and a cost for each look of the search. The searcher's objective is to find a search plan and a time for stopping the search that will maximize his expected return.

The search model assumed for this section is the one of search with discrete effort given in Chapter IV. The target distribution is discrete, so that $p(j)$ gives the probability that the target is in cell j for $j \in J$. Effort must be allocated to cells in discrete looks, and time is considered discrete; that is, time $t = n$ corresponds to the nth look of the search. The target is assumed to have value $0 \leq V(j) < \infty$ if it is found in cell j. There is the cost $\gamma(j, k)$ of making the kth search in cell j and the probability $\beta(j, k)$ of detecting the target for the first time on the kth look in cell j given that the target is in cell j.

At each time $t = 0, 1, 2, \ldots$, the searcher may decide to stop or to proceed by looking in the cell of his choice. If the search stops, no additional cost is incurred but no reward is obtained. If the search continues by looking in cell j, then the cost of the look in that cell is incurred, and if the target is found, a reward $V(j)$ is obtained. Naturally, if the target is found, the search stops. The searcher wishes to find both a search plan and a stopping time s that will maximize expected return for the search (i.e., expected value of the target, if found, minus expected cost of the search).

In this section, conditions are found under which $s = 0$ and $s = \infty$ are the optimal stopping times. If $s = \infty$, then the search continues until the target is detected and the optimal search plan is the one that minimizes expected cost to find the target.

When $V(j) = V_0$ for $j \in J$ and the ratios $p(j)\beta(j, \cdot)/\gamma(j, \cdot)$ are decreasing for $j \in J$, it is shown that the optimal plan follows a locally optimal search as long as the search continues. This means that the search and stop problem is effectively reduced to finding the optimal stopping time. Unfortunately this last problem is difficult to solve in general, so that usually one must settle for an approximation to the optimal stopping time.

For optimal search and stop, the notion of a search plan given in Chapter IV is extended to a search and stop plan. A *search and stop plan* ξ is a sequence $(\xi_1, \xi_2, \ldots, \xi_s)$, where $\xi_i \in J$ for $i = 1, 2, \ldots, s$ and $s \in \{0, 1, 2, \ldots, \infty\}$. The plan ξ tells the searcher to perform the ith look in cell ξ_i if the target has not been detected previously and to stop searching after the sth look if the target has not been found. The class of search and stop plans is denoted by Ξ_s. Let Ξ_∞ be the subclass of plans for which $s = \infty$. Then Ξ_∞ is equal to the class Ξ of search plans defined in Chapter IV.

The state of the search or system given that the target has not been found by time t is specified by the pair (p_t, f_t), where p_t is the posterior probability distribution for the target's location at time t, and f_t the allocation of effort up to time t [i.e., $f_t(j)$ is the number of looks placed in cell j by time t]. Let $\mathbf{R}[p_t, f_t, \xi]$ be the marginal expected return when plan ξ is followed from time t on, the state of the search at time t is given by (p_t, q_t), and the target has not been found by time t. Associated with the plan ξ, there is a sequence of pairs $(\xi_1, k_1), (\xi_2, k_2), \ldots, (\xi_s, k_s)$, where ξ_n gives the cell in which the nth look is made and k_n the number of looks out of the first n that are made in cell j_n.

Let

$$\mathbf{S} = \min\{s, \text{ time at which the target is detected}\}.$$

Then

$$\mathbf{R}[p_t, f_t, \xi] = \sum_{n=1}^{\mathbf{S}} \frac{V(\xi_n) p_t(\xi_n) \beta(\xi_n, f_t(\xi_n) + k_n)}{1 - b(\xi_n, f_t(\xi_n))} - \mathbf{E}\left[\sum_{n=1}^{\mathbf{S}} \gamma(\xi_n, f_t(\xi_n) + k_n)\right],$$

where

$$\Pr\{\mathbf{S} \le n\} = \begin{cases} \sum_{r=1}^{n} \dfrac{p_t(\xi_r) \beta(\xi_r, f_t(\xi_r) + k_r)}{1 - b(\xi_r, f_t(\xi_r))} & \text{for } n < s, \\ 1 & \text{for } n \ge s. \end{cases}$$

It is clear from the above formulas that given knowledge of p_t and f_t, $\mathbf{R}[p_t, f_t, \xi]$ does not depend on t. Henceforth, we shall write simply $\mathbf{R}[p, f, \xi]$. Thus, for the purpose of determining marginal return, the state of the search is completely specified by p and f and is independent of time, given p and f.

For a search in state (p, f), let $(T_j p, T_j f)$ be the state resulting from an unsuccessful look in cell j. In addition, let

$$\tilde{\beta}(j) = \frac{\beta(j, f(j) + 1)}{1 - b(j, f(j))} \quad \text{and} \quad \tilde{\gamma}(j) = \gamma(j, f(j) + 1) \quad \text{for } j \in J.$$

Then $\tilde{\beta}(j)$ is the conditional probability of detecting the target on the next look in cell j given that the target is in cell j and the previous $f(j)$ looks have failed to detect the target. One can now write

$$T_i p(j) = \begin{cases} \dfrac{p(i)[1 - \tilde{\beta}(i)]}{1 - p(i)\tilde{\beta}(i)} & \text{for } j = i, \\ \dfrac{p(j)}{1 - p(i)\tilde{\beta}(i)} & \text{for } j \ne i, \end{cases}$$

$$T_i f(j) = \begin{cases} f(i) + 1 & \text{for } j = i, \\ f(j) & \text{for } j \ne i. \end{cases}$$

5.1 Optimal Search and Stop with Discrete Effort

Let

$$g(p,f) = \sup_{\xi \in \Xi_s} R[p,f,\xi] \quad \text{for any state} \quad (p,f).$$

Then g is called the *optimal return function*. If $\xi^* \in \Xi_s$ satisfies

$$R[p,f,\xi^*] = g(p,f),$$

then ξ^* is called an *optimal search and stop plan for* (p,f).

Existence of Optimal Plans

The following theorem proves the existence of optimal search and stop plans and shows that g satisfies the optimality equation of dynamic programming.

Theorem 5.1.1. *Let J be finite and γ bounded away from 0. Then an optimal search and stop policy ξ^* exists for each (p,f) and the optimal return function g satisfies*

$$g(p,f) = \max\{0, \max_{j \in J}[\tilde{\beta}(j)p(j)V(j) - \tilde{\gamma}(j) + [1 - \tilde{\beta}(j)p(j)]g(T_j p, T_j f)]\}.$$

(5.1.1)

Proof. The proof makes use of Theorems 8.2 and 9.1 of Strauch (1966). In order to use these results, it is necessary to transform the problem in the following way. Let

$$\bar{V} = \max_{j \in J} V(j),$$

and let $\hat{V}(j) = V(j) - \bar{V}$. Consider a modified search and stop problem that is obtained from the original one by replacing V with $\hat{V}$ and imposing a cost $\bar{V}$ if the searcher stops before finding the target. Let $\hat{R}[p,f,\xi]$ be the expected return starting from state (p,f) when using plan ξ in the modified problem. For any $\xi \in \Xi_s$ such that the expected termination time of the search $E[S]$ is finite,

$$R[p,f,\xi] = \hat{R}[p,f,\xi] + \bar{V}.$$

Since γ is bounded away from 0, it follows that if $E[S] = \infty$, then $R[p,f,\xi] = \hat{R}[p,f,\xi] = -\infty$. Thus, any strategy that is optimal for the modified problem is optimal for the original one. The modified problem is what Strauch calls a "negative dynamic programming problem," i.e., all returns are nonpositive.

By Strauch (1966, Theorem 8.2) $\hat{g}(p,f) = \sup_{\xi \in \Xi_s} \hat{R}[p,f,\xi]$ satisfies

$$\hat{g}(p,f) = \max\{-\bar{V}, \max_{j \in J}[\tilde{\beta}(j)p(j)\hat{V}(j) - \tilde{\gamma}(j) + [1 - \tilde{\beta}(j)p(j)]\hat{g}(T_j p, T_j f)]\}.$$

Since $g = \hat{g} + \bar{V}$, it follows that g satisfies (5.1.1). By Strauch (1966, Theorem 9.1), an optimal search and stop plan ξ^* exists for the modified problem. By the above discussion, ξ^* is optimal for the original problem and the theorem is proved.

Example 5.1.2. This example (due to Ross, 1969) shows that one cannot drop the assumption that γ be bounded away from 0 in Theorem 5.1.1.
Suppose $J = \{1, 2\}$ and

$$\beta(1, k) = (\tfrac{1}{2})^k \quad \text{for} \quad k = 1, 2, \ldots,$$

$$\beta(2, k) = \begin{cases} 1 & \text{for} \quad k = 1, \\ 0 & \text{for} \quad k > 1, \end{cases}$$

$$\gamma(1, k) = 0, \quad \gamma(2, k) = 1 \quad \text{for} \quad k = 1, 2, \ldots,$$

$$V(1) = V(2) = 10.$$

If there exists an optimal policy for any state (p, f) such that $p(1) > 0$, then it must be $\xi^1 = (1, 1, \ldots)$. For if $(\xi_1, \xi_2, \ldots, \xi_s) \neq \xi^1$, then

$$\mathbf{R}[p, f, (1, \xi_1, \ldots, \xi_s)] > \mathbf{R}[p, f, (\xi_1, \xi_2, \ldots, \xi_s)]$$

because a search of cell 1 is free. However, suppose $p(1) = \tfrac{1}{10}$, $p(2) = \tfrac{9}{10}$, and $f(1) = f(2) = 0$. Then

$$\mathbf{R}[p, f, \xi^1] = \tfrac{1}{10}(10) = 1$$

is not the maximum return. For if one considers the plan that places the first n looks in cell 1, then a look in cell 2, and returns to looking in cell 1, one obtains for the return of this plan

$$\tfrac{1}{10}\{10[1 - (\tfrac{1}{2})^n] + 9(\tfrac{1}{2})^n\} + \tfrac{9}{10} 9.$$

This expression approaches $9\tfrac{1}{10}$ as n approaches ∞, and thus ξ^1 is not optimal. The result is that there is no optimal plan for this situation.

The reason the existence proof of Theorem 5.1.1 breaks down in this example is that since $\gamma(1, k) = 0$ for $k = 1, 2, \ldots$, it is no longer true that a plan ξ with infinite expected termination time has $\mathbf{R}[p, f, \xi] = -\infty$. Observe that if one replaces $g(p, f)$ by $\mathbf{R}[p, f, \xi^1]$ in (5.1.1), the equation is satisfied. This shows that there may be solutions to Eq. (5.1.1) that do not yield the optimal return. In this case, one may check directly that g also satisfies (5.1.1). This emphasizes the point that even when the optimal return function g satisfies the optimality equation of dynamic programming, it does not follow that any solution of that equation yields the optimal return function and an optimal strategy.

Sufficient Conditions for $s = \infty$ to Be Optimal

Recall that Ξ_∞ is the class of search and stop plans for which $s = \infty$ and that $\mu(\xi)$ is the mean cost to detect the target using plan $\xi \in \Xi = \Xi_\infty$.

5.1 Optimal Search and Stop with Discrete Effort

Theorem 5.1.3. Suppose $\sum_{j\in J} p(j)V(j) < \infty$. If $\xi^* \in \Xi_\infty$ minimizes the expected cost to find the target among all plans in Ξ_∞, then

$$\mathbf{R}[p, f, \xi^*] \geq \mathbf{R}[p, f, \xi] \quad \text{for } \xi \in \Xi_\infty. \tag{5.1.2}$$

Proof. For $\xi \in \Xi_\infty$ such that probability of detecting the target is one,

$$\mathbf{R}[p, f, \xi] = \sum_{j\in J} p(j)V(j) - \mu(\xi).$$

Thus, if $\mu(\xi^*) < \infty$, then (5.1.2) holds because $\mu(\xi^*) \leq \mu(\xi)$ for $\xi \in \Xi$. If $\mu(\xi^*) = \infty$, then $\mathbf{R}[p, f, \xi] = \infty$ for all $\xi \in \Xi_\infty$, and the theorem is proved.

Remark 5.1.4. Recall that if the ratio $p(j)\beta(j, \cdot)/\gamma(j, \cdot)$ is decreasing for $j \in J$ and γ is bounded away from 0, then any locally optimal plan $\xi^* \in \Xi_\infty$ will minimize the expected cost to find the target (see Theorem 4.3.2). Thus, one may find ξ^* to satisfy (5.1.2) in this situation.

The following lemma will be used in Theorem 5.1.6 to find sufficient conditions for the optimal search and stop plan to be one that continues searching until the target is found, i.e., a member of Ξ_∞.

Lemma 5.1.5. Suppose J is finite and γ is bounded away from 0. If for (p, f) it is true that $\tilde{\beta}(j)p(j)V(j) > \tilde{\gamma}(j)$ for some $j \in J$, then no optimal plan stops searching at the state (p, f). If $\tilde{\beta}(j)p(j)V(j) \geq \tilde{\gamma}(j)$ for some $j \in J$, then there is some optimal plan that does not stop at the state (p, f).

Proof. Suppose $\tilde{\beta}(j)p(j)V(j) > \tilde{\gamma}(j)$. From Theorem 5.1.1, we know that an optimal plan $\xi^* \in \Xi_s$ exists for (p, f) and satisfies

$$\mathbf{R}[p, f, \xi^*] \geq \tilde{\beta}(j)p(j)V(j) - \tilde{\gamma}(j) - [1 - \tilde{\beta}(j)p(j)]g(T_j p, T_j f)$$
$$> 0 + [1 - \tilde{\beta}(j)p(j)]g(T_j p, T_j f) \geq 0,$$

where the last inequality follows from $g \geq 0$. Thus, no optimal plan can stop at (p, f) because it would obtain marginal return of zero for doing so.

If $\tilde{\beta}(j)p(j)V(j) \geq \tilde{\gamma}(j)$, then

$$g(p, f) = \mathbf{R}[p, f, \xi^*] \geq [1 - \tilde{\beta}(j)p(j)]g(T_j p, T_j f).$$

If $g(p, f) = 0$, then it follows that $[1 - \tilde{\beta}(j)p(j)]g(T_j p, T_j f) = 0$ and the plan $\xi^* = (j)$ is optimal (i.e., $\mathbf{R}[p, f, \xi^*] = 0$). If $g(p, f) > 0$, then the search must continue at (p, f). This proves the theorem

Let

$$\tilde{\gamma}(j) = \sup\{\gamma(j, k) : k = 1, 2, \ldots\}$$
$$\underline{\beta}(j) = \inf\left\{\frac{\beta(j, k)}{1 - b(j, k-1)} : k = 1, 2, \ldots\right\} \quad \text{for } j \in J.$$

Theorem 5.1.6. Suppose J is finite and γ is bounded away from 0. If

$$\sum_{j \in J} \frac{\tilde{\gamma}(j)}{\underline{\beta}(j) V(j)} \leq 1, \tag{5.1.3}$$

then there is an optimal search and stop plan with stopping time $s = \infty$ for any state (p, f).

Proof. Suppose the conclusion of the theorem does not hold. Then there is a state (p, f) such that all optimal plans stop if they reach this state. By Lemma 5.1.5, this implies

$$\underline{\beta}(j) p(j) V(j) < \tilde{\gamma}(j) \quad \text{for } j \in J,$$

which implies

$$p(j) < \frac{\tilde{\gamma}(j)}{\underline{\beta}(j) V(j)} \quad \text{for } j \in J,$$

which in turn yields

$$1 < \sum_{j \in J} \frac{\tilde{\gamma}(j)}{\underline{\beta}(j) V(j)} \leq \sum_{j \in J} \frac{\bar{\gamma}(j)}{\underline{\beta}(j) V(j)}. \tag{5.1.4}$$

But (5.1.4) contradicts (5.1.3), which is assumed to be true. Thus, the conclusion of the theorem must hold.

Corollary 5.1.7. Suppose J is finite. For $j \in J$, let

$$\gamma(j, k) = \gamma_j > 0 \quad \text{and} \quad \beta(j, k) = \alpha_j (1 - \alpha_j)^{k-1} \quad \text{for } k = 1, 2, \ldots,$$

where $\alpha_j > 0$. If

$$\sum_{j \in J} \frac{\gamma_j}{\alpha_j V(j)} \leq 1, \tag{5.1.5}$$

then any locally optimal plan $\xi^* \in \Xi_\infty$ is an optimal search and stop plan.

Proof. Observe that $\beta(j, k)/[1 - b(j, k - 1)] = \alpha_j$ for $j \in J$, $k = 1, 2, \ldots$. Thus $\underline{\beta}(j) = \alpha_j$ and $\bar{\gamma}_j = \gamma_j$ and (5.1.5) yields (5.1.3). Thus, by Theorem 5.1.6 there is an optimal search and stop plan with $s = \infty$ for any state (p, f). Hence, for any state (p, f),

$$\sup_{\xi \in \Xi_s} \mathbf{R}[p, f, \xi] = \sup_{\xi \in \Xi_\infty} \mathbf{R}[p, f, \xi]. \tag{5.1.6}$$

Observe that the ratios $p(j)\beta(j, k)/\gamma(j, k) = p(j)\alpha_j(1 - \alpha_j)^{k-1}/\gamma_j$ are decreasing in k for $j \in J$. Thus, by Theorem 4.3.2 any locally optimal plan ξ^* mini-

5.1 Optimal Search and Stop with Discrete Effort

mizes the mean cost to find the target. Since there are only a finite number of cells, a locally optimal plan is guaranteed to exist. By Theorem 5.1.3,

$$\mathbf{R}[p, f, \xi^*] = \sup_{\xi \in \Xi_\infty} \mathbf{R}[p, f, \xi],$$

and by (5.1.6) ξ^* is an optimal search and stop plan. This proves the theorem.

Sufficient Conditions for $s = 0$ to Be Optimal

We now show that if the expected return from a single look is always nonpositive, then $s = 0$ is an optimal search and stop plan.

Let ε_j be the probability distribution on J that assigns probability one to cell j.

Theorem 5.1.8. *Suppose J is finite and γ is bounded away from 0. If for $j \in J$*

$$\frac{\beta(j, k)}{1 - b(j, k - 1)} V(j) \leq \gamma(j, k) \quad \text{for} \quad k = 1, 2, \ldots, \quad (5.1.7)$$

then $s = 0$ is an optimal search and stop plan.

Proof. By Theorem 5.1.1, there is an optimal search and stop policy ξ^* for (ε_j, f), where f is an arbitrary member of $\hat{F}(J)$. Since $p(j) = 1$,

$$\mathbf{R}[\varepsilon_j, f, \xi^*] = \Pr\{\text{target found}\} V(j) - [\text{expected cost of search}]. \quad (5.1.8)$$

From (5.1.8), it is clear that ξ^* may be chosen so that it makes all of its looks in cell j, for any look in a cell other than j may be removed from the plan with the result of lowering expected cost without affecting the probability of detecting the target. Thus, (5.1.8) may be written [cf. (4.3.1)],

$$\mathbf{R}[\varepsilon_j, f, \xi^*] = \frac{1}{1 - b(j, f(j))} \left\{ V(j) \sum_{k=1}^{s} \beta(j, k + f(j)) - \sum_{k=1}^{s} \gamma(j, k + f(j))[1 - b(j, k - 1 + f(j))] \right\}. \quad (5.1.9)$$

Combining the two summations in (5.1.9) into a single summation, one sees by virtue of (5.1.7) that each term is negative and that $s = 0$ is optimal for (ε_j, f). Thus, $\mathbf{R}[\varepsilon_j, f, \xi] \leq 0$ for any search and stop plan ξ and

$$g(p, f) = \sup_{\xi \in \Xi_s} \sum_{j \in J} p(j) \mathbf{R}[\varepsilon_j, f, \xi] = 0. \quad (5.1.10)$$

Hence, $s = 0$ is optimal, and the theorem is proved.

Properties of Optimal Search and Stop Plans

Here it is shown that if $V(j)$ does not depend on j [i.e., $V(j) = V_0$ for $j \in J$], then an optimal search and stop plan either performs its next look in a cell with the highest value of $p(j)\tilde{\beta}(j)/\tilde{\gamma}(j)$ or stops searching. In the case where $p(j)\beta(j, \cdot)/\gamma(j, \cdot)$ is decreasing for $j \in J$, it is shown that the optimal search and stop plan is determined solely by the stopping time s, in the sense that it is optimal to follow any locally optimal search plan up to the stopping time s. In this case, the problem is converted into an optimal stopping problem of the type considered by Chow et al. (1971).

For any search and stop plan ξ, let (i, j, ξ) be the plan that first searches cell i, then cell j, and then follows ξ.

Lemma 5.1.9. *For any $\xi \in \Xi_s$ such that $\mathbf{R}[p, f, \xi] < \infty$,*

$$\mathbf{R}[p, f, (i, j, \xi)] \geq \mathbf{R}[p, f, (j, i, \xi)] \tag{5.1.11}$$

if and only if

$$p(i)\tilde{\beta}(i)/\tilde{\gamma}(i) \geq p(j)\tilde{\beta}(j)/\tilde{\gamma}(j). \tag{5.1.12}$$

In addition, strict inequality holds in (5.1.11) *if and only if it holds in* (5.1.12).

Proof. If $i = j$, the result holds trivially. Suppose $i \neq j$:

$$\mathbf{R}[p, f, (i, j, \xi)] = \tilde{\beta}(i)p(i)V(i) - \tilde{\gamma}(i)$$
$$+ [1 - \tilde{\beta}(i)p(i)]\left\{\frac{\tilde{\beta}(j)p(j)V(j)}{1 - \tilde{\beta}(i)p(i)} - \tilde{\gamma}(j)\right.$$
$$+ \left[1 - \frac{\tilde{\beta}(j)p(j)}{1 - \tilde{\beta}(i)p(i)}\right]\mathbf{R}[T_j T_i p, T_j T_i f, \xi]\right\}$$
$$= \tilde{\beta}(i)p(i)V(i) - \tilde{\gamma}(i) + \tilde{\beta}(j)p(j)V(j) - \tilde{\gamma}(j) + \tilde{\beta}(i)p(i)\tilde{\gamma}(j)$$
$$+ [1 - \tilde{\beta}(i)p(i) - \tilde{\beta}(j)p(j)]\mathbf{R}[T_j T_i p, T_j T_i f, \xi].$$

One may obtain $\mathbf{R}[p, f, (j, i, \xi)]$ by simply interchanging i and j in the above formula. Since $T_j T_i p = T_i T_j p$ and $T_j T_i f = T_i T_j f$, it follows that

$$\mathbf{R}[p, f, (i, j, \xi)] - \mathbf{R}[p, f, (j, i, \xi)] = \tilde{\beta}(i)p(i)\tilde{\gamma}(j) - \tilde{\beta}(j)p(j)\tilde{\gamma}(i), \tag{5.1.13}$$

and the lemma is proved.

Theorem 5.1.10. *Let J be finite and γ be bounded away from 0. Assume that $p(j)\beta(j, \cdot)/\gamma(j, \cdot)$ is decreasing for $j \in J$. Suppose (p, f) is the state of the search and*

$$p(i)\tilde{\beta}(i)/\tilde{\gamma}(i) = \max_{j \in J} p(j)\tilde{\beta}(j)/\tilde{\gamma}(j). \tag{5.1.14}$$

5.1 Optimal Search and Stop with Discrete Effort

Then the following hold:

(1) *If* $\tilde{\beta}(i)p(i)V(i) \geq \tilde{\gamma}(i)$, *then there is an optimal plan* $\xi^* = (\xi_1, \xi_2, \ldots, \xi_s)$ *such that* $\xi_1 = i$.

(2) *Either there is an optimal plan* $\xi^* = (\xi_1, \xi_2, \ldots, \xi_s)$ *such that* $\xi_1 = i$ *or no optimal plan ever looks in cell i*.

(3) *If* $p(i)\tilde{\beta}(i)/\tilde{\gamma}(i) > p(j)\tilde{\gamma}(j)/\tilde{\beta}(j)$ *for* $j \neq i$, *then either every optimal plan looks in cell i first or no optimal plan ever looks in cell i*.

Proof. To prove (1), we first show that there is some optimal strategy that looks in cell i. Suppose no such strategy exists. Then by Theorem 5.1.1, there is an optimal strategy ξ^*, and it never looks in cell i. Let p_t denote the posterior target distribution after t looks of plan ξ^* have failed to detect the target. Then $p_t(i) \geq p(i)$ for $t = 1, 2, \ldots$. By Lemma 5.1.5, we may suppose that ξ^* continues searching at each time t so long as the target has not been detected. Thus, $\xi^* \in \Xi_\infty$, and by Theorem 5.1.3 any plan that minimizes the expected cost to find the target is optimal. By Theorem 4.3.2, any locally optimal plan minimizes the expected cost to find the target, and thus there is an optimal plan that looks in i. This contradicts the assumption that no such plan exists.

Let $\xi^* = (\xi_1, \xi_2, \ldots, \xi_s)$ be an optimal search and stop plan that searches cell i for the first time on the nth look. Let f_n be the allocation that results from adding n looks of the plan ξ^* to the initial allocation f. Let $\tilde{\beta}_n$ and $\tilde{\gamma}_n$ denote the functions $\tilde{\beta}$ and $\tilde{\gamma}$ associated with the state (p_n, f_n). Then

$$\tilde{\beta}_n(j) = \frac{\beta(j, f_n(j) + 1)}{1 - b(j, f_n(j))} \quad \text{and} \quad \tilde{\gamma}_n(j) = \gamma(j, f_n(j) + 1) \quad \text{for } j \in J.$$

Let D be the probability that the target is not detected by time n, given that it was not detected by the allocation f. Then

$$p_n(j) = \frac{1}{D} p(j) \frac{1 - b(j, f_n(j))}{1 - b(j, f(j))}.$$

and

$$\frac{p_n(j)\tilde{\beta}_n(j)}{\tilde{\gamma}_n(j)} = \frac{1}{D} \frac{p(j)\beta(j, f_n(j) + 1)}{[1 - b(j, f(j))]\gamma(j, f_n(j) + 1)}.$$

Suppose $n > 1$. Then since $f_{n-2}(i) = f(i)$ and the ratios $p(j)\beta(j, \cdot)/\gamma(j, \cdot)$ are decreasing for $j \in J$, it follows that

$$p_{n-2}(i)\tilde{\beta}_{n-2}(i)/\tilde{\gamma}_{n-2}(i) = \max_{j \in J} p_{n-2}(j)\tilde{\beta}_{n-2}(j)/\tilde{\gamma}_{n-2}(j).$$

Thus, by Lemma 5.1.9, there is an optimal plan that looks in cell i on step $n - 1$. By induction, it follows that there is an optimal plan that looks in cell i on the first look.

To prove (2), we observe that the above argument has shown that if there is an optimal plan that looks in cell i, there is one that looks in cell i first. Finally, (3) follows from the above argument and the fact that in Lemma 5.1.9 strict equality in (5.1.11) holds if and only if strict inequality holds in (5.1.12). This proves the theorem.

Theorem 5.1.11. *Suppose that the assumptions of Theorem* 5.1.10 *hold and, in addition,* $V(j) = V_0 > 0$ *for* $j \in J$. *Then the following hold*:

(1) *Either there is an optimal plan that looks in cell i first or $s = 0$ is the only optimal plan.*

(2) *If $p(j)\tilde{\beta}(j)/\tilde{\gamma}(j) < p(i)\tilde{\beta}(i)/\tilde{\gamma}(i)$, then no optimal plan looks in cell j first.*

Proof. To prove (1), we observe that if there exists an optimal plan that looks in cell i, then by Theorem 5.1.12, there is one that looks in cell i first. Thus, we may prove (1) by showing that if no optimal plan searches cell i, then $s = 0$ is the only optimal plan.

Suppose no optimal plan searches cell i and that $\xi^* = (\xi_1, \ldots, \xi_s)$ is optimal. Then $s < \infty$. For if $s = \infty$ and the plan never looks in cell i, then $\mathbf{R}[p, f, \xi^*] = -\infty$. Suppose $s \neq 0$ and $j = \xi_s$. Since j is the last cell to be looked in, it follows that

$$\tilde{\beta}_{s-1}(j)p_{s-1}(j)V_0 \geq \tilde{\gamma}_{s-1}(j),$$

or else it would be better not to make a last look in cell j. Since the search does not look in cell i and the ratios $p(j)\beta(j, \cdot)/\gamma(j, \cdot)$ are decreasing, an argument similar to the one given in Theorem 5.1.10 shows that

$$\frac{\tilde{\beta}_{s-1}(i)p_{s-1}(i)}{\tilde{\gamma}_{s-1}(i)} \geq \frac{\tilde{\beta}_{s-1}(j)p_{s-1}(j)}{\tilde{\gamma}_{s-1}(j)} \geq \frac{1}{V_0}.$$

However, by (1) of Theorem 5.10 there is an optimal plan that searches cell i at time s, which contradicts our assumption. Thus, we must have $s = 0$. This proves (1).

To prove (2), suppose there is an optimal plan that searches j first. The argument in the preceding paragraph shows that there exists such a plan, which looks in cell i at some time. By use of Lemma 5.1.9 and a series of transpositions, one can show that a strictly larger expected return is obtained by searching cell i first, contradicting the optimality of the plan that searches j first. This proves (2) and the theorem.

Under the conditions of Theorem 5.1.11, we see that as long as one continues to search, he follows a locally optimal plan. Since the sequence of ratios $\{p(j_n)\tilde{\beta}_n(j_n)/\tilde{\gamma}_n(j_n) : n = 1, 2, \ldots\}$ is the same for any locally optimal plan in Ξ_∞, it follows from Lemma 5.1.9 that two locally optimal plans coupled with the same stopping time produce the same expected return. This

5.1 Optimal Search and Stop with Discrete Effort

means that in order to find an optimal search and stop plan, one can choose a fixed locally optimal plan and find the optimal stopping rule for that plan. Thus, in this case the problem of optimal search and stop is reduced to an optimal stopping problem. Unfortunately, the optimal stopping problem is not easy to solve.

Counterexamples

Here counterexamples to two conjectures are given to illustrate some of the complexities of optimal search and stop problems. Both the counterexamples are due to Ross (1969).

Example 5.1.12. Consider the following conjecture:

(1) If $\gamma(1, k) > V(1)$ for $k = 1, 2, \ldots$, then an optimal plan never looks in cell 1.

The following example shows conjecture (1) to be false. Let $J = \{1, 2\}$,

$$\beta(1, 1) = \beta(2, 1) = 1, \quad p(1) = \tfrac{3}{4}, \quad p(2) = \tfrac{1}{4},$$
$$\gamma(1, k) = 5, \quad \gamma(2, k) = 10 \quad \text{for} \quad k = 1, 2, \ldots,$$
$$V(1) = 0, \quad V(2) = 210.$$

If the searcher looks first in cell 2 and then proceeds optimally (i.e., stops searching), his expected return is

$$\tfrac{1}{4}(210) - 10 = 42\tfrac{1}{2}.$$

If the searcher looks first in cell 1 and proceeds optimally (i.e., looks in cell 2 if the target has not been found), then his expected return is

$$\tfrac{1}{4}(200) - 5 = 45.$$

Thus, the optimal plan calls for the first look to be in cell 1.

Example 5.1.13. Consider the following conjecture:

(2) If an optimal strategy does not stop at a state (p, f), then it searches a box with maximal $p(j)\tilde{\beta}(j)/\tilde{\gamma}(j)$.

The following example shows conjecture (2) to be false. Let $f = 0$ and consider the same situation as in Example 5.1.14, but with

$$\gamma(1, k) = \gamma(2, k) = 10 \quad \text{for} \quad k = 1, 2, \ldots.$$

If the searcher first looks in cell 1, his maximum return is $200/4 - 10 = 40$. If the searcher first looks in cell 2, his maximum return is $210/4 - 10 = 42\tfrac{1}{2}$.

Thus, the optimal search plan looks first in cell 2. However,

$$\frac{p(2)\tilde{\beta}(2)}{\tilde{\gamma}(2)} = \frac{1}{40} < \frac{3}{40} = \frac{p(1)\tilde{\beta}(1)}{\tilde{\gamma}(1)}.$$

Observe that when $V(j) = V_0$ for $j \in J$, Theorem 5.1.11 shows conjecture (2) to be true.

Remark 5.1.14. In Ross (1969, Example 3) Ross shows that when there are n cells, it is not true that the optimal n-stage look-ahead strategy is the optimal strategy. At each time t, an optimal n-stage look-ahead strategy proceeds by calculating the strategy that maximizes the expected return over the next n time periods and then performing the first look of this strategy. Chew (1973) finds conditions under which an optimal n-stage look-ahead strategy is optimal for a search with n cells. His conditions pertain to searches in which some cell is unsearchable, i.e., for some $j_0 \in J$ such that $p(j_0) > 0$, either $\beta(j_0, k) = 0$ or $\gamma(j_0, k) = \infty$ for $k = 1, 2, \ldots$.

In the form in which Chew (1967, 1973) states the optimal search and stop problem, there are no rewards for finding the target but instead a penalty D for stopping the search before finding the target. The method of proof of Theorem 5.1.1 shows that no extra generality is achieved by adding a penalty cost D for failing to find the target. For by that method of proof, the optimal stop and search plan for the problem with penalty cost is shown to be the same as the optimal plan with target values $\hat{V}(j) = V(j) + D$ for $j \in J$ and no penalty cost.

In general, one cannot find an optimal search plan ξ^*. However, Ross (1969) and Chew (1973) give methods of approximating the optimal plan ξ^* for searches in which for $j \in J$, $\gamma(j, k) = \gamma_j$ and $\beta(j, k) = \alpha_j(1 - \alpha_j)^{k-1}$ for $k = 1, 2, \ldots$. The approximations in Chew (1973) are designed to deal with searches with an unsearchable cell.

5.2. OPTIMAL SEARCH AND STOP WITH CONTINUOUS EFFORT

This section considers the problem of optimal search and stop when search effort is continuous. In contrast to Section 5.1, the value of the detected target is assumed to be constant over space but allowed to vary in time. Thus, there is a nonnegative function $\mathbf{V}$ such that $\mathbf{V}(t)$ gives the value of the target if it is found at time t. For this section, a substantially simpler cost structure is assumed; that is, a cumulative effort function M is fixed, and it is assumed that for this fixed function M there is a cost function $\mathbf{C}$ such that $\mathbf{C}(t)$ gives the cost of the search up to time t regardless of the plan $\varphi \in \Phi(M)$ that is followed.

A search and stop plan is a pair (φ, s), where $\varphi \in \Phi(M)$ and s is a stopping time such that the plan φ is followed until either the target is detected or time

5.2 Optimal Search and Stop with Continuous Effort

s is reached. The stopping time s may take on any value in $[0, \infty]$. An optimal search and stop plan is one that maximizes expected return. This is defined precisely below.

The following lemmas are used to prove Theorems 5.2.3 and 5.2.4.

Lemma 5.2.1. *Let the search space be X, with M continuous and $b(x, \cdot)$ continuous for $x \in X$. For $\varphi \in \Phi(M)$, let $\mathbf{P}_\varphi(t) = P[\varphi(\cdot, t)]$ for $t \geq 0$. Then $\mathbf{P}_\varphi$ is continuous for $\varphi \in \Phi(M)$.*

Proof. Recall that $\varphi \in \Phi(M)$ implies that $\varphi(x, \cdot)$ is increasing for $x \in X$ and $\int_X \varphi(x, t) \, dx = M(t)$ for $t \geq 0$. Let $\varphi \in \Phi(M)$. Since M is continuous,

$$\lim_{t \to t_0} \int_X \varphi(x, t) \, dx = \int_X \varphi(x, t_0) \, dx. \tag{5.2.1}$$

The monotone nature of $\varphi(x, \cdot)$ and (5.2.1) imply $\lim_{t \to t_0} \varphi(x, t) = \varphi(x, t_0)$ for a.e. $x \in X$. By the dominated convergence theorem,

$$\lim_{t \to t_0} \mathbf{P}_\varphi(t) = \lim_{t \to t_0} \int_X p(x) b(x, \varphi(x, t)) \, dx$$
$$= \int_X p(x) b(x, \varphi(x, t_0)) \, dx$$
$$= \mathbf{P}_\varphi(t_0),$$

and the lemma is proved.

Lemma 5.2.2. *Let the search space be J, with M continuous, and $b(j, \cdot)$ continuous for $j \in J$. For $\varphi \in \Phi(M)$, let $\mathbf{P}_\varphi(t) = P[\varphi(\cdot, t)]$ for $t \geq 0$. Then $\mathbf{P}_\varphi$ is continuous for $\varphi \in \Phi(M)$.*

Proof. The proof is similar to that of Lemma 5.2.1 and is not given.

Theorems 5.2.3 and 5.2.4, proved next, give conditions under which the optimal search and stop problem is reduced to a problem of finding an optimal stopping time; that is, they show that the optimal search and stop plan is obtained by optimally stopping a uniformly optimal search plan. Let $\mathbf{R}_\varphi(s)$ be the expected return obtained from using search plan φ and stopping time s. Then

$$\mathbf{R}_\varphi(s) = \int_0^s V(t) \mathbf{P}_\varphi(dt) - \int_0^s C(t) \mathbf{P}_\varphi(dt) - [1 - \mathbf{P}_\varphi(s)] C(s), \tag{5.2.2}$$

where the above integration is understood in the Lebesgue–Stieltjes sense. The plan (φ^*, s^*) is *optimal within* $\Phi(M)$ if

$$\mathbf{R}_{\varphi^*}(s^*) \geq \mathbf{R}_\varphi(s) \quad \text{for all} \quad (\varphi, s) \text{ such that} \quad \varphi \in \Phi(M), \quad s \in [0, \infty].$$

Theorem 5.2.3. Let the conditions of Lemma 5.2.1 be satisfied. In addition, assume that $b(x, \cdot)$ is increasing for $x \in X$, $\mathbf{V}$ is decreasing, $\mathbf{C}$ is increasing, and $\mathbf{P}_\varphi(0) = 0$ for $\varphi \in \Phi(M)$. Then there is a $\varphi^* \in \Phi(M)$ that is uniformly optimal within $\Phi(M)$ and

$$\mathbf{R}_{\varphi^*}(s) \geq \mathbf{R}_\varphi(s) \quad \text{for} \quad \varphi \in \Phi(M), \quad s \in [0, \infty]. \tag{5.2.3}$$

Proof. By Lemma 5.2.1, $\mathbf{P}_\varphi$ is continuous for $\varphi \in \Phi(M)$. Since $\mathbf{V}$ and $\mathbf{C}$ are monotone, the integration in (5.2.2) may be taken to be Riemann–Stieltjes and the standard integration by parts formula holds. Thus,

$$\mathbf{R}_\varphi(s) = \int_0^s \mathbf{P}_\varphi(t)\, d[-V(s)] + \mathbf{V}(s)\mathbf{P}_\varphi(s) - \int_0^s [1 - \mathbf{P}_\varphi(t)]\, dC(t) - \mathbf{C}(0).$$

By Theorem 2.4.6, there is a φ^* that is uniformly optimal within $\Phi(M)$. Since $\mathbf{P}_{\varphi^*}(t) \geq \mathbf{P}_\varphi(t)$ for $t \geq 0$ and $\varphi \in \Phi(M)$, (5.2.3) follows and the theorem is proved.

The following theorem is a discrete search space counterpart of the above. In distinction to Theorem 5.2.3, $b(j, \cdot)$ is assumed to be concave in Theorem 5.2.4. This is necessary in order to use Theorem 2.4.10 to guarantee the existence of a uniformly optimal plan within $\Phi(M)$.

Theorem 5.2.4. Let the conditions of Lemma 5.2.2 be satisfied. In addition, assume that $b(j, \cdot)$ is concave and increasing for $j \in J$, $\mathbf{V}$ is decreasing, $\mathbf{C}$ is increasing, and $\mathbf{P}_\varphi(0) = 0$ for $\varphi \in \Phi(M)$. Then there is a $\varphi^* \in \Phi(M)$ that is uniformly optimal and

$$\mathbf{R}_{\varphi^*}(s) \geq \mathbf{R}_\varphi(s) \quad \text{for} \quad \varphi \in \Phi(M), \quad s \in [0, \infty].$$

Proof. The proof follows in the same manner as that of Theorem 5.2.3.

If one can find φ^* in the situation of Theorems 5.2.3 and 5.2.4, then the problem of finding an optimal search and stop plan is reduced to finding s^* to maximize $\mathbf{R}_{\varphi^*}$. However, the above theorems do not gurantee the existence of such an s^*. The following theorem gives rather specialized conditions under which one may find s^*, the optimal stopping time.

Theorem 5.2.5. Assume that the conditions of Theorem 5.2.3 or 5.2.4 are satisfied. In addition, suppose that $\mathbf{C}'$, the derivative of $\mathbf{C}$, is positive and increasing and that for a plan φ^* that is uniformly optimal within $\Phi(M)$, $\mathbf{P}_{\varphi^*}'$, the derivative of $\mathbf{P}_{\varphi^*}$, exists and $\mathbf{P}_{\varphi^*}'/(1 - \mathbf{P}_{\varphi^*})$ is decreasing. Let

$$s^* = \inf\left\{ t : \frac{\mathbf{P}_{\varphi^*}'(t)}{1 - \mathbf{P}_{\varphi^*}(t)} \leq \frac{\mathbf{C}'(t)}{\mathbf{V}(t)} \right\}. \tag{5.2.4}$$

Then (φ^*, s^*) is an optimal search and stop plan within $\Phi(M)$.

5.2 Optimal Search and Stop with Continuous Effort

Proof. Under the assumptions of this theorem, for $\varphi = \varphi^*$ one can write (5.2.2) as

$$\mathbf{R}_{\varphi^*}(s) = \int_0^s \{\mathbf{V}(t)\mathbf{P}_{\varphi^*}'(t) - [1 - \mathbf{P}_{\varphi^*}(t)]\mathbf{C}'(t)\} \, dt - \mathbf{C}(0). \quad (5.2.5)$$

Observe that the integrand in (5.2.5) is positive if and only if

$$\frac{\mathbf{P}_{\varphi^*}'(t)}{1 - \mathbf{P}_{\varphi^*}(t)} > \frac{\mathbf{C}'(t)}{\mathbf{V}(t)}.$$

The ratio on the left is decreasing by assumption, and the ratio on the right is increasing by virtue of $\mathbf{V}$ decreasing and $\mathbf{C}'$ increasing. Thus, for $0 \leq t < s^*$ defined in (5.2.4), the integrand is positive and for $s^* < t \leq \infty$, the integrand is less than or equal to zero. Thus, from (5.2.5) it is clear that s^* is an optimal stopping time. This proves the theorem.

Example 5.2.6. In this example, Theorem 5.2.5 is applied to the problem of optimal search and stop when the target distribution is circular normal and the detection function is exponential [i.e., $b(x, z) = 1 - e^{-z}$ for $x \in X, z \geq 0$].

Let the functions $\mathbf{V}$ and $\mathbf{C}' = \mathbf{c}$ be constant and let $M(t) = Wvt$, where W is the sweep width of the sensor and v the constant speed of the search vehicle (see Example 2.2.7). Let $\mathbf{C}(0) = 0$. From Example 2.2.7, we have

$$\mathbf{P}_{\varphi^*}(t) = 1 - (1 + H\sqrt{t}) \exp(-H\sqrt{t}),$$

where

$$H = (Wv/\pi\sigma^2)^{1/2}.$$

Then

$$\frac{\mathbf{P}_{\varphi^*}'(t)}{1 - \mathbf{P}_{\varphi^*}(t)} = \frac{H^2}{2(1 + H\sqrt{t})}.$$

Thus, s^* is the solution to

$$H^2/2(1 + H\sqrt{s^*}) = \mathbf{c}/\mathbf{V},$$

i.e.,

$$s^* = \begin{cases} H^{-2}\left(\dfrac{H^2\mathbf{V}}{2\mathbf{c}} - 1\right)^2 & \text{if } H^2 > 2\mathbf{c}/\mathbf{V}, \\ 0 & \text{otherwise.} \end{cases}$$

In addition, one may calculate

$$\mathbf{R}_{\varphi^*}(s^*) = \begin{cases} \mathbf{V}\left[1 - \dfrac{6\mathbf{c}}{\mathbf{V}H}\left\{1 - \left(\dfrac{\mathbf{V}H^2}{6\mathbf{c}} + \dfrac{1}{3}\right) \exp\left[-\left(\dfrac{H^2\mathbf{V}}{2\mathbf{c}} - 1\right)\right]\right\}\right] & \text{if } H^2 > \dfrac{2\mathbf{c}}{\mathbf{V}}, \\ 0 & \text{otherwise.} \end{cases}$$

One may check (Richardson and Belkin, 1972) for calculations of optimal stopping times and expected returns in some situations involving uncertain sweep width.

Remark 5.2.7. If one could find conditions that guarantee that the ratio $\mathbf{P}_{\varphi*}'/(1 - \mathbf{P}_{\varphi*})$ is decreasing, then by virtue of Theorem 5.2.5 one could assert the existence of optimal search and stop plans under those conditions. The purpose of this remark is to show that the assumption that b is a regular detection function is not enough to guarantee that the above ratio is decreasing.

Let X be the plane and let the target distribution be uniform over a unit square in the plane. Assume

$$b(z) = 1 - \exp\left\{-\int_0^z [1 + \ln(1 + y)] \, dy\right\} \quad \text{for } z \geq 0,$$

where the variable x is suppressed in writing $b(z)$. Then

$$b'(z) = [1 + \ln(1 + z)] \exp\left\{-\int_0^z [1 + \ln(1 + y)] \, dy\right\}.$$

One may check that $b''(z) < 0$ for $z > 0$ and that b is regular.

Suppose $M(t) = t$ for $t \geq 0$. Then

$$\frac{\mathbf{P}_{\varphi*}'(t)}{1 - \mathbf{P}_{\varphi*}(t)} = \frac{b'(t)}{1 - b(t)} = 1 + \ln(1 + t) \quad \text{for } t \geq 0.$$

Thus, $\mathbf{P}_{\varphi*}'/(1 - \mathbf{P}_{\varphi*})$ is increasing even though b is regular.

Remark 5.2.8. It was noted in the beginning of this section that the cost structure assumed here is substantially simpler that the one used in Section 5.1, in that cost does not depend on the search plan. In order to use a cost structure similar to Section 5.1, one could adopt a model similar to the one to be given in Chapter VI; that is, the plans are constrained to be in the class $\Phi(M)$, but the cost of following a plan φ for time t is given by

$$C[\varphi(\cdot, t)] = \int_X c(x, \varphi(x, t)) \, dx,$$

where c is a specified nonnegative Borel function. Then among search and stop plans (φ, s) for which $\varphi \in \Phi(M)$ and $s = \infty$, one can show, just as in Theorem 5.1.3, that the plan φ^* that minimizes expected cost to detect the target will be optimal. Thus, the plan φ^* is the one given in Theorem 6.4.3 for search space X and in Theorem 6.4.6 for search space J.

The possibility of extending the remaining results in Section 5.1 to the case of continuous effort has not been explored yet.

NOTES

The results in Section 5.1 extend those of Ross (1969) who considered the situation where the detection function is geometric, i.e., for $j \in J$, there is an $\alpha_j > 0$ such that $\beta(j, k) = \alpha_j(1 - \alpha_j)^{k-1}$ for $k = 1, 2, \ldots$, and the cost function is independent of k, i.e., $\gamma(j, k) = \gamma_j$ for $j \in J$, $k = 1, 2, \ldots$. In this situation, the state of the system is specified completely by the posterior target distribution. This results from the lack of memory of the geometric detection function, i.e.,

$$\frac{\beta(j, k+1)}{1 - b(j, k)} = \alpha_j \quad \text{for} \quad j \in J, \quad k = 1, 2, \ldots,$$

and the constancy of the cost function. With the exception of Theorem 5.1.8, the proofs in Section 5.1 are extensions of those given by Ross (1969) to the more general situation of Section 5.1.

The development in Section 5.2 is motivated by results in Richardson and Belkin (1972) where restricted versions of Theorems 5.2.3 and 5.2.5 appear. In particular, for the counterpart of Theorem 5.2.3, Richardson and Belkin (1972) assume that b is regular and that $p(x)b'(x, z)$ is bounded. The counterpart of Theorem 5.2.5 is further restricted to target distributions that are either uniform or circular normal.

Prior to Ross (1969), Chew (1967) proved a version of Theorem 5.1.11 for the case of a geometric detection function, $\gamma(j, k) = \gamma_0$ and $V(j) = V_0$. Chew allows one cell to be unsearchable (see Remark 5.1.14). Under these assumptions, Chew also proved the existence of optimal search and stop plans and found methods of approximating optimal plans. Further discussion of approximating optimal plans is given by Ross (1969) and Chew (1973). Chew (1973) also finds conditions under which an optimal n-stage look-ahead policy is optimal (see Remark 5.1.14).

A discussion of the nature of optimal policy regions for search and stop problems is given by Kan (1974).

Chapter VI

Search in the Presence of False Targets

This chapter discusses the problem of optimal search in the presence of objects, false targets, that may be mistaken for the target. Section 6.1 discusses false-target models and models for search in the presence of false targets. Section 6.2 introduces and motivates the use of minimization of mean time as the optimality criterion for searches in the presence of false targets. Section 6.3 gives a concise statement of the search problem in the presence of false targets. Then optimal nonadaptive search plans are found under the restriction that contact investigation be immediate and conclusive. It is also shown that this same plan is optimal in an extended class of plans that allows one to delay contact investigation. The proof of the optimality of these plans is based on optimization theorems given in Section 6.4. These theorems are generalizations of Theorems 2.2.4 and 2.2.5.

In Section 6.5 examples are given to show that the effect of false targets is to slow the rate of expansion of a search and that there is a situation in which there is no uniformly optimal nonadaptive plan even though the search space is X and the detection function is regular. Section 6.6 introduces semiadaptive plans, which combine the optimal nonadapative plan with some feedback to obtain a plan with a smaller mean time than the optimal nonadaptive plan. The mean time for a semiadaptive plan is not as small as the mean time for an optimal adaptive plan. However, optimal adaptive plans have been found only

6.1 *False-Target Models* 137

in special cases, whereas semiadaptive plans may be obtained for a broad range of searches.

Examples are given to show how a semiadaptive plan is constructed and to show that the optimal nonadapative plan coincides with the semiadaptive plan in the case of Poisson-distributed false targets. Finally an application of the notion of semiadaptive plans to searches for multiple targets is outlined.

6.1. FALSE-TARGET MODELS

In the preceding chapters it has been assumed that the sensor or detection device has perfect discrimination, that is, a detection will be called only when the target is present. Of course, most sensors do not have perfect discrimination. When searching with an active sonar for a submarine on the ocean bottom, a large rock of roughly the same dimensions as the submarine may generate a response similar to that generated by a submarine. Magnetic anomalies in the earth's crust may cause a magnetometer to register a response resembling that caused by a submarine. A radar looking for a ship in distress on the ocean surface may detect another ship of similar size and not be able to distinguish it from the one in distress without a close visual inspection. For some sensors, random fluctuations in background noise (possibly electrical or acoustical) may cause a detection to be called when no target is present.

Whenever a detection is called, whether on the target or not, one says that a *contact* is made. In this chapter it is assumed that contacts are caused either by the target or by real stationary objects. Any object that is capable of causing a contact and is not the target is called a *false target*. The search is assumed to take place in two phases. The first phase, called *broad search*, is performed with a sensor capable of detecting the target and also subject to calling a detection on a false target. This sensor is called the *broad-search sensor*. Once a contact is made the searcher has the option of ignoring the contact until later or entering the second phase, i.e., investigation of the contact to determine if it is the target. In order to enter the second phase, broad search must stop. This second phase is called *contact investigation*, and the searcher may choose to interrupt this to return to broad search. The searcher also has the ability to mark the location of contacts so that having investigated a contact once, he will not have to investigate it again should it be redetected.

A search situation such as the above may occur in underwater search. One wishes to search with a sensor having a large sweep width in order to reduce the time necessary to detect the target. However, sensors with large sweep widths such as magnetometers or side-looking sonars are often unable to distinguish between the target, say a sunken submarine, and a wreck or large magnetic rock. Thus, when a contact is made it is necessary to investigate it in order to identify it. This may be done by steering close to the contact to look at it through the viewport of a submersible, by lowering a camera or tv to take a picture of it, or by sending divers down to look at it. Typically, the sensor used for contact identification has a much smaller sweep width than the broad-search sensor.

Possibly one could eliminate false-target problems in the above example by performing a purely visual search, but this will usually require a larger total amount of time to find the target, because underwater viewing ranges are much shorter than detection ranges for sonar or magnetometer. In this chapter, we consider only searches in which one must detect the target as a contact with the broad-search sensor and then perform a contact investigation to identify the contact. In this case, the question is how to divide search effort between broad search and contact investigation and, further, how to allocate effort within each of these divisions. Having found the optimal plan for a search involving a broad-search sensor and contact investigation, one could compare it to the optimal plan for a purely visual search to decide which method of search to use.

Before attacking the problem of allocating effort between broad search and contact investigation, we discuss broad-search plans, contact investigation policies, and false-target models.

Broad-Search Plans

Allocations that refer to broad search are called *broad-search allocations*. Similarly, a function $\varphi: X \times [0, \infty) \to [0, \infty)$ is called a *broad-search plan* if

(i) $\varphi(\cdot, t) \in F(X)$ is an allocation for $t \in [0, \infty)$,
(ii) $\varphi(x, \cdot)$ is increasing for $x \in X$.

Mathematically, a broad-search plan is the same as a search plan defined in Section 2.2. The term broad-search plan is used simply to indicate that broad-search effort rather than contact investigation effort is being allocated. Let $\Phi(M)$ be the class of broad-search plans such that

$$\int_X \varphi(x, t) \, dx = M(t) \quad \text{for} \quad t \geq 0.$$

Note that the class $\Phi(M)$ defined here is identical to that defined in Section 2.2.

Analogous definitions of broad-search plan and the class $\Phi(M)$ are understood to hold for the discrete search space J.

In order to explain the meaning of M, it is convenient to distinguish two types of time, broad-search time and total time. Broad-search time refers to time

6.1 False-Target Models

spent in broad search and total time refers to the cumulative time spent in broad search and contact investigation. Since broad search must stop in order for contact investigation to begin, one might think of broad-search time as being measured by a clock that runs only during the time when broad-search effort is being applied. One now can say that $M(t)$ gives the amount of broad-search effort available by broad-search time t.

Contact Investigation Policies

If a contact investigation policy requires that once a contact investigation is begun it must be continued until the contact is identified as the target or a false target, then it is called a *conclusive contact investigation* policy. When using conclusive contact investigation, the searcher is still allowed to delay the beginning of a contact investigation at his will. However, once begun it must be carried to a conclusion. If, in addition, the search is required to investigate contacts as soon as they are detected, then one is using an *immediate and conclusive contact investigation* policy.

Independent Identically Distributed False-Target Model

Let the search space be X, and assume that there is a known distribution on the number $\mathbf{N}$ of false targets in the search space, that is, $\mathbf{N}$ is a random variable and the probabilities

$$\nu(N) = \Pr\{\mathbf{N} = N\} \quad \text{for} \quad N = 0, 1, 2, \ldots \tag{6.1.1}$$

are assumed to be known. Also assumed known is the joint density function q^N for the distribution of the location of the false targets given $\mathbf{N} = N$; that is, given that there are N false targets,

$$\int_{S_1}\int_{S_2}\cdots\int_{S_N} q^N(x^1, x^2, \ldots, x^N)\, dx^1 \cdots dx^N \tag{6.1.2}$$

is the probability of the joint occurrence of the ith false target being in S_i, where S_i is a subset of X, for $i = 1, 2, \ldots, N$.

The combination of the number distribution given in (6.1.1) and the conditional location distribution given $\mathbf{N} = N$ serves to define the distribution of false targets. However, the above model will not be used in the generality presented. Instead, the following special case will be considered. Let q be a probability density function on X. If $\mathbf{N} = N$, then it is assumed that the N false targets are independently distributed, each with distribution given by the probability density function q. Thus,

$$q^N(x^1, x^2, \ldots, x^N) = q(x^1)q(x^2)\cdots q(x^N). \tag{6.1.3}$$

There are, however, many reasonable situations in which the false targets will not be independently distributed. Dobbie (1973) points out that one might be in the situation where the false targets are wrecks whose earth coordinates are accurately known, while the searcher's position is not accurately known in these coordinates. Then the location of the false targets is not independently or identically distributed with respect to the searcher. For if the searcher detects and identifies one false target, this will provide information about the location of the other false targets. The reason for restricting attention to situations satisfying (6.1.3) is that no substantial results are known for the more general case.

In addition to describing the distribution and number of false targets, one must specify the mechanism by which false targets are detected. This is done by assuming the existence of a function a such that $a(x, z)$ gives the probability of detecting a false target located at point x with effort density z applied at point x. More formally, a function $a: X \times [0, \infty) \to [0, 1]$ such that $a(x, z)$ gives the probability of detecting a false target at x with effort density z is called a *false target detection function* and is assumed to be Borel measurable.

Suppose that the allocation $f \in F(x)$ has been applied to the search space. For a given false target, the probability of detecting that false target is

$$\int_X q(x)a(x, f(x))\, dx,$$

in analogy with the probability of detecting the real target, which is given by

$$\int_X p(x)b(x, f(x))\, dx.$$

If there are N false targets, each with location distribution given by the density q, then the expected number detected with allocation f is

$$N \int_X q(x)a(x, f(x))\, dx.$$

Thus, if the number of false targets $\mathbf{N}$ is a random variable with expectation $\mathbf{E}[\mathbf{N}]$, the expected number of false targets detected by the allocation f is

$$\sum_{N=1}^{\infty} \Pr\{\mathbf{N} = N\} N \int_X q(x)a(x, f(x))\, dx = \mathbf{E}[\mathbf{N}] \int_X q(x)a(x, f(x))\, dx. \quad (6.1.4)$$

Let $\delta(x) = \mathbf{E}[\mathbf{N}]q(x)$ for $x \in X$. Then δ is called the *false-target density* and (6.1.4) becomes

$$\int_X \delta(x)a(x, f(x))\, dx. \quad (6.1.5)$$

6.1 False-Target Models

Mean Time to Investigate False Targets

Suppose that it requires an expected time $\tau(x)$ to investigate and identify a false target detected at a point x. Since only the mean is specified, the time to investigate a false target can be random or deterministic. The assumption is made here that expected time to investigate a contact depends only on location. If the policy of immediate and conclusive contact investigation is used, then using an argument similar to the one leading to (6.1.4) and (6.1.5), the expected time spent investigating false targets when the allocation f is applied to X is found to be

$$\mathrm{E}[N] \int_X \tau(x) q(x) a(x, f(x))\, dx = \int_X \tau(x) \delta(x) a(x, f(x))\, dx. \quad (6.1.6)$$

Poisson False-Target Model

For the second approach to defining a false-target distribution, a Poisson model for the location and detection of false targets is discussed. In this model, one need specify only the false target density δ, which must be a Borel function. Then for any allocation $f \in F(X)$, the expected number of false targets detected in the Borel set S using that allocation is assumed to be

$$d(S, f) = \int_S \delta(x) a(x, f(x))\, dx. \quad (6.1.7)$$

Let $\Delta(S, f)$ be the random variable that gives the number of false targets detected in S using allocation f. If $d(S, f) < \infty$, it is assumed that

$$\Pr\{\Delta(S, f) = k\} = \frac{1}{k!} [d(S, f)]^k e^{-d(S, f)} \quad \text{for} \quad k = 0, 1, 2, \ldots, \quad (6.1.8)$$

that is, $\Delta(S, f)$ has a Poisson distribution with mean $\mathrm{E}[\Delta(S, f)] = d(S, f)$.

Moreover, it is assumed that the additional false targets detected by an incremental allocation h also have a Poisson distribution, namely,

$$\Pr\{\Delta(S, f + h) - \Delta(S, f) = k\}$$
$$= \frac{1}{k!} [d(S, f + h) - d(S, f)]^k \exp\{-[d(S, f + h) - d(S, f)]\}$$
$$\text{for} \quad k = 0, 1, 2, \ldots. \quad (6.1.9)$$

Finally, if S_1 and S_2 are disjoint Borel subsets of X, then $\Delta(S_1, f)$ is independent of $\Delta(S_2, f)$ for any $f \in F(X)$, i.e., the number of false targets detected in S_1 is independent of the number detected in S_2.

An understanding of the nature of the Poisson assumption may be obtained by considering conditions (a)–(d) below. When these conditions are satisfied, it can be shown (see Karlin, 1968, p. 337) that the false-target distribution is Poisson.

For any Borel subset $S \subset X$ and allocation $f \in F(X)$,

(a) $0 < \Pr\{\Delta(S,f) = 0\} < 1$ if $0 < d(S,f) < \infty$;
(b) The probability distribution of $\Delta(S,f)$ depends only on $d(S,f)$ and

$$\Pr\{\Delta(S,f) \geq 1\} \to 0 \quad \text{as} \quad d(S,f) \to 0;$$

(c) If $S_1, \ldots, S_n$ are disjoint Borel sets, then $\Delta(S_1, f), \ldots, \Delta(S_n, f)$ are mutually independent random variables;
(d) $\lim_{d(S,f) \to 0} \Pr\{\Delta(S, h) \geq 1\}/\Pr\{\Delta(S, h) = 1\} = 1$.

Assumption (a) states that if some broad search has been performed in an area where $\delta > 0$, then there is positive (but not equal to one) probability of contacting no false targets. The first part of (b) says essentially that the number of false contacts does not depend on the shape of the region but only on the area of the region, the false-target density, and the thoroughness of the search. The motivation for the second part of (b) is obvious after one notes that $d(S,f)$ approaches zero whenever the area of S approaches 0, the false contact density in S approaches 0, or the search effort density function approaches 0 in S. The assumption in (c) says that the contacting of false targets in one area does not affect the contacting of them in any disjoint area. This assumption requires, for example, that the operators of the broad-search sensor do not improve their ability to distinguish false targets from the real target as the search progresses. Assumption (d) states that as $d(S, f) \to 0$, the probability of contacting more than one false target becomes small compared to the probability of contacting exactly one.

Note that if one knows only the false-target density δ and if one believes (c), then the Poisson false-target model is a reasonable one to use.

*False-Target Detection as a Poisson Process

Let φ be a broad-search allocation and S a Borel subset of X. Then $\Delta(S, \varphi(\cdot, t))$ is the number of false targets detected in S by broad-search time t, and $\Delta(S, \varphi)$ may be thought of as a nonhomogeneous Poisson process in time. Similarly, for a fixed broad-search time t and broad-search plan φ, one may think of the number of false targets detected as a nonhomogeneous Poisson process in space (see Karlin, 1968, p. 337, for a discussion of such processes in the homogeneous case).

Relation between Poisson and Independent Identically Distributed False-Target Models

The Poisson model of false-target detection implies a spatial distribution of false targets as follows. Let $a(x, f(x)) = 1$ for $x \in X$ in (6.1.7), and let S be

6.1 False-Target Models

a Borel subset of X. Since $a(x, f(x)) = 1$ for $x \in S$, all false targets located in S will be detected and

$$\bar{d}(S) = \int_S \delta(x)\, dx \qquad (6.1.10)$$

gives the expected number of false targets in S. If $\bar{d}(S) < \infty$, then (6.1.8) implies that

$$\frac{1}{k!}[\bar{d}(S)]^k e^{-\bar{d}(S)} \qquad (6.1.11)$$

is the probability of there being k false targets in S for $k = 0, 1, 2, \ldots$. Again, if S_1 and S_2 are disjoint Borel sets, the number of false targets in S_1 is independent of the number in S_2.

If $\bar{d}(X) < \infty$, then the Poisson model may be easily stated in terms of the independent identically distributed model. As before, let $\mathbf{N}$ be the random variable that gives the total number of false targets in X. The proof by Karlin (1968, Theorem 1.2, p. 343) may be easily modified to show that, conditioned on $\mathbf{N} = N$, the distribution of the location of the N false targets is the same as that obtained by making N independent draws from the probability distribution given by the density

$$q(x) = \delta(x)/\bar{d}(X) \qquad \text{for} \quad x \in X. \qquad (6.1.12)$$

Thus, the joint density function for the location of the N false targets is given by (6.1.3), where q is defined by (6.1.12). From (6.1.11), it follows that $\mathbf{N}$ is Poisson distributed with mean $E[\mathbf{N}] = \bar{d}(X)$. Thus, $E[\mathbf{N}]q(x) = \delta(x)$, which coincides with the definition of δ for the independent identically distributed false target model. If $\bar{d}(X) = \infty$, as is the case in Example 6.5.1, then the correspondence between the two models is not so easily made.

As in the independent identically distributed model, it is assumed that the mean time to investigate a false target located at x is $\tau(x)$. It is desired to compute the expected time spent investigating the false targets detected by a broad-search allocation f if immediate and conclusive contact investigation is pursued. To do this, one first partitions X into a sequence $S_1, S_2, \ldots$ of disjoint Borel sets whose union is equal to X and for which $\bar{d}(S_i) < \infty$ for $i = 1, 2, \ldots$.[1] By the same reasoning that led to (6.1.12), one finds that conditioned on there being $\mathbf{N} = N$ false targets in S_i, the distribution of the location of the N false targets is the same as N independent draws from the probability

---

[1] It is always possible to find such a sequence. First one partitions X into a countable number of sets with finite measure. Each of these sets is further partitioned into subsets defined by the points x such that $k \leq \delta(x) < k + 1$ for $k = 0, 1, 2, \ldots$.

distribution defined on S_i by the density $\delta(x)/\bar{d}(S_i)$ for $x \in S_i$. Following the same argument as led to (6.1.6), one obtains

$$\bar{d}(S_i) \int_{S_i} \tau(x) \frac{\delta(x)}{\bar{d}(S_i)} a(x, f(x)) \, dx = \int_{S_i} \tau(x) \delta(x) a(x, f(x)) \, dx, \quad (6.1.13)$$

which is the expected time spent investigating the false targets detected in S_i by allocation f. Thus, the expected time spent identifying the false targets detected in X using the allocation f is obtained by summing over i in (6.1.13) to obtain

$$\int_X \tau(x) \delta(x) a(x, f(x)) \, dx, \quad (6.1.14)$$

which is identical to (6.1.6).

Discrete False-Target Distributions

As with the continuous false-target distributions, two models are presented, the independent identically distributed and Poisson models.

For the independent identically distributed false-target model, it is assumed that the distribution of the number of false targets $\mathbf{N}$ is specified, that is,

$$\nu(N) = \Pr\{\mathbf{N} = N\} \quad \text{for} \quad N = 0, 1, 2, \ldots$$

is given. The location of each false target has the distribution specified by $q(j)$, the probability that the false target is located in cell j for $j \in J$. The location of one false target is assumed to be independent of the location of all other false targets.

Let $a(j, z)$ give the probability of detecting a false target located in cell j with z effort placed in cell j. Again, detection of one false target is assumed to be independent of the detection of any other false target. Define

$$\delta(j) = \mathbf{E}[\mathbf{N}]q(j) \quad \text{for} \quad j \in J.$$

Then δ is the discrete analog of the false-target density defined above. Let $\tau(j)$ be the expected time to identify a false target detected in cell j. An argument similar to the one that led to (6.1.6) shows that

$$\sum_{j \in J} \tau(j) \delta(j) a(j, f(j)) \quad (6.1.15)$$

is the expected amount of time to identify the false targets detected with broad-search allocation f.

For a Poisson model, one specifies only δ_j, the expected number of false

6.2 Criterion for Optimality

targets in cell j for $j \in J$. The distribution of the number of false targets in cell j is assumed to be Poisson, i.e.,

$$\Pr\{N \text{ false targets in cell } j\} = \delta_j^N e^{-\delta_j}/N! \quad \text{for} \quad N = 0, 1, 2, \ldots.$$

The number of false targets in one cell is independent of the number in any other cell.

In addition, the probability of detecting one false target is assumed to be independent of the probability of detecting any other false target. One can show that (6.1.15) also gives the expected time to investigate the false targets detected with broad-search allocation f for the Poisson false-target model.

6.2. CRITERION FOR OPTIMALITY

In this section we discuss the criterion of optimality, which will be used for evaluating search plans in the presence of false targets. We also discuss the class of plans over which one is allowed to optimize. The second discussion leads to the definition of optimal nonadaptive and optimal adaptive plans. A hybrid or semiadaptive plan is also introduced, which combines some of the virtues of adaptive plans with the simplicity and availability of the optimal nonadaptive plan.

In order to avoid confusion, it will be said that the target is *detected* when it appears as a contact and that it has been *identified* when contact investigation shows that contact to be the target. Finally, the target is *found* when it is both detected and identified.

In previous chapters, where there are no false targets to complicate the search, the question of which optimality criterion to use is made almost moot by the existence for most searches (see Section 2.4) of uniformly optimal plans. Such a plan maximizes the probability of detection (i.e., finding the target) at each time t. Thus, it minimizes the time to reach a given probability of detecting the target and the mean time to detect the target.

In this chapter, minimization of the mean time to find the target is the optimization criterion. One would also like to find plans that maximize the probability of finding the target by a given time and uniformly optimal plans. However, finding a plan to maximize the probability of finding the target by a given time appears to be a difficult problem not approachable by the techniques presented in this book. As for uniformly optimal plans, it is shown in Example 6.5.2 that there are situations in which no uniformly optimal plan exists in the class of nonadaptive plans.

The possibility of detecting false targets introduces in a substantial way the possibility of feedback during a search. When there are no false targets, the only feedback is whether the target is detected or not. If the target is detected, the search stops; if not, it continues as planned, in the sense that one can say in a deterministic fashion how the search effort will be allocated, if the target is not detected by a certain time. In the presence of false targets, the detection and identification of a false target may yield information about the distribution of the remaining false targets. It is possible that this information can be used to improve the search plan.

An *adaptive* search plan is one that is allowed to make use of all the feedback obtained from the search, i.e., location and time of detection of contacts, which contacts have been identified as false targets, and whether or not the target has been found. Of course, a nonadaptive plan is a special case of an adaptive plan. Mathematically, a nonadaptive search plan is one whose broad-search plan and contact investigation policy are a function of broad-search time and location only. Of course, the search stops when the target is found.

Ideally, one would like to find an optimal adaptive plan, i.e., a plan that minimizes the mean time to find the target among all adaptive plans. Unfortunately, this is a very difficult problem. The only optimal adaptive search plan in the presence of false targets to be found in the search literature at the present time is given by Dobbie (1973). However, this plan is for a very simple search situation in which there are two cells and exactly one false target known to be in the second cell. In addition, the contact investigation policy is required to be immediate and conclusive. It does not appear that the methods used there can be extended to allow more cells or more false targets. As a result, this chapter will be concerned with finding optimal nonadaptive plans. Having done this, it will be possible to consider semiadaptive plans, which make some use of feedback and thereby yield a smaller mean time to find the target than nonadaptive plans.

6.3. OPTIMAL NONADAPTIVE PLANS

In this section, optimal nonadaptive search plans are found when contact investigation is required to be immediate and conclusive, that is, the optimal broad-search plan φ^* is found for use with immediate and conclusive contact investigation. It is then shown that φ^* plus immediate contact investigation is also optimal in an extended class of plans in which contact investigation may be delayed. All contact investigation policies are required to be conclusive (i.e., once a contact investigation is begun it must be continued until the contact is identified).

Statement of Problem

Before finding optimal nonadaptive plans, a concise statement of the search problem is made for the case where the search space is X. The statement for a discrete search space J will be clear by analogy.

6.3 Optimal Nonadaptive Plans

The target location distribution is given by a density function p defined on X such that

$$\int_X p(x)\,dx = 1,$$

i.e., the target distribution cannot be defective. There is a Borel-measurable broad-search target detection function $b: X \times [0, \infty) \to [0, 1]$ such that if $f \in F(X)$ is a broad-search allocation,

$$P[f] = \int_X p(x)b(x, f(x))\,dx \tag{6.3.1}$$

gives the probability of detecting the target with allocation f.

Correspondingly, for false targets there is given a Borel-measurable false-target density function δ such that for any Borel set $S \subset X$

$$\int_S \delta(x)\,dx \tag{6.3.2}$$

is the expected number of false targets contained in S. There is a Borel-measurable false-target detection function $a: X \times [0, \infty) \to [0, 1]$ such that for $f \in F(X)$

$$\int_X \delta(x)a(x, f(x))\,dx \tag{6.3.3}$$

is the expected number of false targets detected with broad-search allocation f. Recall that false targets are assumed to be real objects that cannot be distinguished from the target by the broad-search sensor. In addition, it is assumed that when a false target is detected and identified, it can be marked or eliminated in such a way that it will not be investigated again.

There is a Borel function τ such that $\tau(x)$ is the mean time to identify a false target detected at $x \in X$. The mean time to investigate the false targets detected by broad-search allocation $f \in F(X)$ is assumed to be

$$\int_X \tau(x)\delta(x)a(x, f(x))\,dx. \tag{6.3.4}$$

Observe that only the false-target density δ has been specified and that it is simply assumed that (6.3.4) gives the mean time to investigate the false targets detected by the broad-search allocation f. It has been shown in Section 6.1 that either a Poisson false-target model or an independent identically distributed false-target model will yield (6.3.4) as the correct expression for mean time to investigate the false targets detected by broad-search allocation f.

For a specified cumulative effort function M, a *search plan* consists of a broad-search plan $\varphi \in \Phi(M)$ and a contact investigation policy. Since only nonadaptive plans are considered, both φ and the contact investigation, policy are allowed to depend only on location and broad-search time. Thus $\varphi: X \times [0, \infty) \to [0, \infty)$ and, since $\varphi \in \Phi(M)$,

(i) $\varphi(\cdot, t) \in F(X)$ for $t \geq 0$,
(ii) $\varphi(x, \cdot)$ is increasing   for $x \in X$,   (6.3.5)
(iii) $\int_X \varphi(x, t) \, dx = M(t)$ for $t \geq 0$.

Recall that $M(t)$ gives the amount of broad-search effort that can be exerted by broad-search time t. The decision to investigate a contact or not must depend only on the location of the contact and the broad-search time. Also, since conclusive contact investigation is assumed, once a contact investigation is begun it must be continued until the contact is identified.

The problem addressed in this section is to find the nonadaptive search plan that minimizes the mean time to find the target. It is assumed that the mean time to investigate the contact that is the target depends only on the location of the target and not on the search plan. Thus, minimizing the mean time to find the target is equivalent to minimizing the mean time until the beginning of the investigation of the contact that is the target.

Immediate and Conclusive Contact Investigation

Here we find the optimal nonadaptive plan among those that use immediate and conclusive contact investigation. Since the contact investigation policy is specified, the problem is reduced to finding $\varphi^* \in \Phi(M)$ to minimize the mean time to find the target. As we just noted, this problem is equivalent to finding φ^* to minimize $\bar{\mu}(\varphi^*)$, the mean time to detect the target. In order to accomplish this, an expression for $\bar{\mu}(\varphi)$ is derived.

To compute $\bar{\mu}(\varphi)$ it is assumed that broad-search effort is applied at the rate v, so that

$$M(t) = vt \quad \text{for } t \geq 0$$

and that a broad-search allocation f requires broad search time

$$\frac{1}{v} \int_X f(x) \, dx$$

to be applied. Let

$$c(x, z) = \frac{z}{v} + \tau(x)\delta(x)a(x, z) \quad \text{for } x \in X, \ z \geq 0. \quad (6.3.6)$$

Then

$$C[f] = \int_X c(x, f(x)) \, dx = \frac{1}{v} \int_X f(x) \, dx + \int_X \tau(x)\delta(x)a(x, f(x)) \, dx \quad (6.3.7)$$

is the mean time spent in broad search and investigating false targets when the broad-search allocation f is applied. This is easily seen, since $(1/v) \int_X f(x) \, dx$

6.3 Optimal Nonadaptive Plans

is the time spent in broad search and $\int_X \tau(x)\delta(x)a(x, f(x))\,dx$ the mean time spent investigating the false targets detected by f.

Let $\varphi \in \Phi(M)$. If the target is detected at broad-search time t, then $C[\varphi(\cdot, t)]$ gives the expected time spent in broad search and contact investigation before the target is detected. Since $P[\varphi(\cdot, t)]$ is the probability of detecting the target by broad-search time t,

$$\bar{\mu}(\varphi) = \int_0^\infty C[\varphi(\cdot, t)] P[\varphi(\cdot, dt)]. \tag{6.3.8}$$

Thus, we wish to find $\varphi^* \in \Phi(M)$ such that

$$\bar{\mu}(\varphi^*) \leq \bar{\mu}(\varphi) \quad \text{for} \quad \varphi \in \Phi(M).$$

Such φ^* is found for the case where the false- and real-target detection functions are the same (i.e., $a = b$) in the following theorem. The proof of this theorem is based on the optimization theorems proved in Section 6.4.

Theorem 6.3.1. *Let $a = b$, and assume that b is a regular detection function. Let*

$$c(x, z) = \frac{z}{v} + \tau(x)\delta(x)b(x, z) \quad \text{for} \quad x \in X, \; z \geq 0, \tag{6.3.9}$$

where v is a positive constant, and assume that τ, δ, and $b'(\cdot, 0)$ are bounded above. For $x \in X$, let

$$\rho_x(z) = \frac{p(x)b'(x, z)}{c'(x, z)} \quad \text{for} \quad z \geq 0, \tag{6.3.10}$$

$$\rho_x^{-1}(\lambda) = \begin{cases} \text{inverse of } \rho_x \text{ evaluated at } \lambda & \text{for } 0 < \lambda \leq \rho_x(0), \\ 0 & \text{for } \lambda > \rho_x(0). \end{cases} \tag{6.3.11}$$

In addition, let

$$U(\lambda) = \int_X \rho_x^{-1}(\lambda)\,dx \quad \text{for} \quad \lambda > 0 \tag{6.3.12}$$

and $M(t) = vt$ for $t \geq 0$. Then φ^* defined by

$$\varphi^*(x, t) = \rho_x^{-1}(U^{-1}(M(t))) \quad \text{for} \quad x \in X, \; t \geq 0, \tag{6.3.13}$$

is a member of $\Phi(M)$ and satisfies

$$P[\varphi^*(\cdot, t)] = \max\{P[f] : f \in F(X) \text{ and } C[f] \leq C[\varphi^*(\cdot, t)]\} \quad \text{for} \quad t \geq 0, \tag{6.3.14}$$

and

$$\bar{\mu}(\varphi^*) \leq \bar{\mu}(\varphi) \quad \text{for} \quad \varphi \in \Phi(M). \tag{6.3.15}$$

Proof. The theorem is proved by verifying that the hypotheses of Lemma 6.4.1 and Theorem 6.4.3 hold. For Lemma 6.4.1, one observes that since b is regular and

$$\rho_x(z) = \frac{p(x)}{[vb'(x, z)]^{-1} + \tau(x)\delta(x)} \quad \text{for} \quad z \geq 0,$$

then ρ_x is positive continuous and strictly decreasing to 0 for $x \in X$ such that $p(x) > 0$. Since $c'(x, z) \geq 1/v$ for $x \in X$, $z \geq 0$, the remaining hypotheses of Lemma 6.4.1 hold. Since b is regular, it is absolutely continuous (see discussion at the beginning of Section 6.4). Thus $c(x, \cdot)$ is absolutely continuous and $c(x, 0) = 0$ for $x \in X$. Let $\bar{\tau}$, $\bar{\delta}$, and $\bar{b}$ be finite constants that bound, respectively, τ, δ, and $b'(\cdot, 0)$ above. Then, for $\varphi \in \Phi(M)$,

$$C[\varphi(\cdot, t)] = \frac{1}{v}\int_X \varphi(x, t)\, dx + \int_X \tau(x)\delta(x)b(x, \varphi(x, t))\, dx$$

$$\leq t + \bar{\tau}\bar{\delta}\bar{b}\int_X \varphi(x, t)\, dx$$

$$= t + \bar{\tau}\bar{\delta}\bar{b}M(t) < \infty \quad \text{for} \quad t \geq 0.$$

The remaining hypotheses of Theorem 6.4.3 are clearly satisfied, and thus (6.3.14) and (6.3.15) hold. This proves the theorem.

The following theorem extends Theorem 6.3.1 to the case where $a \neq b$. In the following, $a'(x, \cdot)$ denotes the derivative of $a(x, \cdot)$.

Theorem 6.3.2. *Assume that v is constant and τ, δ, and $a'(\cdot, 0)$ are bounded above. Let*

$$c(x, z) = \frac{z}{v} + \tau(x)\delta(x)a(x, z) \quad \text{for} \quad x \in X, \quad z \geq 0.$$

Let ρ_x, ρ_x^{-1}, M, and U be defined as in Theorem 6.3.1. Assume that ρ_x is positive continuous and strictly decreasing to zero for $x \in X$ such that $p(x) > 0$. Assume that $a(x, \cdot)$ and $b(x, \cdot)$ are absolutely continuous and $a(x, 0) = b(x, 0) = 0$ for $x \in X$. Then φ^* defined by (6.3.13) satisfies (6.3.14) and (6.3.15).

Proof. The proof follows by verifying that the hypotheses of Lemma 6.4.1 and Theorem 6.4.3 are satisfied.

For completeness, a version of Theorem 6.3.1 is stated for the case of a discrete target distribution. A version of Theorem 6.3.2 could also be given for discrete target distributions.

6.3 Optimal Nonadaptive Plans

Theorem 6.3.3. *Let $a = b$ and assume that b is a regular detection function on J. Let*

$$c(j, z) = \frac{z}{v} + \tau(j)\delta(j)b(j, z) \quad \text{for } j \in J, \ z \geq 0, \quad (6.3.16)$$

where v is a positive constant, and assume that τ, δ, and $b'(\cdot, 0)$ are bounded above. For $j \in J$, let

$$\rho_j(z) = p(j)b'(j, z)/c'(j, z) \quad \text{for } z \geq 0, \quad (6.3.17)$$

$$\rho_j^{-1}(\lambda) = \begin{cases} \text{inverse of } \rho_j \text{ evaluated at } \lambda & \text{for } 0 < \lambda \leq \rho_j(0), \\ 0 & \text{for } \lambda > \rho_j(0). \end{cases}$$

In addition, let

$$U(\lambda) = \sum_{j \in J} \rho_j^{-1}(\lambda) \quad \text{for } \lambda > 0 \quad (6.3.18)$$

and $M(t) = vt$ for $t > 0$. Then φ^* defined by

$$\varphi^*(j, t) = \rho_j^{-1}(U^{-1}(M(t))) \quad \text{for } j \in J, \ t \geq 0, \quad (6.3.19)$$

is a member of $\Phi(M)$ and satisfies

$$P[\varphi^*(\cdot, t)] = \max\{P[f] : f \in F(J) \text{ and } C[f] \leq C[\varphi^*(\cdot, t)]\} \quad (6.3.20)$$

and

$$\bar{\mu}(\varphi^*) \leq \bar{\mu}(\varphi) \quad \text{for } \varphi \in \Phi(M). \quad (6.3.21)$$

Proof. The theorem follows from Lemma 6.4.2 and Theorem 6.4.6.

*Immediate Contact Investigation Is Optimal among Delay Policies with Breathers

In Theorems 6.3.1 and 6.3.2, the optimal nonadaptive broad-search density φ^* is found for plans that use immediate and conclusive contact investigation. Here the class of nonadaptive search plans considered is broadened to allow one to delay contact investigation, although contact investigation is still required to be conclusive. In this expanded class, it is shown that φ^* coupled with immediate contact investigation is still optimal. This result agrees with the intuitive feeling that there is no point to broad searching in an area if contacts are not investigated.

A *nonadaptive investigation delay policy* is a Borel function κ mapping $X \times [0, \infty)$ into the doubleton $\{0, 1\}$. The function κ has the following interpretation. If $\kappa(x, t) = 1$, then a contact made at point x by broad-search time t will be investigated immediately if it has not already been investigated. If $\kappa(x, t) = 0$, then a contact located at x will not be investigated at broad-search time t. For $x \in X$, the function $\kappa(x, \cdot)$ is further restricted so that if

$\kappa(x, t_0) = 1$ for some t_0, then $\kappa(x, t) = 1$ for all $t \geq t_0$, that is, once $\kappa(x, \cdot)$ reaches 1, it must stay there. In effect, $\kappa(x, \cdot)$ is a delay function telling how long the searcher will delay before beginning to investigate contacts at point x.

As an example, consider the following contact investigation policy. No contacts will be investigated until the (prior) probability of contacting the target has reached 0.9. After that, contacts will be investigated immediately. When combined with a broad-search plan m, the contact investigation policy κ can be expressed as follows. Let t_0 be the broad-search time at which $P[\varphi(\cdot, t_0)] = 0.9$. Then for $x \in X$,

$$\kappa(x, t) = \begin{cases} 0 & \text{for } 0 \leq t < t_0, \\ 1 & \text{for } t \geq t_0. \end{cases} \qquad (6.3.22)$$

Note that the investigation policy depends only on broad-search time and location and is thus a nonadaptive policy. If more than one contact is made by broad-search time t_0, there is still a question as to the order in which contacts are to be investigated. If the target distribution has a density that is highest near the origin, then a reasonable policy is to investigate the contacts closest to the origin first.

In order to apply the Lagrange multiplier techniques presented in this book to find optimal nonadaptive plans, it is necessary to require that contact investigation policies never result in the investigation of two contacts without an intervening period of broad search. In addition, it is assumed that for all $t \geq 0$

$$\lim_{s \to t} \kappa(x, s) = \kappa(x, t) \qquad \text{for a.e.} \quad x \in X. \qquad (6.3.23)$$

A delay policy that satifies (6.3.23) for all $t \geq 0$ and that always has an intervening period of broad search between any two contact investigations is called a *delay policy with breathers*. The requirement that (6.3.23) be satisfied imposes no essential restriction beyond the requirement of intervening periods of broad search. For if (6.3.23) were not satisfied for some t, then it would be possible to have more than one contact investigation scheduled for the same broad-search time as with the policy in (6.3.22). The detailed reason for this restriction will be made clear when the expression for the mean time to find the target is developed below.

While the restriction to delay policies with breathers appears severe, one can still approximate (in mean time) the policy given in (6.3.22) by delay policies with breathers as follows. Let $\|x\|$ denote the distance of the point x from the origin in Euclidean n-space, i.e., if $x = (x_1, x_2, \ldots, x_n)$, then $\|x\| = (x_1^2 + x_2^2 + \cdots + x_n^2)^{1/2}$. Let t_0 be defined as above, and for $x \in X$, let

$$\kappa_\varepsilon(x, t) = \begin{cases} 0 & \text{for } 0 \geq t < t_0 + \varepsilon\|x\|/(1 + \|x\|), \\ 1 & \text{for } t \geq t_0 + \varepsilon\|x\|/(1 + \|x\|). \end{cases}$$

6.3 Optimal Nonadaptive Plans

Using plan κ_ε, at most ε additional broad-search time is spent before the contacts made by broad-search time t_0 have been investigated. Since there is probability 0 of obtaining two contacts the same distance from the origin, κ_ε specifies (with probability 1) that some broad search take place between each contact investigation. As ε approaches 0, the probability of detecting additional contacts in the additional broad-search time approaches 0. Thus for a fixed broad-search plan φ, one can see that the mean time to find the target when using policy κ_ε approaches the mean time when using κ as $\varepsilon \to 0$. Thus, if it is shown that immediate contact investigation is optimal in the class of nonadaptive policies with breathers, it will follow that it is optimal compared to policies such as the one defined in (6.3.22). The same result will hold for any contact investigation policy that can be approximated in mean time by policies with breathers.

Let $\mathcal{N}$ be the class of nonadaptive search plans having contact investigation delay policies with breathers and broad-search plans in $\Phi(M)$, where $M(t) = vt$ for $t \geq 0$. Then $\mathcal{N}$ consists of pairs (φ, κ), where $\varphi \in \Phi(M)$ and κ is a delay policy with breathers, which depends only on location and broad-search time.

Let κ^* be the policy that calls for immediate contact investigation. Then

$$\kappa^*(x, s) = 1 \quad \text{for} \quad x \in X, \quad s \geq 0.$$

Observe that the conditions of Theorem 6.3.1. imply that M is continuous and that the false target detection function a is continuous in the second variable, i.e., $a(x, \cdot)$ is continuous for $x \in X$. The result of these two assumptions is that the probability of making two contacts at the same broad-search time is 0, and thus κ^* is a policy with breathers.

In order to show that (φ^*, κ^*) is optimal in $\mathcal{N}$, we find an expression for $\bar{\mu}(\varphi^*, \kappa^*)$, the mean time to the beginning of the investigation of the contact that is the target. As before, since contact investigation is conclusive and the mean time to investigate the contact that is the target does not depend on the search plan, minimizing $\bar{\mu}$ is equivalent to minimizing the mean time to find the target.

Let

$$\mathbf{p}(x, y, z) = p(x)yb(x, z), \tag{6.3.24}$$

$$\mathbf{c}(x, y, z) = \frac{z}{v} + \tau(x)\delta(x)ya(x, z) \quad \text{for} \quad x \in X, \; 0 \leq y \leq 1, \; z \geq 0. \tag{6.3.25}$$

For $t \geq 0$, define

$$\mathbf{P}_t[\varphi, \kappa] = \int_X \mathbf{p}(x, \kappa(x, t), \varphi(x, t)) \, dx \tag{6.3.26}$$

$$\mathbf{C}_t[\varphi, \kappa] = \int_X \mathbf{c}(x, \kappa(x, t), \varphi(x, t)) \, dx \tag{6.3.27}$$

for $(\varphi, \kappa) \in \mathcal{N}$.

Then $\mathbf{P}_t[\varphi, \kappa]$ is the probability of having detected the target and begun the investigation of the contact that is the target by broad-search time t.

If $\varphi \in \Phi(M)$, where $M(t) = vt$ for $t \geq 0$, then

$$\mathbf{C}_t[\varphi, \kappa] = t + \int_X \tau(x)\delta(x)\kappa(x, t)a(x, \varphi(x, t)) \, dx. \qquad (6.3.28)$$

It is assumed the second term on the right-hand side of (6.3.28) gives the expected time spent investigating false targets by broad-search time t. Again if the false targets satisfy either the Poisson or independent identically distributed model, then an argument similar to the one leading to (6.1.6) shows that the second term on the right-hand side of (6.3.28) gives the expected time spent investigating false targets by broad-search time t provided that there is always a breather between investigating contacts.

If more than one contact were investigated without an intervening period of broad search, then one of the contacts investigated might be the target, in which case the investigation of the remaining contacts would not take place. Since all contacts detected by time t for which $\kappa(x, t) = 1$ are included in the integral (6.3.28), it will overestimate the time spent investigating false targets unless contact investigation policies are required to have breathers.

Let

$$g(t) = \mathbf{P}_t[\varphi, \kappa] \quad \text{and} \quad h(t) = \mathbf{C}_t[\varphi, \kappa] \quad \text{for} \quad t \geq 0.$$

Then

$$\bar{\mu}(\varphi, \kappa) = \int_0^\infty h(t)g(dt). \qquad (6.3.29)$$

Theorem 6.3.4. *Suppose that the conditions of Theorem 6.3.1 are satisfied and that φ^* is given by (6.3.13). Let κ^* be the contact investigation policy that calls for immediate contact investigation. Then*

$$\bar{\mu}(\varphi^*, \kappa^*) \leq \bar{\mu}(\varphi, \kappa) \quad \text{for} \quad (\varphi, \kappa) \in \mathcal{N}, \qquad (6.3.30)$$

i.e., (φ^*, κ^*) minimizes the mean time to find the target among all plans in $\mathcal{N}$.

Proof. Let $(\varphi, \kappa) \in \mathcal{N}$. By using (6.3.23) and the fact that $a(x, \cdot) = b(x, \cdot)$ is continuous, one can show that $\mathbf{C}_t[\varphi, \kappa]$ is a continuous and strictly increasing function of t. Let

$$g_2(t) = \mathbf{P}_t[\varphi, \kappa], \quad h_2(t) = \mathbf{C}_t[\varphi, \kappa] \quad \text{for} \quad t \geq 0.$$

Since $h_2(t) \geq t$, it follows that if $\lim_{t \to \infty} g_2(t) < 1$, then $\bar{\mu}(\varphi, \kappa) = \infty$ and

6.4 Optimization Theorems

(6.3.30) holds trivially. Thus, we may assume that g_2 and h_2 satisfy, respectively, the hypotheses on g and h of Lemma 6.4.4. Let

$$g_1(t) = \mathbf{P}_t[\varphi^*, \kappa^*] = P[\varphi^*(\cdot, t)]$$
$$h_1(t) = \mathbf{C}_t[\varphi^*, \kappa^*] = C[\varphi^*(\cdot, t)]$$

for $t \geq 0$.

It is shown in Theorem 6.4.3 that g_1 and h_1 satisfy the hypotheses of Lemma 6.4.4. Thus, to prove (6.3.30) one may use Lemma 6.4.5, provided one shows that for s and $t \geq 0$

$$g_2(s) > g_1(t) \quad \text{implies} \quad h_2(s) > h_1(t). \tag{6.3.31}$$

Fix s and $t \geq 0$. For $\lambda = U^{-1}(M(t))$, $(\lambda, \varphi^*(\cdot, t))$ maximizes the pointwise Lagrangian. Thus for $x \in X$, $0 \leq y \leq 1$, $z \geq 0$,

$$\ell(x, \lambda, \varphi^*(x, t)) \geq \ell(x, \lambda, z)$$
$$\geq y\ell(x, \lambda, z)$$
$$= p(x)yb(z) - \lambda\left(\frac{yz}{v} + \tau(x)\delta(x)yb(z)\right)$$
$$\geq p(x)yb(z) - \lambda\left(\frac{z}{v} + \tau(x)\delta(x)yb(z)\right)$$
$$= \mathbf{p}(x, y, z) - \lambda\mathbf{c}(x, y, z).$$

Thus,

$$g_1(t) - \lambda h_1(t) = P[\varphi^*(\cdot, t)] - \lambda C[\varphi^*(\cdot, t)]$$
$$\geq \mathbf{P}_s[\varphi, \kappa] - \lambda \mathbf{C}_s[\varphi, \kappa]$$
$$= g_2(s) - \lambda h_2(s),$$

i.e.,

$$g_2(s) - g_1(t) \leq \lambda[h_2(s) - h_1(t)]. \tag{6.3.32}$$

Since $\lambda > 0$, (6.3.32) implies (6.3.31). By Lemma 6.4.5,

$$\bar{\mu}(\varphi^*, \kappa^*) = \int_0^\infty h_1(t)g_1(dt) \leq \int_0^\infty h_2(t)g_2(dt) = \bar{\mu}(\varphi, \kappa),$$

and the theorem is proved.

6.4. OPTIMIZATION THEOREMS

In this section, the optimization theorems that are used to find optimal nonadaptive plans are proved. These theorems generalize the results of

Theorems 2.2.4 and 2.2.5 in a somewhat different direction than was pursued in Theorems 2.4.3 and 2.4.4. Here as in Section 2.4 the general cost functional

$$C[f] = \int_X c(x, f(x))\, dx$$

is used. However, Theorems 2.4.3 and 2.4.4 deal with uniformly optimal plans within the class $\tilde{\Phi}(M)$. Recall that, if $\varphi \in \tilde{\Phi}(M)$,

$$\int_X c(x, \varphi(x, t))\, dx = M(t) \quad \text{for } t \geq 0.$$

In this section, the object is to find plans $\varphi^* \in \Phi(M)$ that are optimal for their cost at each time t, i.e.,

$$\int_X \varphi^*(x, t)\, dx = M(t),$$

and

$$P[\varphi^*(\cdot, t)] = \max\{P[f] : f \in F(X) \text{ and } C[f] \leq C[\varphi^*(\cdot, t)]\} \quad \text{for } t \geq 0.$$

The following lemmas guarantee the existence of the inverse of the function U, which appears here in a somewhat more general form than in Theorems 2.2.4 and 2.2.5. The inverse of U will be used to find optimal broad-search allocations. In Lemma 6.4.1, $b(x, \cdot)$ is assumed to be absolutely continuous in order to allow one to integrate $b'(x, \cdot)$, the derivative of $b(x, \cdot)$, to obtain $b(x, \cdot)$. Recall that $b(x, \cdot)$ is absolutely continuous if $b'(x, z)$ exists for a.e. $z \in [0, \infty)$ and

$$b(x, z_2) - b(x, z_1) = \int_{z_1}^{z_2} b'(x, z)\, dz \quad \text{for } z_1, z_2 \in [0, \infty).$$

Similar comments apply to the assumption that $b(j, \cdot)$ is absolutely continuous in Lemma 6.4.2. As before, $c'(x, \cdot)$ indicates the derivative of $c(x, \cdot)$.

Lemma 6.4.1. *Let*

$$\rho_x(z) = p(x)b'(x, z)/c'(x, z) \quad \text{for } x \in X, \ z \geq 0,$$

be such that ρ_x is positive, continuous, and strictly decreasing to zero for x such that $p(x) > 0$. Assume that $b(x, \cdot)$ is absolutely continuous, $b(x, 0) = 0$, and $c'(x, \cdot) \geq A > 0$ for $x \in X$, where A is a constant. Let

$$\rho_x^{-1}(\lambda) = \begin{cases} \text{inverse of } \rho_x \text{ evaluated at } \lambda & \text{for } 0 < \lambda \leq \rho(x, 0), \\ 0 & \text{for } \lambda > \rho(x, 0), \end{cases} \quad (6.4.1)$$

$$U(\lambda) = \int_X \rho_x^{-1}(\lambda)\, dx \quad \text{for } \lambda > 0. \quad (6.4.2)$$

6.4 Optimization Theorems

Then an inverse function U^{-1} may be defined on $[0, \infty)$ such that $U(U^{-1}(K)) = K$ for $0 \leq K < \infty$.

Proof. This lemma is proved by showing that U is continuous and strictly decreasing for $\lambda_\ell < \lambda \leq \lambda_u$, where

$$\lambda_\ell = \sup\{\lambda : U(\lambda) = \infty\}, \qquad \lambda_u = \inf\{\lambda : U(\lambda) = 0\}.$$

Since ρ_x is decreasing, $\rho_x(z) \geq \lambda$ for $0 \leq z \leq \rho_x^{-1}(\lambda)$. Thus, for $x \in X$ and $\lambda > 0$,

$$\frac{p(x)}{A} \geq \frac{p(x)b(x, \rho_x^{-1}(\lambda))}{A} = \int_0^{\rho_x^{-1}(\lambda)} \frac{p(x)b'(x, z)}{A}\, dz$$

$$\geq \int_0^{\rho_x^{-1}(\lambda)} \rho_x(z)\, dz$$

$$\geq \lambda \rho_x^{-1}(\lambda).$$

It follows that

$$\rho_x^{-1}(\lambda) \leq p(x)/A\lambda \qquad \text{for } x \in X, \ \lambda > 0, \tag{6.4.3}$$

and

$$U(\lambda) = \int_X \rho_x^{-1}(\lambda)\, dx \leq 1/A\lambda \qquad \text{for } \lambda > 0. \tag{6.4.4}$$

The continuous and strictly decreasing nature of ρ_x guarantees that ρ_x^{-1} is continuous and strictly decreasing for $0 < \lambda < \rho_x(0)$. The dominated convergence theorem (Theorem A.1) and (6.4.3) may be used to show that U is continuous. That $\lim_{\lambda \to \infty} U(\lambda) = 0$ follows from (6.4.4). Since $\rho_x^{-1}(\lambda)$ increases to ∞ as λ decreases to zero for x such that $p(x) > 0$, one may invoke the monotone convergence theorem (Theorem A.2) to show that $\lim_{\lambda \to 0} U(\lambda) = \infty$. The strictly decreasing nature of ρ_x^{-1} for x such that $p(x) > 0$ guarantees that U is strictly decreasing for $\lambda_\ell < \lambda \leq \lambda_u$. Thus, U^{-1} exists and the lemma is proved.

Lemma 6.4.2 is the discrete counterpart of Lemma 6.4.1.

Lemma 6.4.2. *Let*

$$\rho_j(z) = p(j)b'(j, z)/c'(j, z) \qquad \text{for } j \in J, \ z \geq 0,$$

be such that ρ_j is positive, continuous, and strictly decreasing to zero for j such that $p(j) > 0$. Assume that $b(j, \cdot)$ is absolutely continuous, $b(j, 0) = 0$, and

VI Search in the Presence of False Targets

$c'(x, \cdot) \geq A > 0$ for $j \in J$, where A is a constant. Let

$$\rho_j^{-1}(\lambda) = \begin{cases} \text{inverse of } \rho_j \text{ evaluated at } \lambda & \text{for } 0 < \lambda \leq \rho_j(0), \\ 0 & \text{for } \lambda > \rho_j(0), \end{cases}$$
(6.4.5)

$$U(\lambda) = \sum_{j \in J} \rho_j^{-1}(\lambda) \quad \text{for} \quad \lambda > 0. \tag{6.4.6}$$

Then an inverse function U^{-1} may be defined on $[0, \infty)$ such that $U(U^{-1}(K)) = K$ for $0 \leq K < \infty$.

Proof. The proof of this lemma proceeds in the same manner as Lemma 6.3.1. One may use the dominated and monotone convergence theorems applied to discrete measures to obtain the strictly decreasing and continuous nature of U from the continuous, strictly decreasing behavior of ρ_j^{-1}. Alternatively, one can use Theorems A.3 and A.4 in the manner of Theorem 2.2.3.

If $\varphi \in \Phi(M)$, then one can think of $C[\varphi(\cdot, t)]$ as being the expected cost of a search given that it terminates at time t. Correspondingly, $P[\varphi(\cdot, t)]$ is the probability of the search terminating by time t. Define

$$\bar{\mu}(\varphi) = \int_0^\infty C[\varphi(\cdot, t)] P[\varphi(\cdot, dt)] \quad \text{for} \quad \varphi \in \Phi(M). \tag{6.4.7}$$

The integration in (6.4.7) is understood to be Lesbesgue–Stieltjes integration (see, for example, Feller, 1966). If $\lim_{t \to \infty} P[\varphi(\cdot, t)] = 1$, then $\bar{\mu}(\varphi)$ gives the expected cost of detecting the target. In the following theorem, $\varphi^* \in \Phi(M)$ is found such that

$$\bar{\mu}(\varphi^*) \leq \bar{\mu}(\varphi) \quad \text{for} \quad \varphi \in \Phi(M).$$

Theorem 6.4.3. *Assume that the hypotheses of Lemma 6.4.1 hold. Let the cumulative effort function M be continuous, strictly increasing, and satisfy $\lim_{t \to \infty} M(t) = \infty$ and $M(0) = 0$. Let*

$$\varphi^*(x, t) = \rho_x^{-1}(U^{-1}(M(t))) \quad \text{for} \quad x \in X, \ t \geq 0. \tag{6.4.8}$$

Then $\varphi^* \in \Phi(M)$. If, in addition, $c(x, \cdot)$ is absolutely continuous, $c(x, 0) = 0$ for $x \in X$, and $C[\varphi^*(\cdot, t)] < \infty$ for $t > 0$ and $\varphi^* \in \Phi(M)$, then letting $\lambda = U^{-1}(M(t))$, the pair $(\lambda, \varphi^*(\cdot, t))$ maximizes the pointwise Lagrangian and

$$P[\varphi^*(\cdot, t)] = \max\{P[f] : f \in F(X) \text{ and } C[f] \leq C[\varphi^*(\cdot, t)]\} \quad \text{for} \quad t \geq 0. \tag{6.4.9}$$

In addition,

$$\bar{\mu}(\varphi^*) \leq \bar{\mu}(\varphi) \quad \text{for} \quad \varphi \in \Phi(M). \tag{6.4.10}$$

6.4 Optimization Theorems

Proof. To show that $\varphi^* \in \Phi(M)$, one observes that $\rho_x^{-1}(\lambda) \geq 0$ for $x \in X$, $\lambda \geq 0$, so that $\varphi^*(\cdot, t) \in F(X)$ for $t \geq 0$. Since ρ_x^{-1} is decreasing and $U^{-1}(M(t))$ decreases as t increases, $\varphi^*(x, \cdot)$ is increasing for $x \in X$. Thus, φ^* is a search plan. Finally, by Lemma 6.4.1,

$$\int_X \varphi^*(x, t)\, dx = U(U^{-1}(M(t))) = M(t) \quad \text{for } t \geq 0,$$

and it follows that $\varphi^* \in \Phi(M)$.

To prove (6.4.9), fix t such that $M(t) > 0$. [If $M(t) = 0$, then $\varphi(x, t) = 0$ for a.e. $x \in X$ for $\varphi \in \Phi(M)$.] Let $\lambda = U^{-1}(M(t))$. Since ρ_x is decreasing, it follows from (6.4.8) that for $x \in X$

$$\begin{aligned} \rho_x(z) &\geq \lambda &&\text{for} && 0 < z < \varphi^*(x, t), \\ &\leq \lambda &&\text{for} && \varphi^*(x, t) < z < \infty. \end{aligned} \quad (6.4.11)$$

Since $c(x, \cdot) > 0$, (6.4.11) implies for $x \in X$

$$\begin{aligned} p(x)b'(x, z) - \lambda c'(x, z) &\geq 0 &&\text{for} && 0 < z < \varphi^*(x, t), \\ &\leq 0 &&\text{for} && \varphi^*(x, t) < z < \infty. \end{aligned} \quad (6.4.12)$$

Then (6.4.12) implies that $(\lambda, \varphi^*(\cdot, t))$ maximizes the pointwise Lagrangian. Thus, by Theorem 2.1.3, (6.4.9) holds.

To prove (6.4.10), two cases are considered. For case 1, suppose that $\lim_{t \to \infty} P[\varphi^*(\cdot, t)] < 1$. Since $c'(x, z) \geq A$, $c(x, z) \geq Az$ and

$$C[\varphi^*(\cdot, t)] \geq A \int_X \varphi^*(x, t)\, dx = AM(t).$$

As a result, $\lim_{t \to \infty} C[\varphi^*(\cdot, t)] = \infty$. Thus, by (6.4.9) $\lim_{t \to \infty} P[\varphi(\cdot, t)] < 1$ and $\bar{\mu}(\varphi) = \infty$ for $\varphi \in \Phi(M)$. For this case, (6.4.10) holds trivially.

For case 2, suppose that $\lim_{t \to \infty} P[\varphi^*(\cdot, t)] = 1$. In order to prove (6.4.10) in this case, we use Lemma 6.4.5 (proved below). Fix $\varphi \in \Phi(M)$ and let

$$g(t) = P[\varphi(\cdot, t)], \quad h(t) = C[\varphi(\cdot, t)] \quad \text{for } t \geq 0.$$

Then (6.4.7) becomes

$$\bar{\mu}(\varphi) = \int_0^\infty h(t)g(dt). \quad (6.4.13)$$

One may show that h is a continuous function in two steps:

Step 1: $\lim_{s \to 0} \varphi(x, t + s) = \varphi(x, t)$ *for a.e.* $x \in X$. Since M is continuous,

$$0 = \lim_{s \to 0^+} [M(t + s) - M(t)] = \lim_{s \to 0^+} \int_X [\varphi(x, t + s) - \varphi(x, t)]\, dx$$

$$= \int_X [\lim_{s \to 0^+} \varphi(x, t + s) - \varphi(x, t)]\, dx, \quad (6.4.14)$$

where the interchange of limit and integration is justified by the dominated convergence theorem. Since $\varphi(x, \cdot)$ is increasing for $x \in X$, $\lim_{s \to 0^+} \varphi(x, t + s) \geq \varphi(x, t)$. Thus, by (6.4.14) $\lim_{s \to 0^+} \varphi(x, t + s) = \varphi(x, t)$ for a.e. $x \in X$. A similar argument for the left-handed limit finishes the proof of step 1.

Step 2: $\lim_{s \to 0} h(t + s) = h(t)$. Since $c(x, \cdot)$ is assumed to be absolutely continuous, it is of course continuous for $x \in X$. Thus, by the dominated convergence theorem and the result in step 1,

$$\lim_{s \to 0} h(t + s) = \lim_{s \to 0} \int_X c(x, \varphi(x, t + s))\, dx$$
$$= \int_X c(x, \varphi(x, t))\, dx = h(t),$$

which proves that h is continuous.

Since M is strictly increasing and $c'(x, \cdot) \geq A > 0$, it follows that h is strictly increasing. The assumption that $\rho(x, \cdot)$ is positive for $p(x) > 0$ implies that $b'(x, \cdot) > 0$ for x and that g is increasing. Let

$$g_1(t) = P[\varphi^*(\cdot, t)], \quad h_1(t) = C[\varphi^*(\cdot, t)]$$
$$g_2(t) = P[\varphi(\cdot, t)], \quad h_2(t) = C[\varphi(\cdot, t)] \quad \text{for } t \geq 0.$$

Then by virtue of (6.4.9) the hypotheses of Lemma 6.4.5 are satisfied, and (6.4.10) follows from (6.4.13). This proves the theorem.

Lemma 6.4.4. *Let g be a probability distribution function on $[0, \infty)$ such that $g(0) = 0$ and $\lim_{t \to \infty} g(t) = 1$. Let $h: [0, \infty) \to [0, \infty)$ be continuous and strictly increasing such that $h(0) = 0$. Let $h(\infty) = \lim_{t \to \infty} h(t)$. Then*

$$\int_0^\infty h(t) g(dt) = \int_0^\infty [1 - g(t)] h(dt). \tag{6.4.15}$$

Furthermore, let $u(y) = g(h^{-1}(y))$ for $0 \leq y \leq h(\infty)$; then

$$\int_0^\infty h(t) g(dt) = \int_0^{h(\infty)} [1 - u(y)]\, dy. \tag{6.4.16}$$

Proof. Since g is increasing and h continuous, the integration in (6.4.15) becomes standard Riemann–Stieltjes integration. To see that (6.4.15) holds, observe that

$$\int_0^y h(t) g(dt) = -[1 - g(y)]h(y) + [1 - g(0)]h(0) + \int_0^y [1 - g(t)] p(dt). \tag{6.4.17}$$

6.4 Optimization Theorems

Since $h(0) = 0$, $[1 - g(0)]h(0) = 0$. If

$$\lim_{y \to \infty} \int_0^y h(t)g(dt) < \infty, \tag{6.4.18}$$

then

$$0 = \lim_{y \to \infty} \int_y^\infty h(t)g(dt) \geq \lim_{y \to \infty} h(y)[1 - g(y)] = 0.$$

In this case, passing to the limit in (6.4.17) yields (6.4.15). If the limit in (6.4.18) is infinite, then the left-hand side of (6.4.15) is infinite, and by (6.4.17)

$$\int_0^y h(t)g(dt) \leq \int_0^y [1 - g(t)]h(dt) \quad \text{for } y > 0.$$

Thus, the right-hand side of (6.4.15) is also infinite, and (6.4.15) holds in all cases.

Since h is continuous and strictly increasing, one can define an inverse function h^{-1} on $[0, h(\infty))$. Equation (6.4.16) follows from result **a** in Loeve (1963, p. 342) by observing that h creates a measure on $[0, \infty)$ and that the measure induced by the mapping $h: [0, \infty) \to [0, \infty)$ is simply Lebesgue measure on $[0, h(\infty))$. Since $u(h(s)) = g(s)$, (6.4.16) follows directly from that result and (6.4.15). This proves the lemma.

Lemma 6.4.5. *Suppose that g_1 and h_1 (g_2 and h_2) satisfy the hypotheses on g and h, respectively, in Lemma 6.4.4. If for any s and $t \geq 0$*

$$g_2(s) > g_1(t) \quad \text{implies} \quad h_2(s) > h_1(t), \tag{6.4.19}$$

then

$$\int_0^\infty h_1(t)g_1(dt) \leq \int_0^\infty h_2(t)g_2(dt). \tag{6.4.20}$$

Proof. Let $u_1(y) = g_1(h_1^{-1}(y))$ for $0 \leq y < h_1(\infty)$ and $u_2(y) = g_2(h_2^{-1}(y))$ for $0 \leq y < h_2(\infty)$. By Lemma 6.4.4,

$$\int_0^\infty h_1(t)g_1(dt) = \int_0^{h_1(\infty)} [1 - u_1(y)] \, dy, \tag{6.4.21}$$

$$\int_0^\infty h_2(t)g_2(dt) = \int_0^{h_2(\infty)} [1 - u_2(y)] \, dy. \tag{6.4.22}$$

Define $t_1 = \inf(\{\infty\} \cup \{t : g_1(t) = 1\})$. Let $0 \leq y < h_1(t_1)$, and let t and s be such that $h_1(t) = y$ and $h_2(s) = y$. Then

$$u_1(y) = g_1(t) \quad \text{and} \quad u_2(y) = g_2(s).$$

By (6.4.19), $u_1(y) = g_1(t) \geq g_2(s) = u_2(y)$, and (6.4.20) follows from (6.4.21) and (6.4.22). This proves the lemma.

The following theorem gives the counterpart of Theorem 6.4.3 for discrete target distributions. The proof of this theorem is analogous to that of Theorem 6.4.3 and is not given.

Theorem 6.4.6. *Assume that the hypotheses of Lemma 6.4.2 hold. Let the cumulative effort function M be strictly increasing, continuous, and satisfy* $\lim_{t \to \infty} M(t) = \infty$ *and* $M(0) = 0$. *Let*

$$\varphi^*(j, t) = \rho_j^{-1}(U^{-1}(M(t))) \quad \text{for} \quad j \in J, \quad t \geq 0. \quad (6.4.23)$$

Then $\varphi^* \in \Phi(M)$. *If, in addition,* $c(j, \cdot)$ *is absolutely continuous,* $c(j, 0) = 0$ *for* $j \in J$, *and* $C[\varphi^*(\cdot, t)] < \infty$ *for* $t > 0$, *then*

$$P[\varphi^*(\cdot, t)] = \max\{P[f] : f \in F(J) \text{ and } C[f] \leq C[\varphi^*(\cdot, t)]\}, \quad (6.4.24)$$

and

$$\bar{\mu}(\varphi^*) \leq \bar{\mu}(\varphi) \quad \text{for} \quad \varphi \in \Phi(M). \quad (6.4.25)$$

6.5. EXAMPLES

This section considers two examples that reveal interesting properties of the optimal nonadaptive broad-search plan φ^* for searches that use immediate and conclusive contact investigation. Example 6.5.1 shows that one effect of false targets is to slow the rate at which search expands away from areas of high prior probability density. In Example 6.5.2, it is shown that there are situations involving false targets in which no uniformly optimal plan exists even when the detection function is regular and the search space continuous.

Example 6.5.1. In this example, it is shown that the presence of false targets slows the rate at which the optimal nonadaptive plan using immediate and conclusive contact investigation expands away from the areas of high prior probability density.

Let X be Euclidean 2-space, the plane, and let

$$p(x) = \frac{1}{2\pi\sigma^2} \exp\left(-\frac{x_1^2 + x_2^2}{2\sigma^2}\right) \quad \text{for} \quad x \in X,$$

$$a(x, z) = b(x, z) = 1 - e^{-z} \quad \text{for} \quad x \in X, \quad z \geq 0.$$

The false-target distribution is assumed to be Poisson with density δ, which is taken to be a fixed positive constant. The contact investigation time τ is also assumed to be constant. Since p is circularly symmetric, it is convenient

6.5 Examples

to express p in terms of polar coordinates, i.e.,

$$p(r) = \frac{1}{2\pi\sigma^2} \exp\left(-\frac{r^2}{2\sigma^2}\right) \quad \text{for} \quad 0 \leq \theta \leq 2\pi, \ r \geq 0. \quad (6.5.1)$$

Since p does not depend on θ, this variable has been suppressed. A similar device is employed for ρ, ρ^{-1}, and φ^*.

Contact investigation is assumed to be immediate and conclusive. Thus, one is interested in finding φ^*, the optimal nonadaptive broad-search plan. By Theorem 6.3.1, φ^* is found as follows. Calculate

$$\rho_r(z) = \frac{p(r)b'(r, z)}{c'(r, z)} = \frac{p(r)e^{-z}}{1/v + \tau\delta e^{-z}} \quad \text{for} \quad r \geq 0, \ z \geq 0, \quad (6.5.2)$$

and

$$\rho_r^{-1}(\lambda) = \begin{cases} \ln(vp(r)/\lambda - \tau\delta v) & \text{for} \ r < h(\lambda), \\ 0 & \text{for} \ r \geq h(\lambda), \end{cases} \quad (6.5.3)$$

where

$$h(\lambda) = \sigma\{2 \ln v - 2 \ln[2\pi\sigma^2\lambda(1 + \tau\delta v)]\}^{1/2} \quad \text{for} \ \lambda > 0, \quad (6.5.4)$$

and

$$U(\lambda) = 2\pi \int_0^{h(\lambda)} \ln\left(\frac{v \exp(-r^2/2\sigma^2)}{2\pi\sigma^2\lambda} - \tau\delta v\right) r \, dr \quad \text{for} \ \lambda > 0. \quad (6.5.5)$$

Then

$$\varphi^*(r, t) = \rho_r^{-1}(U^{-1}(vt)) \quad \text{for} \ r \geq 0, \ t \geq 0. \quad (6.5.6)$$

Unfortunately, we are not able to calculate U and U^{-1} in closed form, so that we cannot display φ^* in a more explicit fashion than in (6.5.6). However, it is a simple matter to calculate U and U^{-1} numerically on a computer for any given values of the parameters. Thus, it is straightforward to find φ^* by numerical methods. In spite of the lack of a closed-form expression for φ^*, it is possible to show that increasing δ has the effect of slowing the rate at which search expands away from the origin.

Fix $r_1 > 0$. Then effort is first exerted at a point a distance r_1 from the origin at the broad-search time t such that

$$U^{-1}(vt) = \rho(r_1, 0) = \frac{v \exp(-r_1^2/2\sigma^2)}{2\pi\sigma^2(1 + \tau\delta v)}.$$

Let $0 \leq r_0 < r_1$. By (6.5.6), the search density applied to a point at distance r_0 from the origin by time t is

$$\rho^{-1}(r_0, \rho(r_1, 0)) = \ln\left\{\exp\left(\frac{r_1^2 - r_0^2}{2\sigma^2}\right) + \tau\delta v\left[\exp\left(\frac{r_1^2 - r_0^2}{2\sigma^2}\right) - 1\right]\right\}.$$

Since $r_1 > r_0$, this density increases as δ increases. Thus, as the false-target density δ increases, more effort is expended and more broad-search time passes before the search expands to a circle of radius r_1.

The same argument can be made for arbitrary regular detection functions and circularly symmetric target distributions with central tendencies.

Example 6.5.2. In this example, a situation involving false targets is presented in which there is no nonadaptive plan using immediate and conclusive contact investigation, which maximizes the probability of detecting the target at each total time T. Recall that total time includes both contact investigation time and broad-search time. The example includes a regular detection function and a continuous target distribution. Thus, the situation for searches involving false targets is inherently less satisfactory than for searches without them. For in the case where there are no false targets, Theorems 2.2.4 and 2.2.5 show how to find uniformly optimal plans when the detection function is regular.

Let X be the plane. Consider two regions R_1 and R_2 in the plane each having unit area. Let

$$p(x) = \begin{cases} 0.1 & \text{for } x \in R_1, \\ 0.9 & \text{for } x \in R_2, \\ 0 & \text{otherwise,} \end{cases} \qquad \delta(x) = \begin{cases} 1 & \text{for } x \in R_1, \\ 89 & \text{for } x \in R_2, \\ 0 & \text{otherwise.} \end{cases}$$

The false-target distribution is assumed to be Poisson. Assume $v = 1$, and for $x \in X$ let

$$\tau(x) = 1 \qquad \text{and} \qquad b(x, z) = a(x, z) = 1 - e^{-z} \qquad \text{for } z \geq 0.$$

Consider two broad-search plans φ_1 and φ_2 defined for $t \geq 0$ as follows:

$$\varphi_1(x, t) = \begin{cases} t & \text{for } x \in R_1, \\ 0 & \text{for } x \in R_2, \end{cases} \qquad \varphi_2(x, t) = \begin{cases} 0 & \text{for } x \in R_1, \\ t & \text{for } x \in R_2. \end{cases}$$

Since $M(t) = t$, both plans are in $\Phi(M)$. In addition, both plans are coupled with immediate and conclusive contact investigation.

To find the target by total time $1 + h$, it must be contacted by total time h. Let $g_1(T)$ and $g_2(T)$ be the probabilities of finding the target by total time T for plans 1 and 2, respectively. Then

$$g_1(1 + h) \leq \Pr\{\text{contacting target by broad-search time } h\}$$
$$= 0.1(1 - e^{-h}),$$

and

$$g_2(1 + h) \geq \Pr\left\{\begin{matrix}\text{contacting target by broad-search time } h \\ \text{and not detecting any false targets by} \\ \text{broad-search time } h\end{matrix}\right\}$$
$$= 0.9(1 - e^{-h})e^{-89h}.$$

6.6 Semiadaptive Plans

Hence
$$\lim_{h \to 0} g_1(1 + h)/g_2(1 + h) \leq \tfrac{1}{9},$$

and $g_2(1 + h) > g_1(1 + h)$ for all h sufficiently small. Recall that

$$\rho(x, z) = \frac{p(x)e^{-z}}{1 + \delta(x)e^{-z}} \quad \text{for} \quad x \in X, \quad z \geq 0,$$

so that

$$\rho(x, 0) = \begin{cases} 0.05 & \text{for} \quad x \in R_1, \\ 0.01 & \text{for} \quad x \in R_2. \end{cases}$$

Thus, the optimal nonadaptive broad-search plan φ^*, begins searching in R_1. As a result, for some $h^* > 0$, φ^* coincides with φ_1 for t in the interval $[0, h^*]$. Let $g^*(T)$ be the probability of finding the target by total time T when following φ^*. Thus, there is an $h_0 > 0$ such that

$$g^*(1 + h) = g_1(1 + h) < g_2(1 + h) \quad \text{for} \quad 0 \leq h \leq h_0.$$

It now follows that no plan in $\Phi(M)$ can maximize probability of detection as a function of total time for all T. For if φ were such a plan and $g(T)$ the probability of finding the target by total time T when using this plan, then

$$g(T) \geq g_2(T) > g^*(T) \quad \text{for} \quad 1 \leq T \leq 1 + h_0,$$

and

$$g(T) \geq g^*(T) \quad \text{for} \quad T \notin [1, 1 + h_0].$$

Thus,

$$\bar{\mu}(\varphi) = \int_0^\infty [1 - g(T)] \, dt < \int_0^\infty [1 - g^*(T)] \, dT = \bar{\mu}(\varphi^*). \quad (6.5.7)$$

But (6.5.7) contradicts Theorem 6.3.1, which yields that $\bar{\mu}(\varphi^*) \leq \bar{\mu}(\varphi)$ for $\varphi \in \Phi(M)$. Thus, there is no plan φ in $\Phi(M)$ that maximizes the probability of detection for all total time T for this search.

6.6. SEMIADAPTIVE PLANS

This section introduces the concept of semiadaptive search plans. It is shown that a semiadaptive plan yields a mean time to find the target that lies between that of the optimal nonadaptive and optimal adaptive plans. The advantage of a semiadaptive plan is that it performs better than the optimal nonadaptive plan and is substantially easier to find than the optimal adaptive plan in the presence of false targets. This section also discusses the use of semiadaptive plans in multiple-target situations.

Nonadaptive, Adaptive, and Semiadaptive Plans

Recall that a member of $\mathcal{N}$, the class of nonadaptive search plans, is a function of broad-search time and location only. A nonadaptive plan makes no use of feedback generated by the search concerning the number, location, and detection times of contacts. The only feedback used is to stop the search when the target is found. When the class of search plans is expanded to allow the use of all the information gained during the search, one obtains the class $\mathcal{A}$ of adaptive search plans. Of course, $\mathcal{N} \subset \mathcal{A}$, so that the optimal adaptive plan produces a mean time to find the target that is less than or equal to that produced by the optimal nonadaptive plan.

Let μ_n denote the mean time to find the target when using an optimal nonadaptive plan. For the search situations considered in this section, an optimal nonadaptive plan will always exist. Let μ_a denote the mean time to find the target when using the optimal adaptive plan. If there is no optimal adaptive plan, then μ_a is understood to be the infimum over all adaptive plans of the mean time to find the target. By the above discussion, $\mu_a \leq \mu_n$. Thus, if one has a choice, he would prefer to use the optimal adaptive plan. However, it appears to be very hard to find the optimal adaptive plan for searches involving false targets. Optimal adaptive plans have been found only for the situation of a search with two cells and one false target known to be in the second cell (see Dobbie, 1973). A discussion of the methods and difficulties of finding optimal adaptive plans in the context of control theory may be found in Sworder (1966). In contrast, Theorems 6.3.1–6.3.3 give one a method of finding optimal nonadaptive plans for a wide class of searches using immediate and conclusive contact investigation.

A semiadaptive plan as defined below makes use of feedback but is simple enough to be calculated in the same fashion as the nonadaptive plan on which it is based.

Definition of Semiadaptive Plan

In defining semiadaptive search plans, only immediate and conclusive contact investigation will be considered. Let the search situation be such that the hypotheses of either Theorem 6.3.1 or 6.3.3 are satisfied. This allows one to calculate the optimal nonadaptive plan. Since contact investigation is immediate and conclusive, specification of a search plan amounts to specifying the broad-search allocation.

A *semiadaptive search plan* is specified by a sequence of update times $0 < t_1 < t_2 < t_3 \cdots$. This sequence is required to be finite or to satisfy $\lim_{k \to \infty} t_k = \infty$. At time 0, the search begins by following the optimal nonadaptive plan. If the target is not found by broad-search time t_1, then information gained up to that time (such as detection and identification of false

6.6 Semiadaptive Plans

targets) is used to revise the target distribution and the false-target distribution in a Bayesian manner. Detection functions are similarly revised. For the broad-search time interval $(t_1, t_2]$, the optimal nonadaptive search plan is followed for the revised distributions and detection functions obtained at time t_1. If the target is not found by broad-search time t_2, the process is repeated again. This procedure continues until the target is found.

Intuitively, semiadaptive plans are better than nonadaptive plans, since they periodically assess the information gained by the search and use this information to plan for the next interval of search. However, they are less efficient than the optimal adaptive plan because they do not plan ahead to take advantage of information as it is generated.

Comparison of Mean Times for Nonadaptive, Semiadaptive, and Adaptive Plans

It will now be shown that $\mu_a \leq \mu_s \leq \mu_n$ where μ_a, μ_s, and μ_n are the mean times to find the target for the optimal adaptive, semiadaptive, and optimal nonadaptive plans respectively.

Theorem 6.6.1. *Assume that the conditions of either Theorem 6.3.1 or 6.3.3 hold and that the false targets are either Poisson or independent identically distributed. Let μ_s be the mean time to find the target when using the semiadaptive plan based on the sequence $\{t_1, t_2, \ldots\}$ of update times. Then*

$$\mu_a \leq \mu_s \leq \mu_n \tag{6.6.1}$$

if either (a) or (b) holds:

(a) *the sequence of update points is finite,*
(b) $\lim_{k \to \infty} t_k = \infty$ *and* $\mu_s < \infty$.

Proof. One can show, under the conditions of either Theorem 6.3.1 or 6.3.3. and with false targets Poisson or independent identically distributed, that an optimal nonadaptive policy can be found after each update time, so that a semiadaptive plan does exist. This is true because the Bayesian update of the distributions and detection functions does not destroy the validity of the hypotheses of those theorems. For example, if b is a regular detection function on X and the allocation f is applied to X without detecting the target, then the updated detection function $\tilde{b}$ given by

$$\tilde{b}(x, z) = \frac{b(x, z + f(x)) - b(x, f(x))}{1 - b(x, f(x))} \quad \text{for} \quad x \in X, \quad z \geq 0,$$

is still regular. This will be discussed in more detail later in this section when explicit formulas for the posterior detection and distribution functions are found.

Since the semiadaptive plan is a member of the class $\mathscr{A}$ of adaptive plans, $\mu_a \leq \mu_s$ holds automatically.

Suppose (a) holds. Then (6.6.1) will be shown to hold by mathematical induction. Suppose there is one update time t_1. Let $\mathbf{T}_n$ and $\mathbf{T}_s$ be the random times at which the target is found in the nonadaptive and semiadaptive plans, respectively. In addition, let $\mathbf{S}_n$ and $\mathbf{S}_s$ be the random broad-search times at which the target is found in the nonadaptive and semiadaptive plans, respectively. Since the nonadaptive and semiadaptive plans coincide up to broad-search time t_1, it follows that[1]

$$E[\mathbf{T}_n \mid \mathbf{S}_n \leq t_1] = E[\mathbf{T}_s \mid \mathbf{S}_s \leq t_1],$$

and

$$\Pr\{\mathbf{S}_n \leq t_1\} = \Pr\{\mathbf{S}_s \leq t_1\}.$$

Since the (t_1, ∞) segment of the semiadaptive plan follows an optimal nonadaptive plan for the updated distributions at broad-search time t_1,

$$E[\mathbf{T}_n \mid \mathbf{S}_n > t_1] \geq E[\mathbf{T}_s \mid \mathbf{S}_s > t_1].$$

Thus,

$$\begin{aligned}\mu_s &= E[\mathbf{T}_s \mid \mathbf{S}_s \leq t_1]\Pr\{\mathbf{S}_s \leq t_1\} + E[\mathbf{T}_s \mid \mathbf{S}_s > t_1]\Pr\{\mathbf{S}_s > t_1\} \\ &\leq E[\mathbf{T}_n \mid \mathbf{S}_n \leq t_1]\Pr\{\mathbf{S}_n \leq t_1\} + E[\mathbf{T}_n \mid \mathbf{S}_n > t_1]\Pr\{\mathbf{S}_n > t_1\} \\ &= \mu_n.\end{aligned}$$

This proves (6.6.1) for the case of one update point.

Suppose that (6.6.1) holds when there are k update points. Consider a sequence $\{t_1, t_2, \ldots, t_{k+1}\}$ of update points. Let $\mathbf{T}_s^k$ and $\mathbf{T}_s^{k+1}$ be the times to find the target using the semiadaptive plans associated with $\{t_1, \ldots, t_k\}$ and $\{t_1, \ldots, t_{k+1}\}$, respectively. Let $\mathbf{S}_s^k$ and $\mathbf{S}_s^{k+1}$ be the corresponding broad-search times. Then

$$E[\mathbf{T}_s^{k+1} \mid \mathbf{S}_s^{k+1} \leq t_{k+1}] = E[\mathbf{T}_s^k \mid \mathbf{S}_s^k \leq t_{k+1}],$$

and

$$\Pr\{\mathbf{S}_s^{k+1} \leq t_{k+1}\} \leq \Pr\{\mathbf{S}_s^k \leq t_{k+1}\}.$$

Since the (t_{k+1}, ∞) segment of the semiadaptive plan with $k+1$ update points follows the optimal nonadaptive plan for the updated distributions at broad-search time t_{k+1}, it follows that

$$E[\mathbf{T}_s^{k+1} \mid \mathbf{S}_s^{k+1} > t_{k+1}] \leq E[\mathbf{T}_s^k \mid \mathbf{S}_s^k > t_{k+1}].$$

Thus, $E[\mathbf{T}_s^{k+1}] \leq E[\mathbf{T}_s^k]$. By the induction hypothesis, $E[\mathbf{T}_s^k] \leq \mu_n$. Thus, $E[\mathbf{T}_s^{k+1}] \leq \mu_n$ and (6.6.1) holds for case (a).

---
[1] The notation $E[X \mid A]$ denotes the expectation of X conditioned on the event A.

6.6 Semiadaptive Plans

Suppose (b) holds. Let $\mathbf{T}_s$ and $\mathbf{S}_s$ be the total and broad-search times, respectively, at which the target is found when using the semiadaptive plan. Define

$$\varepsilon_k = E[\mathbf{T}_s \mid \mathbf{S}_s > t_k] \Pr\{\mathbf{S}_s > t_k\} \qquad \text{for} \quad k = 1, 2, \ldots.$$

Let $g(t) = \Pr\{\mathbf{S}_s \leq t\}$ for $t \geq 0$. Since $\mu_s < \infty$,

$$\lim_{k \to \infty} \varepsilon_k = \lim_{k \to \infty} \int_{t_k}^{\infty} E[\mathbf{T}_s \mid \mathbf{S}_s = t] g(dt) = 0.$$

Let $\mathbf{T}_s^k$ and $\mathbf{S}_s^k$ denote the total and broad-search times, respectively, to find the target for the semiadaptive plan that uses only the first k update points of $\{t_1, t_2, \ldots\}$. Then

$$E[\mathbf{T}_s \mid \mathbf{S}_s \leq t_{k+1}] \Pr\{\mathbf{S}_s \leq t_{k+1}\} = E[\mathbf{T}_s^k \mid \mathbf{S}_s^k \leq t_{k+1}] \Pr\{\mathbf{S}_s^k \leq t_{k+1}\}$$
$$\leq E[\mathbf{T}_s^k] \leq \mu_n \qquad \text{for} \quad k = 1, 2, \ldots,$$

where the last inequality follows from the fact that (a) implies (6.6.1) holds. As a result,

$$\mu_s = E[\mathbf{T}_s] \leq \mu_n + \varepsilon_{k+1} \qquad \text{for} \quad k = 1, 2, \ldots.$$

Since $\lim_{k \to \infty} \varepsilon_k = 0$, $\mu_s \leq \mu_n$ and (6.6.1) follows. This proves the theorem.

Remark 6.6.2. Theorem 6.6.1 may be extended to allow the update times to be nonnegative random variables $\mathbf{t}_1 \leq \mathbf{t}_2 \leq \cdots$. If there are a finite number of update times, then the proof proceeds exactly as in Theorem 6.6.1. In case the number is infinite, condition (b) is replaced by

(b) $\lim_{k \to \infty} \mathbf{t}_k = \infty$ with probability one and $\mu_s < \infty$.

Consider the semiadaptive plan corresponding to a given set of update points. If an additional update time is added to this set, it is reasonable to suppose that the resulting semiadaptive plan would have a smaller mean time to find the target than the original one. Whether this is true or not is still an open question. Going one step further, one might consider semiadaptive plans in which the updating is done continuously. Again, one would hope that semiadaptive plans with continuous updating would produce smaller mean times than plans with discrete updating, or at least that they would produce mean times smaller than the optimal nonadaptive plans. The latter has been shown for a particular class of searches involving false targets (Stone, 1973c).

Calculating Posterior Distributions and Detection Functions

In order to apply Theorem 6.6.1, one has to calculate the posterior target and false-target distributions and detection functions at the update times. Here formulas are given that allow one to perform these calculations.

Let $\tilde{b}_k$ denote the posterior or updated detection function for the interval between the kth and $(k+1)$st update times. Let $f_k(x)$ be the effort density accumulated at point $x \in X$ by the update time t_k. [A similar definition holds for $f_k(j)$ for $j \in J$.] Then

$$\tilde{b}_k(x, z) = \frac{b(x, z + f_k(x)) - b(x, f_k(x))}{1 - b(x, f_k(x))} \quad \text{for} \quad x \in X, \quad z \geq 0, \quad (6.6.2)$$

and for the search space J,

$$\tilde{b}_k(j, z) = \frac{b(j, z + f_k(j)) - b(j, f_k(j))}{1 - b(j, f_k(j))} \quad \text{for} \quad j \in J, \quad z \geq 0. \quad (6.6.3)$$

Let $\tilde{p}_k$ represent the posterior target distribution given that the target is not detected by the time of the kth update, that is, for the search space X,

$$\tilde{p}_k(x) = \frac{p(x)[1 - b(x, f_k(x))]}{1 - P[f_k]} \quad \text{for} \quad x \in X, \quad (6.6.4)$$

and for the search space J,

$$\tilde{p}_k(j) = \frac{p(j)[1 - b(j, f_k(j))]}{1 - P[f_k]} \quad \text{for} \quad j \in J. \quad (6.6.5)$$

Suppose that the false targets are independently and identically distributed; that is, each false target has its distribution given by the function q and the distribution of each false target is independent of every other false target and the real target. In addition, it is assumed that $a = b$ and that detection of one false target is independent of the detection of any other false target and the real target. The number of false targets $\mathbf{N}$ is a random variable with

$$\nu(N) = \Pr\{\mathbf{N} = N\} \quad \text{for} \quad N = 0, 1, 2, \ldots.$$

Let

$$Q[f] = \int_X q(x) b(x, f(x))\, dx \quad \text{for} \quad f \in F(X),$$

and

$$Q[f] = \sum_{j \in J} q(j) b(j, f(j)) \quad \text{for} \quad f \in F(J).$$

Then $Q[f]$ is the probability of detecting a given false target with allocation f.

If a false target remains undetected by the update time t_k, then its posterior distribution is given by

$$\tilde{q}_k(x) = \frac{q(x)[1 - b(x, f_k(x))]}{1 - Q[f_k]} \quad \text{for} \quad x \in X, \quad (6.6.6)$$

6.6 Semiadaptive Plans

or, in the case of the search space J,

$$\tilde{q}_k(j) = \frac{q(j)[1 - b(j, f_k(j))]}{1 - Q[f_k]} \quad \text{for} \quad j \in J. \tag{6.6.7}$$

Thus the false targets remaining undetected are independent and identically distributed.

To determine the posterior false target distribution, it remains to find $\tilde{v}_k(N, i)$, the probability of having N false targets remaining undetected by broad-search time t_k given that i false targets have been detected and identified by broad-search time t_k. Since immediate and conclusive contact investigation is employed, all false targets that are detected are identified and no contacts remain unidentified. By Bayes' rule,

$$\tilde{v}_k(N, i) = \frac{\Pr\{\text{detecting } i \text{ false targets} \mid \mathbf{N} = N + i\}v(N + i)}{\sum_{L=0}^{\infty} \Pr\{\text{detecting } i \text{ false targets} \mid \mathbf{N} = L + i\}v(L + i)}.$$

The probability of detecting a given false target is simply $Q[f_k]$. Since the detection of one false target is independent of detecting any other false target,

$$\Pr\{\text{detecting } i \text{ false targets} \mid \mathbf{N} = L + i\} = \binom{L + i}{i} Q^i[f_k](1 - Q[f_k])^L,$$

where

$$\binom{L + i}{i} = \frac{(L + i)!}{L! \, i!}.$$

Thus,

$$\tilde{v}_k(N, i) = \frac{(N + i)(N + i - 1) \cdots (N + 1)(1 - Q[f_k])^N v(N + i)}{\sum_{L=0}^{\infty} (L + i)(L + i - 1) \cdots (L + 1)(1 - Q[f_k])^L v(L + i)} \quad \text{if } i \geq 1, \tag{6.6.8}$$

and

$$\tilde{v}_k(N, 0) = \frac{(1 - Q[f_k])^N v(N)}{\sum_{L=0}^{\infty} (1 - Q[f_k])^L v(L)}. \tag{6.6.9}$$

It is now clear that if the conditions of Theorem 6.3.1 or 6.3.3 are satisfied for the prior detection and distribution functions and the false targets are independently and identically distributed, then the conditions of Theorem 6.3.1 or 6.3.3 are satisfied for the posterior functions calculated for the update at time t_k. In particular, if b is regular, it is clear that $\tilde{b}_k$ given by (6.6.2) or (6.6.3) is regular. The false targets are independently and identically distributed, with $\tilde{q}_k$ specifying their common distribution. The distribution of the number of false targets is now given by $\tilde{v}_k$ in (6.6.8) and (6.6.9). Notice that

$\tilde{v}_k$ is the only function that depends on the number of false targets detected by time t_k.

Let $N_k(i)$ be the number of false targets remaining undetected by time t_k given that i were detected by t_k. The posterior false target density $\tilde{\delta}_k$ is obtained by computing

$$E[N_k(i)] = \sum_{N=1}^{\infty} N v_k(N, i)$$

and setting

$$\tilde{\delta}_k(x) = E[N_k(i)]\tilde{q}_k(x) \quad \text{for} \quad x \in X. \quad (6.6.10)$$

Suppose the false targets are Poisson distributed. If $\bar{d}(X) = \int_X \delta(x)\,dx < \infty$, then the Poisson-distributed case is a special case of the false targets being independently and identically distributed, with

$$q(x) = \delta(x)/\bar{d}(X) \quad \text{for} \quad x \in X,$$

and

$$v(N) = \frac{1}{N!}[\bar{d}(X)]^N e^{-\bar{d}(X)} \quad \text{for} \quad N = 0, 1, 2, \ldots.$$

As a result,

$$\tilde{v}_k(N, i) = \frac{(1/N!)[(1 - Q[f_k])\,\bar{d}(X)]^N}{\sum_{L=0}^{\infty}(1/L!)[(1 - Q[f_k])\,\bar{d}(X)]^L}$$

$$= \frac{1}{N!}[(1 - Q[f_k])\,\bar{d}(X)]^N \exp\{-(1 - Q[f_k])\,\bar{d}(X)\}$$

$$\text{for} \quad N = 0, 1, 2, \ldots. \quad (6.6.11)$$

Observe that in the Poisson case $\tilde{v}_k(N, i)$ does not depend on i, the number of false targets detected by broad-search time t_k.

Since

$$E[N_k(i)] = (1 - Q[f_k])\,\bar{d}(X),$$

and

$$\tilde{q}_k(x) = \frac{\delta(x)[1 - b(x, f_k(x))]}{\bar{d}(X)(1 - Q[f_k])},$$

it follows that

$$\tilde{\delta}_k(x) = E[N_k(i)]\tilde{q}_k(x) = \delta(x)[1 - b(x, f_k(x))] \quad \text{for} \quad x \in X. \quad (6.6.12)$$

For Poisson false-target distributions where $\bar{d}(X) = \infty$, one may show that

6.6 Semiadaptive Plans

(6.6.12) still holds by partitioning X into a countable number of disjoint sets $S_1, S_2, \ldots$, such that

$$\bigcup_{r=1}^{\infty} S_r = X, \quad \bar{d}(S_r) < \infty \quad \text{for} \quad r = 1, 2, \ldots.$$

One can then show that (6.6.12) holds on S_r for $r = 1, 2, \ldots$. Alternatively, one can obtain (6.6.12) by viewing the number of false targets detected as nonhomogeneous Poisson process in time in the manner described in Section 6.1.

These results suggest that the optimal nonadaptive and the semiadaptive plans coincide when the false-target distribution is Poisson. This is shown to be the case in Example 6.6.4.

Examples

Example 6.6.3. This example shows how one can construct a semiadaptive search plan for a two-cell search involving immediate and conclusive contact investigation.

Let
$$p(j) = \tfrac{1}{2} \quad \text{for} \quad j = 1, 2,$$

$$q(j) = \begin{cases} \tfrac{1}{4} & \text{for } j = 1, \\ \tfrac{3}{4} & \text{for } j = 2, \end{cases}$$

$$\nu(N) = \Pr\{N = N\} = (\tfrac{1}{2})^{N+1} \quad \text{for} \quad N = 0, 1, 2, \ldots.$$

Then $E[N] = 1$, and

$$\delta(j) = E[N]q(j) = \begin{cases} \tfrac{1}{4} & \text{for } j = 1, \\ \tfrac{3}{4} & \text{for } j = 2. \end{cases}$$

Assume that for $j = 1, 2$,

$$\tau(j) = 1, \quad a(j, z) = b(j, z) = 1 - e^{-z} \quad \text{for} \quad z \geq 0.$$

Observe that for $z \geq 0$,

$$\rho_j(z) = \begin{cases} \dfrac{2e^{-z}}{4 + e^{-z}} & \text{for } j = 1, \\[2mm] \dfrac{2e^{-z}}{4 + 3e^{-z}} & \text{for } j = 2. \end{cases}$$

Let $v = 1$, so that the cumulative effort function is given by $M(t) = t$ for $t \geq 0$. Applying the discussion preceding Theorem 3.1.1, one can develop the optimal nonadaptive broad-search plan φ^* from the marginal rate of return function ρ by placing the next small increment of effort in the cell having the highest value of $\rho_j(\varphi^*(j, t))$ at each time t.

Since $\rho_1(0) > \rho_2(0)$, the optimal nonadaptive plan commences search in R_1. Search is concentrated in R_1 until broad-search effort z_1 is accumulated there such that

$$\rho_1(z_1) = \rho_2(0),$$

i.e.,

$$\frac{2e^{-z_1}}{4 + e^{-z_1}} = \frac{2}{7}.$$

Solving for z_1, one finds that

$$z_1 = \ln(\tfrac{3}{2}).$$

Since $M(t) = t$ for $t \geq 0$, one concentrates broad search in cell 1 until broad-search time $t = \ln(\tfrac{3}{2})$. At this time, the optimal nonadaptive plan would continue by splitting broad-search effort between cells 1 and 2 in such a way that the marginal rates of return in the two cells would remain equal.

For the semiadaptive plan it is assumed that there is a single update point at broad-search time $t_1 = \ln(\tfrac{3}{2})$. At broad-search time t_1, the posterior detection and distribution functions are calculated. The plan continues by following the optimal nonadaptive plan for these posterior functions. These functions, and therefore the plan, will depend on the number of false targets found by broad-search time t_1. Of course, if the target has been found, the search stops.

Suppose that one false target is found up to broad-search time t_1. Then

$$\tilde{q}_1(1) = \frac{\tfrac{1}{4}\exp[-\ln(\tfrac{3}{2})]}{\tfrac{3}{4} + \tfrac{1}{4}\exp[-\ln(\tfrac{3}{2})]} = \tfrac{2}{11},$$

$$\tilde{q}_1(2) = \tfrac{9}{11},$$

$$\tilde{p}_1(1) = \frac{\tfrac{1}{2}\exp[-\ln(\tfrac{3}{2})]}{\tfrac{1}{2} + \tfrac{1}{2}\exp[-\ln(\tfrac{3}{2})]} = \tfrac{2}{5},$$

$$\tilde{p}_1(2) = \tfrac{3}{5},$$

$$\tilde{b}_1(1, z) = \frac{\exp[-\ln(\tfrac{3}{2})] - \exp[-\ln(\tfrac{3}{2}) - z]}{\exp[-\ln(\tfrac{3}{2})]} = 1 - e^{-z} \quad \text{for} \quad z \geq 0,$$

$$\tilde{b}_1(2, z) = 1 - e^{-z} \quad \text{for} \quad z \geq 0.$$

In order to calculate $\tilde{\delta}_1$, one must calculate $\tilde{v}_1(N, 1)$ by (6.6.8). Observe that

$$1 - Q(f_1) = \tfrac{3}{4} + \tfrac{1}{4}\exp[-\ln(\tfrac{3}{2})] = \tfrac{11}{12},$$

6.6 Semiadaptive Plans

so that

$$\tilde{v}_1(N, 1) = \frac{(N + 1)(\frac{11}{12})^N(\frac{1}{2})^{N+2}}{\sum_{L=0}^{\infty} (L + 1)(\frac{11}{12})^L(\frac{1}{2})^{L+2}} \quad \text{for} \quad N = 0, 1, 2, \ldots. \quad (6.6.13)$$

Observe that the summation in the denominator of (6.6.13) is equal to

$$\left(\frac{1}{2}\right)\left(\frac{12}{11}\right)\left(\frac{24}{13}\right) \sum_{L=0}^{\infty} (L + 1)\left(\frac{13}{24}\right)\left(\frac{11}{24}\right)^{L+1} = \frac{(12)^2}{(11)(13)} \frac{11}{13} = \left(\frac{12}{13}\right)^2,$$

where the summation in the above equality is recognized as the mean of a geometric distribution. Thus,

$$\tilde{v}_1(N, 1) = \left(\frac{13}{24}\right)^2 (N + 1)\left(\frac{11}{24}\right)^N \quad \text{for} \quad N = 0, 1, 2, \ldots, \quad (6.6.14)$$

and

$$E[N_1(1)] = \left(\frac{24}{11}\right)\left(\frac{13}{24}\right) \sum_{N=1}^{\infty} N(N + 1)\left(\frac{11}{24}\right)^{N+1}\left(\frac{13}{24}\right) = 2\left(\frac{13}{11}\right)\left(\frac{11}{13}\right)^2 = \frac{22}{13},$$

where the summation is recognized as the second factorial moment of a geometric distribution. This may be evaluated by use of generating functions (see Feller, 1957, Chapter XI, Section 1). Thus,

$$\tilde{\delta}_1(j) = \begin{cases} \dfrac{4}{13} & \text{for } j = 1. \\[6pt] \dfrac{18}{13} & \text{for } j = 2. \end{cases}$$

Since $v = 1$, $\tau(j) = 1$, and $\tilde{b}(j, z) = 1 - e^{-z}$ for $z \geq 0$, $j = 1, 2$, the marginal rate of return for the posterior detection and distribution functions is given by

$$\tilde{\rho}_1(j, z) = \frac{\tilde{p}_1(j)e^{-z}}{1 + \tilde{\delta}_1(j)e^{-z}} \quad \text{for} \quad j \in J, \ z \geq 0.$$

Thus,

$$\tilde{\rho}_1(1, z) = \frac{\frac{2}{5}e^{-z}}{1 + \frac{4}{13}e^{-z}}, \qquad \tilde{\rho}_1(2, z) = \frac{\frac{3}{5}e^{-z}}{1 + \frac{18}{13}e^{-z}} \quad \text{for} \ z \geq 0.$$

The semiadaptive plan now follows the optimal nonadaptive plan based on $\tilde{\rho}_1$ given above. Observe that

$$\tilde{\rho}_1(1, 0) = 26/85 \simeq 0.30, \qquad \tilde{\rho}_1(2, 0) = 39/155 \simeq 0.25.$$

Thus, the semiadaptive plan (when one false target is detected before broad-search time t_1) calls for placing additional effort in cell 1 beyond broad-search time $\ln(\frac{3}{2})$ before splitting effort between cells 1 and 2. Recall that the optimal nonadapative plan begins to split effort between cells 1 and 2 at broad-search time $\ln(\frac{3}{2})$.

Example 6.6.4. In this example it is shown that when the false-target distribution is Poisson, the optimal nonadaptive and the semiadaptive plans coincide. This will be shown for the case of a single update time. From this it will be clear that the result holds for any number of update times.

Let b be a regular detection function on X. Assume that $a = b$ and that contact investigation is immediate and conclusive. Let the update time be t_1 and let f_1 be the broad-search allocation applied by broad-search time t_1 in a given search. From (6.6.12), we have

$$\tilde{\delta}_1(x) = \delta(x)[1 - b(x, f_1(x))] \quad \text{for} \quad x \in X.$$

Observe that the derivative of $\tilde{b}_1(x, \cdot)$ evaluated at z is just $b'(x, f_1(x) + z)/[1 - b(x, f_1(x))]$ and recall that

$$\tilde{p}_1(x) = \frac{p(x)[1 - b(x, f_1(x))]}{1 - P[f]} \quad \text{for} \quad x \in X.$$

Thus

$$\tilde{\rho}_1(x, z) = \frac{1}{1 - P[f_1]} \frac{p(x)b'(x, f_1(x) + z)}{v^{-1} + \tau(x)\delta(x)b'(x, f_1(x) + z)} \quad \text{for} \quad x \in X, \; z \geq 0,$$

and we see that

$$\tilde{\rho}_1(x, z) = \frac{\rho(x, f_1(x) + z)}{1 - P[f_1]} \quad \text{for} \quad x \in X, \; z \geq 0.$$

As a result, the optimal nonadaptive plan based on $\tilde{\rho}$ is the same as the continuation of the optimal nonadaptive plan based on ρ. Thus the semiadaptive plan merely continues the optimal nonadaptive plan beyond broad-search time t_1.

The same result holds for discrete target distributions. In fact, it is shown in Stone and Stanshine (1971, Section 4), that this result also extends to delayed contact investigations of the type discussed in Section 6.3. One added difficulty here is calculating posterior-target and false-target distributions when some contacts have not been investigated.

Applications to Multiple Targets

Suppose there are N real targets and one wishes to maximize the expected number of these targets detected within a given cost constraint. The ith target is assumed to have a probability distribution given by p_i for $i = 1, 2, \ldots, N$. These distributions are assumed to be mutually independent and the targets are assumed to be distinguishable when they are detected, that is, one can identify which target has been detected.

The detection function d is assumed to be the same for all targets, and the probability of detecting one target is independent of detecting the others. Let

$$e(x, z) = \sum_{i=1}^{N} p_i(x)b(x, z), \quad c(x, z) = z \quad \text{for} \quad x \in X, \quad z \geq 0.$$

$$e(j, z) = \sum_{i=1}^{N} p_i(j)b(j, z), \quad c(j, z) = z \quad \text{for} \quad j \in J, \quad z \geq 0.$$

Let $E[f]$ be defined as in (2.4.1) or (2.4.2) for $f \in F$. Suppose $M(t) = vt$ for $t \geq 0$. Then if the conditions of Theorem 2.4.3 or 2.4.4 are satisfied, one can find a uniformly optimal search plan φ^* such that $\varphi^* \in \Phi(M)$ and

$$E[\varphi^*(\cdot, t)] = \max\{E[\varphi(\cdot, t)] : \varphi \in \Phi(M)\} \quad \text{for} \quad t \geq 0.$$

[Since $c(x, z) = z$ or $c(j, z) = z$, $\tilde{\Phi}(M) = \Phi(M)$.] In the terminology of this chapter, φ^* is the uniformly optimal nonadaptive search plan; that is, φ^* maximizes the expected number of targets detected by time t for $t \geq 0$ among all nonadaptive plans in $\Phi(M)$.

A semiadaptive plan may be constructed as follows: Let $\{\mathbf{t}_1, \mathbf{t}_2, \ldots, \mathbf{t}_N\}$ be the set of update points, where $\mathbf{t}_i$ is the broad-search time at which the ith target detection occurs for $i = 1, 2, \ldots, N$. At each update time, one computes the posterior target distributions $\tilde{p}_i$ for the targets that remain undetected. One then computes

$$\tilde{e}(x, z) = \sum_{\{i : i\text{th target not detected}\}} \tilde{p}_i(x)\tilde{b}(x, z),$$

where $\tilde{b}$ is the posterior detection function at the time of the update. A similar definition holds for $\tilde{e}(j, z)$. Until the next update time (i.e., until the next target is detected), one follows the nonadaptive plan based on $\tilde{e}$.

As with false targets, one may show that the semiadaptive plan produces a larger expected number of targets detected by time t than the optimal nonadaptive plan for $t \geq 0$.

NOTES

The Poisson-distributed false-target model is taken from Stone and Stanshine (1971), while the independent identically distributed model is based on the work of Dobbie (1973). The theorems in Section 6.3 are based on the results in Stone and Stanshine (1971) and are restricted to conclusive contact investigation policies. In the case where one is allowed to interrupt contact investigation, optimal nonadaptive search plans are found for a class of search problems in Stone *et al.* (1972).

In Dobbie (1973) an optimal adaptive search plan is found for a two-cell search in which there is one false target, which is known to be in the second cell. Contact investigation is assumed to be immediate and conclusive and both the false- and real-target detection functions are equal, homogeneous, and exponential. For this search problem, Stone (1973c) compares μ_a, the mean time for the optimal adaptive plan, to μ_n, the mean time for the optimal nonadaptive plan. For $p(1) = 0, 0.1, \ldots, 0.9, 1$, he finds the maximum difference between μ_a and μ_n to be less than 1%. Of course, the difference between μ_a and the mean time for any semiadaptive plan would be even less. This raises the question as to whether optimal adaptive plans or semiadaptive plans perform significantly better in mean time than optimal nonadaptive plans in the presence of false targets.

Simulation results reported in Stone (1973c) indicate that semiadaptive plans with continuous updating can produce significantly (e.g., as much as 30%) smaller mean times than the optimal nonadaptive plan in more complicated search situations (e.g., 4-cell target distribution with multiple false targets). The results indicate that the semiadaptive plan has the greatest advantage over the nonadaptive plan when the target distribution is uniform and there are a small number of false targets in each cell but the time to investigate a contact is large. Unfortunately, it was not possible to compare the mean time for the optimal adaptive plan to that for the semiadaptive plan because the optimal adaptive plan is not known for these more complicated situations.

The idea of a semiadaptive plan is an adaptation of a similar idea in the theory of discrete-time optimal control. In particular, semiadaptive plans were motivated by the class of open-loop optimal feedback plans discussed by Dreyfus (1965).

In Dobbie (1975b) a problem of involving false targets and a moving real target is considered. A search problem involving false alarms, as opposed to real objects such as false targets, is considered by Smith and Walsh (1971).

Chapter VII

Approximation of Optimal Plans

This chapter considers two types of approximations to optimal search plans for stationary targets. The first type of approximation yields an allocation that is optimal for a given cost within a restricted class $\hat{F}$ of allocations. This approximation is for searches with no false targets. The second type of approximation is an incremental approximation. This type yields a search plan that approximates the optimal plan in mean time to find the target. This approximation is easy to use even for complicated searches involving false targets.

7.1. APPROXIMATIONS THAT ARE OPTIMAL IN A RESTRICTED CLASS

By way of introduction, consider the allocation f^* that is optimal for cost K for a search involving a circular normal target distribution and exponential detection function. From (2.2.5) we have

$$f^*(x) = \begin{cases} \left(\dfrac{K}{\pi\sigma^2}\right)^{1/2} - \dfrac{x_1^2 + x_2^2}{2\sigma^2} & \text{for } x_1^2 + x_2^2 \leq 2\sigma^2\left(\dfrac{K}{\pi\sigma^2}\right)^{1/2}, \\ 0 & \text{for } x_1^2 + x_2^2 > 2\sigma^2\left(\dfrac{K}{\pi\sigma^2}\right)^{1/2}. \end{cases}$$

Figure 7.1.1 shows a cross section of the graph of f^* taken along one of the coordinate axes. The cross section is simply a parabola that opens downward and is truncated at $z = 0$.

Consider the problem of trying to allocate effort so that the effort density decreases to zero in a smooth fashion from the origin out to the circle of radius $(2\sigma)^{1/2}(K/\pi)^{1/4}$. In most searches this would be very difficult if not impossible. To avoid this problem one might consider allocations that are obtained by using layers of effort density all having a uniform height h. The resulting allocation might look like the one given by the dashed line in Fig. 7.1.1.

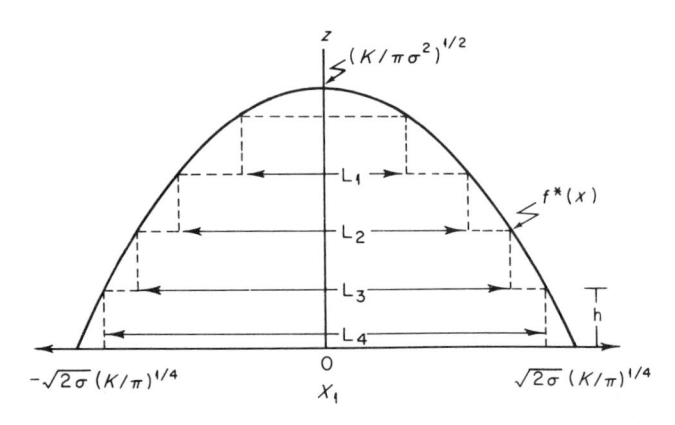

Fig. 7.1.1. Cross section of optimal allocation.

In the case of underwater search, it is often most practical to perform a parallel-path search of the type described in Section 1.2. Suppose that one chose to space the paths S distance apart and to measure effort in swept area. If the sweep width of the sensor is W, then the increment h in effort density resulting from a single parallel-path search is

$$h = W/S.$$

The allocation shown by the dashed line in Fig. 7.1.1 could then be realized by performing successive parallel-path searches in squares parallel to the axes having side lengths L_1, L_2, L_3, and L_4 as shown in Fig. 7.1.1. The resulting allocation in the plane would not have the circular symmetry that f^* does, but it would have the cross section shown in Fig. 7.1.1.

Assume that the value of h has been fixed and that effort density can be allocated to a region only in integer multiples of h. Under this restriction, what is the optimum allocation of a given amount K of effort? The remainder

7.1 Approximations in a Restricted Class

of the section addresses this question for the general case of search for a stationary target in n-space with no false targets.

Theorem 7.1.1. Fix $h > 0$, and let the search space be X. Let
$$Z = \{0, h, 2h, \ldots\}, \qquad \hat{F} = [\text{set of Borel functions } f: X \to Z].$$
Let $0 < K < \infty$. If $C[f^*] = K$, then a necessary and sufficient condition for f^* to be optimal within $\hat{F}$ for cost K is the existence of a finite $\lambda \geq 0$ such that
$$\ell(x, \lambda, f^*(x)) = \max\{\ell(x, \lambda, kh) : k = 0, 1, 2, \ldots\} \qquad \text{for a.e. } x \in X. \quad (7.1.1)$$
If in addition $b(x, \cdot)$ is increasing and right continuous for $x \in X$, then there exists f^* such that $C[f^*] = K$ and f^* is optimal within $\hat{F}$ for cost K.

Proof. That (7.1.1) is a necessary and sufficient condition for f^* to be optimal within $\hat{F}$ for cost K follows directly from Theorem 2.1.5.

Suppose $b(x, \cdot)$ is increasing and right continuous for $x \in X$. Let
$$b_h(x, z) = b(x, kh) \qquad \text{for} \quad kh \leq z < (k+1)h, \quad k = 0, 1, 2, \ldots. \quad (7.1.2)$$
Then $b_h(x, kh) = b(x, kh)$ for $k = 0, 1, 2, \ldots$, $x \in X$, and $b_h(x, \cdot)$ is increasing and right continuous for $x \in X$. By Theorem 2.4.6, there is an $f_h \in F(X)$ such that $C[f_h] = K$ and
$$\int_X p(x) b_h(x, f_h(x))\, dx$$
$$= \max\left\{\int_X p(x) b_h(x, f(x))\, dx : f \in F(X) \text{ and } C[f] \leq k\right\}. \quad (7.1.3)$$

Define $[f(x)/h]$ to be the greatest integer in $f(x)/h$, and let
$$\hat{f}(x) = h[f(x)/h] \qquad \text{for} \quad x \in X.$$
Then $\hat{f} \in \hat{F}$. From the definition of $\hat{f}$ and b_h, it follows that $b_h(x, f_h(x)) = b(x, \hat{f}(x))$ for $x \in X$ and
$$P[\hat{f}] = \int_X p(x) b_h(x, f_h(x))\, dx. \quad (7.1.4)$$

Since $\hat{f}(x) \leq f_h(x)$ for $x \in X$, $C[\hat{f}] \leq C[f_h] = K$. Thus, from (7.1.3), (7.1.4), and the fact that $\hat{F} \subset F(X)$, it follows that $\hat{f}$ is optimal within $\hat{F}$ for cost K.

If $C[\hat{f}] = K$, let $f^* = \hat{f}$. If $C[\hat{f}] < K$, then let R be any region in X with area $(K - C[\hat{f}])/h$, and let
$$f^*(x) = \begin{cases} \hat{f}(x) + h & \text{for } x \in R, \\ \hat{f}(x) & \text{otherwise.} \end{cases}$$
In either case, $f^* \in \hat{F}$ and $C[f^*] = K$. Since $b(x, \cdot)$ is increasing for $x \in X$, $P[f^*] \geq P[\hat{f}]$ and f^* is optimal within $\hat{F}$ for cost K. This proves the theorem.

Remark 7.1.2. Theorem 7.1.2 gurantees that whenever $b(x, \cdot)$ is increasing and right continuous there exists an f^* that is optimal within $\hat{F}$ for cost K. It also says that one must look among the allocations f^* that satisfy (7.1.1) for some λ to find f^*. Since (7.1.1) is also a sufficient condition, the following iterative method for finding f^* is suggested:

(a) Arbitrarily choose $\lambda > 0$.
(b) Find f_λ such that

$$\ell(x, \lambda, f_\lambda(x)) = \max\{\ell(x, \lambda, kh) : k = 0, 1, 2, \ldots\} \quad \text{for} \quad x \in X. \quad (7.1.5)$$

(c) If $f_\lambda(x)$ is not unique, let $\bar{f}_\lambda(x)$ and $\underline{f}_\lambda(x)$ be the largest and smallest values within the set $\{0, h, 2h, \ldots\}$ that satisfy (7.1.5).

(d) If $C[\underline{f}_\lambda] \le K \le C[\bar{f}_\lambda]$, then by choosing an appropriate set $R \subset X$, one can find f^* such that

$$f^*(x) = \begin{cases} \bar{f}_\lambda(x) & \text{for } x \in R, \\ \underline{f}_\lambda(x) & \text{otherwise,} \end{cases}$$

and $C[f^*] = K$. This f^* will satisfy (7.1.1) and be optimal within $\hat{F}$ for cost K.

(e) If $K < C[\underline{f}_\lambda]$, select a value for λ larger than the previous one and return to step (b).

(f) If $K > C[\bar{f}_\lambda]$, select a value for λ smaller than the previous one and return to step (b).

Since $C[\underline{f}_\lambda]$ and $C[\bar{f}_\lambda]$ are decreasing functions of λ, one may use a binary search technique in steps (e) and (f) to choose the new λ. If this is done, the sequence of λ's will converge on a λ that yields $C[f_\lambda] = K$. Thus, one can find an allocation that is optimal within $\hat{F}$ for a cost as close to K as desired.

Example 7.1.3. In this example, Theorem 7.1.1 is used to find an allocation within the class $\hat{F}$ that approximates the optimal allocation f^* for the circular normal target distribution and exponential detection function discussed at the beginning of this section.

Let X be the plane,

$$p(x) = \frac{1}{2\pi\sigma^2} \exp\left(-\frac{x_1^2 + x_2^2}{2\sigma^2}\right) \quad \text{for} \quad x = (x_1, x_2) \in X,$$

$$b(x, z) = 1 - e^{-z} \quad \text{for} \quad x \in X, \quad z \ge 0.$$

Fix $h > 0$, and let $\hat{F}$ be defined as in Theorem 7.1.1. For a given cost K, an allocation $\hat{f}$ is sought that is optimal within $\hat{F}$ for cost K. This allocation will be the best approximation to f^* under the restriction that effort density must be applied in layers of thickness h.

7.1 Approximations in a Restricted Class

Optimal Allocation

In order to use Theorem 7.1.1, one must find for $\lambda > 0$, an f_λ such that

$$\ell(x, \lambda, f_\lambda(x)) = \max\{\ell(x, \lambda, kh) : k = 0, 1, 2, \ldots\} \quad \text{for} \quad x \in X. \quad (7.1.6)$$

Claim. For $\lambda > 0$, f_λ is given by

$$f_\lambda(x) = \begin{cases} 0 & \text{for } x \text{ such that } \quad p(x) \leq \dfrac{\lambda h}{1 - e^{-h}}, \\ nh & \text{for } x \text{ such that } \dfrac{\lambda h e^{(n-1)h}}{1 - e^{-h}} < p(x) \leq \dfrac{\lambda h e^{nh}}{1 - e^{-h}}. \end{cases} \quad (7.1.7)$$

To prove the claim, observe that if $p(x) = \lambda e^{nh}$, then

$$\ell'(x, \lambda, nh) = p(x) e^{-nh} - \lambda = 0,$$

and $f_\lambda(x) = nh$. Since $f_\lambda(x)$ increases as $p(x)$ increases, it follows that for $\lambda e^{nh} \leq p(x) \leq \lambda e^{(n+1)h}$, $f_\lambda(x)$ equals either nh or $(n+1)h$. Suppose $p(x)$ satisfies

$$p(x)(1 - e^{-nh}) - \lambda nh = p(x)(1 - e^{-(n+1)h}) - \lambda(n+1)h. \quad (7.1.8)$$

Then

$$p(x) = \lambda h e^{nh}/(1 - e^{-h}).$$

Let

$$g_n = \lambda h e^{nh}/(1 - e^{-h}) \quad \text{for} \quad n = 0, 1, 2, \ldots.$$

Note that $\lambda e^{nh} \leq g_n \leq \lambda e^{(n+1)h}$.

From (7.1.8), it is clear that for $p(x) > g_n$, the right-hand side of (7.1.8) is larger than the left. Thus $f_\lambda(x) = (n+1)h$ for $g_n < p(x) \leq \lambda e^{(n+1)h}$. Since $f_\lambda(x) = nh$ for $p(x) = \lambda e^{nh}$, the left-hand side is larger than the right for $p(x) = \lambda e^{nh}$. Because the equality in (7.1.8) occurs at only one point, it follows that the left-hand side is greater than the right for $\lambda e^{nh} \leq p(x) \leq g_n$. Thus, $f_\lambda(x) = nh$ for $\lambda e^{nh} \leq p(x) \leq g_n$ and $f_\lambda(x) = (n+1)h$ for $g_n < p(x) \leq \lambda e^{(n+1)h}$. The claim now follows.

Switching to polar coordinates, the density for the target distribution becomes

$$p(r, \theta) = \frac{1}{2\pi\sigma^2} \exp\left(-\frac{r^2}{2\sigma^2}\right) \quad \text{for} \quad 0 \leq \theta \leq 2\pi, \; r \geq 0.$$

Let $A(\lambda) = \ln[(1 - e^{-h})/(2\pi\sigma^2 \lambda h)]$ for $\lambda > 0$. For a given λ, let N be the integer such that $(N - 1)h < A(\lambda) \leq Nh$. Then (7.1.7) becomes

$$f_\lambda(r, \theta) = \begin{cases} 0 & \text{for } \dfrac{r^2}{2\sigma^2} \geq A(\lambda), \\ nh & \text{for } A(\lambda) - nh \leq \dfrac{r^2}{2\sigma^2} \leq A(\lambda) - (n - 1)h, \quad n = 1, \ldots, N - 1, \\ Nh & \text{for } 0 \leq \dfrac{r^2}{2\sigma^2} \leq A(\lambda) - (N - 1)h. \end{cases} \quad (7.1.9)$$

Thus, for λ such that $(N - 1)h < A(\lambda) \leq Nh$,

$$\begin{aligned} U(\lambda) &= \int_0^{2\pi} \int_0^\infty f_\lambda(r, \theta) r \, dr \, d\theta \\ &= 2\pi\sigma^2 Nh[A(\lambda) - (N - 1)h] \\ &\quad + 2\pi\sigma^2 \sum_{n=1}^{N-1} nh[A(\lambda) - (n - 1)h - (A(\lambda) - nh)] \\ &= 2\pi\sigma^2 h\{NA(\lambda) - \tfrac{1}{2}N(N - 1)h\}. \end{aligned} \quad (7.1.10)$$

Now to find $\hat{f}$ it is only necessary to find λ such that $U(\lambda) = K$ and set $\hat{f} = f_\lambda$. Observe that when $A(\lambda) = (N - 1)h$, $U(\lambda) = \pi\sigma^2 h^2 N(N - 1)$, and when $A(\lambda) = Nh$, $U(\lambda) = \pi\sigma^2 h^2 N(N + 1)$. Thus, for a given amount of effort K, choose N so that

$$\pi\sigma^2 h^2 N(N - 1) \leq K \leq \pi\sigma^2 h^2 N(N + 1).$$

Then solve the following equation for $A(\lambda)$:

$$K = 2\pi\sigma^2 h\{NA(\lambda) - \tfrac{1}{2}N(N - 1)h\},$$

i.e.,

$$A(\lambda) = \frac{K}{2\pi\sigma^2 Nh} + \frac{(N - 1)h}{2}. \quad (7.1.11)$$

The λ that satisfies (7.1.11) satisfies $U(\lambda) = K$. Let

$$r_n^2 = 2\sigma^2 \left[\frac{K}{2\pi\sigma^2 Nh} + \frac{(N + 1)h}{2} - (n - 1)h\right] \quad \text{for } n = 1, 2, \ldots, N.$$

Substituting for $A(\lambda)$ in (7.1.9) and letting $\hat{f} = f_\lambda$, one obtains

$$\hat{f}(r, \theta) = \begin{cases} 0 & \text{for } r^2 \geq r_1^2, \\ n & \text{for } r_{n+1}^2 \leq r^2 \leq r_n^2, \quad n = 1, 2, \ldots, N - 1, \\ N & \text{for } 0 \leq r^2 \leq r_N^2. \end{cases} \quad (7.1.12)$$

7.1 Approximations in a Restricted Class

Observe that $\hat{f} \in \hat{F}$ and $C[\hat{f}] = K$. Let λ be such that the equality holds in (7.1.11). Then $f_\lambda = \hat{f}$ satisfies (7.1.6) and, by Theorem 7.1.1, $\hat{f}$ is optimal within $\hat{F}$ for cost K.

Probability of Detection

The probability of detection resulting from the allocation $\hat{f}$ is calculated and compared to the probability produced by f^*, the allocation that is optimal within F for cost K. It is shown that as h approaches 0, $P[\hat{f}]$ approaches $P[f^*]$.

Using (7.1.12), one can calculate

$$1 - P[\hat{f}] = e^{-Nh} \int_0^{2\pi} \int_0^{r_N} \frac{1}{2\pi\sigma^2} \exp\left(\frac{-r^2}{2\sigma^2}\right) r \, dr \, d\theta$$

$$+ \sum_{n=1}^{N-1} e^{-hn} \int_0^{2\pi} \int_{r_{n+1}}^{r_n} \frac{1}{2\pi\sigma^2} \exp\left(\frac{-r^2}{2\sigma^2}\right) r \, dr \, d\theta$$

$$+ \int_0^{2\pi} \int_{r_1}^{\infty} \frac{1}{2\pi\sigma^2} \exp\left(\frac{-r^2}{2\sigma^2}\right) r \, dr \, d\theta$$

$$= e^{-Nh} \left[1 - \exp\left(\frac{-r_N^2}{2\sigma^2}\right)\right] + \exp\left(-\frac{r_1^2}{2\sigma^2}\right)$$

$$+ \sum_{n=1}^{N-1} e^{-nh} \left[\exp\left(\frac{-r_{n+1}^2}{2\sigma^2}\right) - \exp\left(-\frac{r_n^2}{2\sigma^2}\right)\right]$$

$$= e^{-Nh} + N(1 - e^{-h}) \exp\left(\frac{-K}{2\pi\sigma^2 Nh} - \frac{(N+1)h}{2}\right).$$

Thus

$$P[\hat{f}] = 1 - e^{-Nh} - N(1 - e^{-h}) \exp\left(\frac{-K}{2\pi\sigma^2 Nh} - \frac{(N+1)h}{2}\right), \quad (7.1.13)$$

where N is the unique integer that satisfies

$$\pi\sigma^2 h^2 N(N-1) < K \le \pi\sigma^2 h^2 N(N+1). \quad (7.1.14)$$

In Example 2.2.7, Eq. (2.2.21) gives the probability of detecting the target by time t using the allocation f^* that is optimal within $F(X)$ for cost Wvt. Setting $K = Wvt$ in (2.2.21), one obtains

$$P[f^*] = 1 - [1 + (K/\pi\sigma^2)^{1/2}] \exp[-(K/\pi\sigma^2)^{1/2}].$$

Naturally $P[f^*] \ge P[\hat{f}]$. However, one can show that as $h \to 0$, $P[\hat{f}]$ approaches $P[f^*]$. To see this, observe from (7.1.14) that as $h \to 0$, $N^2 h^2 \to K/(\pi\sigma^2)$. Using this and the fact that $\lim_{h \to 0} N(1 - e^{-h}) = \lim_{h \to 0} Nh$, it follows from (7.1.13) that

$$\lim_{h \to 0} P[\hat{f}] = P[f^*].$$

Remark 7.1.4. The approximation obtained in Example 7.1.3 may still be difficult to use because of the circular regions in which layers of effort must be placed. It is often more convenient to place effort over square regions. Thus one might ask: What is the best allocation of effort within $\hat{F}$ for which the layers of effort can be placed in square regions? In the case of the circular normal distribution a method of answering this question is presented in this remark.

Consider the following distribution on $[0, \infty)$. Let I be the cumulative distribution function for a normal random variable with mean zero and unit variance. Let σ be the parameter of the circular normal target distribution considered in Example 7.1.3. Define

$$g(r) = [I(r/2\sigma) - I(-r/2\sigma)]^2 \quad \text{for} \quad r \geq 0.$$

Then $g(r)$ gives the probability that the target is located inside the square of side r that is centered at the origin. By extending the definition of g so that $g(r) = 0$ for $r < 0$, one obtains a probability distribution on the line. Let

$$p(r) = g'(r) = \begin{cases} \dfrac{\sqrt{2}}{\sigma\sqrt{\pi}} \exp\left(\dfrac{-r^2}{8\sigma^2}\right)\left[I\left(\dfrac{r}{2\sigma}\right) - I\left(\dfrac{-r}{2\sigma}\right)\right] & \text{for} \quad r > 0, \\ 0 & \text{for} \quad r < 0. \end{cases} \quad (7.1.15)$$

Let $b(r, z) = 1 - e^{-z}$ for $-\infty < r < \infty$; then for $\lambda > 0$, one may find f_λ to satisfy (7.1.6) (replacing X by the real line) by using (7.1.7) and p as defined in (7.1.15); that is,

$$f_\lambda(r) = \begin{cases} 0 & \text{for } r \text{ such that} \quad p(r) \leq \dfrac{\lambda h}{1 - e^{-h}}, \\ nh & \text{for } r \text{ such that} \quad \dfrac{\lambda h e^{(n-1)h}}{1 - e^{-h}} < p(r) \leq \dfrac{\lambda h e^{nh}}{1 - e^{-h}}. \end{cases} \quad (7.1.16)$$

The allocation of an effort density nh to an $r \geq 0$ is equivalent to allocating effort density nh around the square of side r centered at the origin in the case of the circular normal target distribution. Thus, to find an allocation $\hat{f}$ on the plane that is optimal for cost K within the subclass of allocations $\hat{F}$ in which layers of effort are placed within squares, one must find λ such that

$$\int_{-\infty}^{\infty} 4rf_\lambda(r)\, dr = K. \quad (7.1.17)$$

Since $p(r) = 0$ for $r < 0$, $f_\lambda(r) = 0$ for $r < 0$ and (7.1.17) may be written as

$$\int_0^\infty 4rf_\lambda(r)\, dr = K. \quad (7.1.18)$$

7.1 Approximations in a Restricted Class

Having found f_λ to satisfy (7.1.18), one obtains $\hat{f}$ by letting

$$\hat{f}(x) = f_\lambda(r) \quad \text{for } x \text{ on the square of side } r \text{ centered at the origin in the plane.} \quad (7.1.19)$$

Unfortunately, the form of p given in (7.1.15) prevents one from expressing (7.1.16) in a convenient analytic form and thus obtaining a closed-form expression for the integral in (7.1.18). However, the integral may be readily calculated in a numerical fashion so that a λ to satisfy (7.1.18) may be found by use of numerical techniques. This in turn would yield $\hat{f}$.

Example 7.1.5. One might suppose that as h decreases in Theorem 7.1.1, one would obtain a better approximation to the optimal allocation of K amount of effort. This example shows that this need not be the case.

Let X be the plane and let the target location distribution be uniform over a rectangle of unit area in X. Assume

$$b(x, z) = 1 - e^{-z} \quad \text{for } x \in X, \; z \geq 0.$$

Let $\Pi_h(K)$ be the probability of detection resulting from the allocation of Theorem 7.1.1, which is optimal within $\hat{F}$ for cost K. Let $\Pi_0(K)$ be the probability of detection resulting from the optimal plan when there is no restriction requiring effort to be allocated in multiples of h:

$$\Pi_0(K) = 1 - e^{-K} \quad \text{for } K > 0.$$

For $h > 0$, one can show that Π_h is the function obtained by linear interpolation between the points $(nh, 1 - e^{-nh})$ for $n = 0, 1, \ldots$. For $h = 1$ and $h = 0.75$, this function is shown in Fig. 7.1.2. Observe that

$$\Pi_1(1) > \Pi_{3/4}(1).$$

Thus, for the case where there is one unit of effort to allocate, reducing h from 1 to $\tfrac{3}{4}$ actually reduces the resulting probability of detection rather than increasing it.

This example illustrates that h should not be chosen simply to equal h_0, the smallest operationally reasonable value of h. In fact, to obtain the best choice of h one has to solve the additional optimization problem of finding $h^* \geq h_0$ such that

$$\Pi_{h^*}(K) \geq \Pi_h(K) \quad \text{for } h \geq h_0.$$

Remark 7.1.6. Approximations to the optimal allocation for a circular normal target distribution that require searching only rectangular regions have been investigated in Koopman (1946, Section 7.1) and Reber (1956, 1957). The approximation considered in Koopman (1946) is developed for

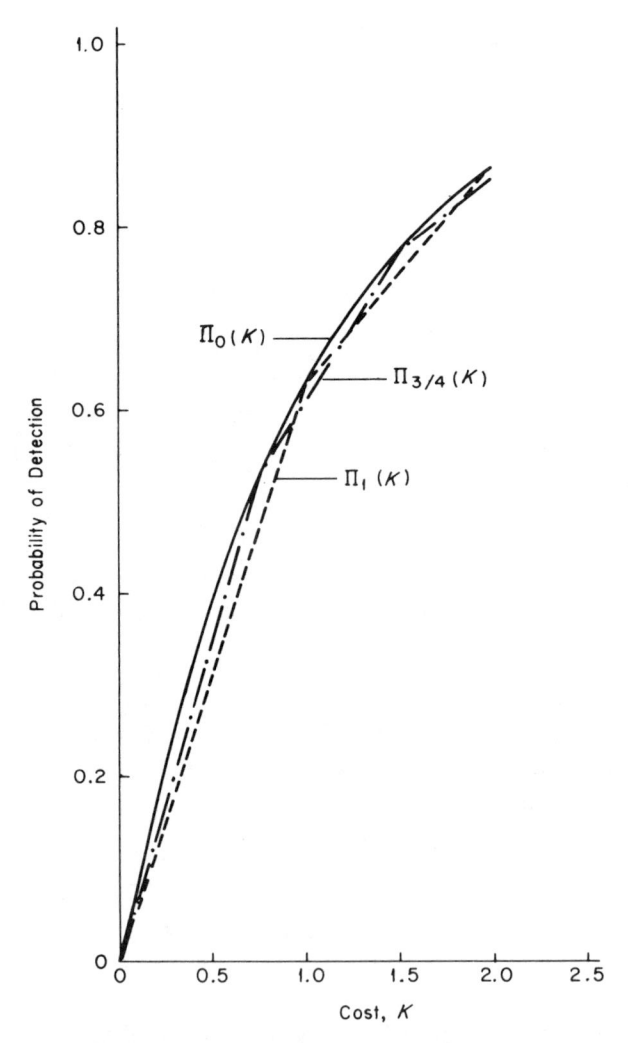

Fig. 7.1.2. Probability of detection for allocations that are optimal within $\hat{F}$ for cost K. [*Note:* Target location distribution is uniform over a rectangle of area 1; $b(x, z) = 1 - e^{-z}$ for $x \in X$, $z \geq 0$.]

the inverse-cube detection function [see (1.2.5)] but may be adapted for use with an exponential detection function. In Reber (1956, 1957) approximating plans are found for general bivariate normal distributions [see (1.1.1)] and a class of detection functions that include the exponential. Plans are also given to achieve probability 0.9 or 0.95 of detecting the target. These plans are also described in Richardson *et al.* (1971).

7.2. INCREMENTAL APPROXIMATIONS

In Chapter II uniformly optimal search plans are found for searches involving a stationary target and a regular detection function. In the presence of false targets, optimal nonadaptive plans are given in Chapter VI. Typically, optimal plans have a complicated form even when one is able to overcome the analytic difficulties involved in finding their explicit expressions. This complicated form makes it difficult to provide planning advice without the aid of computers. Even in cases where the functional form of the plan is simple, such as the plan for the circular normal distribution and exponential detection function given in (2.2.6), the plans usually call for spreading small amounts of effort over ever expanding areas.

As a method of overcoming some of these difficulties, a class of plans is presented that are based on an increment Δ of effort. These plans are called Δ-*plans* and are used for searches where the search space J is composed of a finite number N of cells. The plans develop in the following way:

An increment of effort Δ is fixed throughout the search. Before allocating each increment, one calculates

$$\rho(j, z_j) = \frac{p(j) b'(j, z_j)}{c'(j, z_j)} \quad \text{for } j = 1, 2, \ldots, N,$$

where z_j is the total effort already placed in cell j. The next increment Δ is placed in the cell having the highest value of $\rho(j, z_j)$. If there is more than one of these highest cells, then one may allocate Δ to any one of them.

Note that each increment Δ is allocated to a single cell and that the only computation required is to calculate $\rho(j, z_j)$ for $j = 1, 2, \ldots, N$. Usually this can be easily done with the aid of a calculator and some basic mathematical tables. In addition, after an increment Δ has been placed in cell j only the rate $\rho(j, z_j + \Delta)$ for that cell need be recalculated. The rates in the other cells remain the same as before. The application of Δ-plans is restricted to search spaces having only a finite number of cells. However, in practice, a target distribution over X is often approximated by one with a finite number of cells in the manner discussed in Section 1.1. As an example of this, see Richardson and Stone (1971).

In the case of no false targets, $c(j, z) = z$ and a Δ-plan may be thought of as a discrete approximation to the locally optimal plan given in Section 3.1. In case of search in the presence of false targets, Δ is understood to be an increment of broad-search time; that is, if cell j receives the next increment Δ, then Δ broad-search time is placed in cell j. Any contacts that are generated are investigated immediately. The functions τ, δ, and a are assumed to have the meanings given to them in Chapter VI. In addition $v = 1$ since broad-search effort is measured in time. It is assumed that $a = b$ and that

$$c(j, z) = z + \tau(j)\delta(j)b(j, z) \quad \text{for} \quad j = 1, \ldots, N, \quad z \geq 0.$$

The false targets are assumed to be either Poisson or independent identically distributed. Note that the case of no false targets may be considered as a special case of the false-target situation in which $\delta(j) = 0$ for $j = 1, \ldots, N$.

Let μ_Δ denote the mean time to detect the target when using a Δ-plan. In order to refer to searches with and without false targets, the term *optimal search plan* is used to mean either a uniformly optimal search plan in the case of no false targets or an optimal nonadaptive broad-search plan for immediate and conclusive contact investigation when false targets are present. Let μ^* be the mean time to detect the target when the optimal search plan is used. Note that when false targets are present, both μ_Δ and μ^* do not include the time to identify the contact that is the target.

The main result of this section is to show that if

$$D = \max_{1 \leq j \leq N} [1 + \tau(j)\delta(j)b'(j, 0)],$$

then

$$\mu^* \leq \mu_\Delta \leq \mu^* + 2(N + 1)D\Delta. \tag{7.2.1}$$

This implies that $\mu_\Delta \to \mu^*$ as $\Delta \to 0$ when $D < \infty$. This is the sense in which a Δ-plan approximates an optimal plan.

Let $h(j, n)$ be the cumulative effort placed in cell j after n increments have been applied to the search in a Δ-plan. The following lemma is used to prove the main result of this section.

Lemma 7.2.1. *Suppose the search space J has a finite number N of cells and that b is a regular detection function on J. Let $M(t) = t$ for $t > 0$ and φ^* be an optimal search plan within $\Phi(M)$. Then for $n = 1, 2, \ldots,$*

$$\varphi^*(j, (n - N)\Delta) \leq h(j, n) \leq \varphi^*(j, n\Delta) + \Delta \quad \text{for} \quad j = 1, \ldots, N, \tag{7.2.2}$$

where $\varphi^*(j, (n - N)\Delta) = 0$ when $n < N$.

Proof. Fix n and let $\lambda_n^+ = \max \rho(j, h(j, n))$, where max without an indicated range is understood to run over $j = 1, \ldots, N$. Recall that $\rho_j(z) = \rho(j, z)$ for $j \in J$, $z \geq 0$, and that

$$\rho_j^{-1}(\lambda) = \begin{cases} \text{inverse of } \rho_j \text{ evaluated at } \lambda & \text{for} \quad 0 < \lambda \leq \rho_j(0), \\ 0 & \text{for} \quad \lambda > \rho_j(0). \end{cases}$$

It is claimed that

$$\rho_j^{-1}(\lambda_n^+) \leq h(j, n) \leq \rho_j^{-1}(\lambda_n^+) + \Delta. \tag{7.2.3}$$

7.2 Incremental Approximations

To see that the left-hand side of (7.2.3) holds, observe that $\lambda_n^+ \geq \rho(j, h(j, n))$ by definition and apply ρ_j^{-1} to both sides of this inequality. Since ρ_j^{-1} is decreasing, the left-hand side of (7.2.3) follows.

If $h(j, n) = 0$, the right-hand side of (7.2.3) holds trivially. Suppose $h(j, n) = i\Delta$ for some positive integer i. It follows that

$$\rho_j((i - 1)\Delta) \geq \lambda_n^+, \tag{7.2.4}$$

for if (7.2.4) does not hold, the last increment Δ placed in cell j was not added according to a Δ-plan; that is, this increment was added to cell j at a time when $\rho(j, (i - 1)\Delta)$ was not the highest rate. Thus, by (7.2.4)

$$h(j, n) = \Delta + (i - 1)\Delta \leq \rho_j^{-1}(\lambda_n^+) + \Delta,$$

and (7.2.3) holds.

Since $\sum_{j=1}^N h(j, n) = n\Delta$, summing (7.2.3) yields

$$\sum_{j=1}^N \rho_j^{-1}(\lambda_n^+) \leq n\Delta \leq \sum_{j=1}^N \rho_j^{-1}(\lambda_n^+) + N\Delta. \tag{7.2.5}$$

Thus,

$$(n - N)\Delta \leq \sum_{j=1}^N \rho_j^{-1}(\lambda_n^+). \tag{7.2.6}$$

When $n < N$, the left-hand side of (7.2.2) holds by convention. If $n \geq N$, let $t = (n - N)\Delta$, and let $\lambda(t) = U^{-1}(t)$, where,

$$U(\lambda) = \sum_{j=1}^N \rho_j^{-1}(\lambda).$$

Then by Theorem 2.2.5 or 6.3.3,

$$\varphi^*(j, (n - N)\Delta) = \rho_j^{-1}(\lambda(t)) \quad \text{for } j = 1, \ldots, N,$$

and

$$U(\lambda(t)) = \sum_{j=1}^N \rho_j^{-1}(\lambda(t)) = t = (n - N)\Delta.$$

By (7.2.6),

$$U(\lambda(t)) = (n - N)\Delta \leq \sum_{j=1}^N \rho_j^{-1}(\lambda_n^+) = U(\lambda_n^+).$$

Since U is strictly decreasing, $\lambda_n^+ \leq \lambda(t)$ and

$$\varphi^*(j, (n - N)\Delta) = \rho_j^{-1}(\lambda(t)) \leq \rho_j^{-1}(\lambda_n^+) \leq h(j, n),$$

where the last inequality follows from (7.2.3). This proves the left-hand side of (7.2.2).

By (7.2.5), $\sum_{j=1}^{N} \rho_j^{-1}(\lambda_n^+) \leq n\Delta$, which implies that $\rho_j^{-1}(\lambda_n^+) \leq \varphi^*(j, n\Delta)$. The right-hand side of (7.2.2) now follows from (7.2.3) and the lemma is proved.

Since a Δ-plan specifies the allocation of effort only at integer multiples of Δ, it is convenient to choose a search plan φ_Δ such that $\varphi_\Delta \in \Phi(M)$, where $M(t) = t$ for $t \geq 0$, and

$$\varphi_\Delta(j, n\Delta) = h(j, n) \quad \text{for} \quad j = 1, \ldots, N, \quad n = 1, 2, \ldots. \quad (7.2.7)$$

Clearly such a plan exists. Let φ^* be the optimal search plan in $\Phi(M)$ and define

$$P_\Delta(t) = P[\varphi_\Delta(\cdot, t)], \quad C_\Delta(t) = C[\varphi_\Delta(\cdot, t)]$$
$$P^*(t) = P[\varphi^*(\cdot, t)], \quad C^*(t) = C[\varphi^*(\cdot, t)] \quad \text{for} \quad t \geq 0.$$

Let

$$\mu_\Delta = \int_0^\infty [1 - P_\Delta(t)] C_\Delta(dt),$$

$$\mu^* = \int_0^\infty [1 - P^*(t)] C^*(dt).$$

Then by the discussion in Section 2.2 or 6.4, μ_Δ and μ^* are the mean times to detect the target when using φ_Δ and φ^*, respectively. In Theorem 7.2.2, a bound is found for μ_Δ. Since the proof holds for any plan φ_Δ that satisfies (7.2.7), it gives a bound on mean time to detect when using a Δ-plan.

Theorem 7.2.2. *Suppose the search space J has a finite number N of cells and that b is a regular detection function on J. If*

$$D = \max_{1 \leq j \leq N} [1 + \delta(j)\tau(j)b'(j, 0)],$$

then

$$\mu_\Delta \leq \mu^* + 2(N + 1)D\Delta. \quad (7.2.8)$$

Proof. From the left half of (7.2.2) it follows that

$$P_\Delta(n\Delta) \geq P^*((n - N)\Delta) \quad \text{for} \quad n = 1, 2, \ldots,$$

where for convenience it is understood that $P^*(t) = P_\Delta(t) = 0$ for $t \leq 0$. Thus,

$$P_\Delta(t) \geq P^*((n - N - 1)\Delta) \geq P^*(t - (N + 1)\Delta) \quad \text{for} \quad (n - 1)\Delta \leq t \leq n\Delta.$$

7.2 Incremental Approximations

Since the left- and right-most members of the above inequality do not depend on n, it follows that
$$P_\Delta(t) \geq P^*(t - (N+1)\Delta) \quad \text{for } t \geq 0. \tag{7.2.9}$$
From the right half of (7.2.2), one obtains
$$C_\Delta(n\Delta) \leq \sum_{j=1}^{N} [\varphi^*(j, n\Delta) + \Delta + \tau(j)\delta(j)b(j, \varphi^*(j, n\Delta) + \Delta)]. \tag{7.2.10}$$
Since b is regular, $b'(j, \cdot)$ is decreasing for $j \in J$ and
$$b(j, \varphi^*(j, n\Delta) + \Delta) \leq b(j, \varphi^*(j, n\Delta)) + b'(j, 0)\Delta \quad \text{for } j \in J.$$
From this and (7.2.10), it follows that
$$C_\Delta(n\Delta) \leq C[\varphi^*(\cdot, n\Delta)] + ND\Delta = C^*(n\Delta) + ND\Delta.$$
Thus, for $(n-1) \leq t \leq n\Delta$,
$$C_\Delta(t) \leq C^*(n\Delta) + ND\Delta \leq C^*((n-1)\Delta) + (N+1)D\Delta \leq C^*(t) + (N+1)D\Delta.$$
Since the first and last members of this sequence of inequalities do not depend on n, one can write
$$C_\Delta(t) \leq C^*(t) + (N+1)D\Delta \quad \text{for } t \geq 0. \tag{7.2.11}$$
From (7.2.9), it follows that
$$\mu_\Delta = \int_0^\infty [1 - P_\Delta(t)]C_\Delta(dt)$$
$$\leq \int_0^\infty [1 - P^*(t - (N+1)\Delta)]C_\Delta(dt).$$
Let $g(t) = P^*(t - (N+1)\Delta)$ for $t \geq 0$. Then an integration by parts and use of (7.2.11) yields
$$\mu_\Delta \leq \int_0^\infty [1 - P^*(t - (N+1)\Delta)]C_\Delta(dt) = \int_0^\infty C_\Delta(t)g(dt)$$
$$\leq (N+1)D\Delta + \int_0^\infty C^*(t)g(dt).$$
Since $C^*(t) \leq C^*(t - (N+1)\Delta) + (N+1)D\Delta$,
$$\int_0^\infty C^*(t)g(dt) \leq (N+1)D\Delta + \int_0^\infty C^*(t - (N+1)\Delta)g(dt)$$
$$= (N+1)D\Delta + \int_0^\infty C^*(t)P^*(dt)$$
$$= (N+1)D\Delta + \mu^*.$$

It now follows that

$$\mu_\Delta \leq \mu^* + 2(N + 1)D\Delta,$$

and the theorem is proved.

In the case where there are no false targets [i.e., $c(j, z) = z$ for $j \in J, z \geq 0$], inequality (7.2.8) may be sharpened.

Theorem 7.2.3. *Suppose the search space J has a finite number N of cells and that b is a regular detection function on J. If $c(j, z) = z$ for $j \in J, z \geq 0$, then*

$$\mu_\Delta \leq \mu^* + (N + 1)\Delta. \tag{7.2.12}$$

Proof. The proof proceeds in the same manner as that of Theorem 7.2.2 to obtain (7.2.9). Since $C_\Delta(t) = t$ and $C^*(t) = t$, it follows that

$$\mu_\Delta = \int_0^\infty [1 - P_\Delta(t)] \, dt \leq \int_0^\infty [1 - P^*(t - (N + 1)\Delta)] \, dt \leq (N + 1)\Delta + \mu^*,$$

and the theorem is proved.

Example 7.2.4. There are situations in which Δ-plans perform substantially better than is indicated by the bounds in Theorems 7.2.2 and 7.2.3. The Δ-plan considered here provides an example of such a situation.

Consider a search in which the target distribution is uniform over a square of unit area. Suppose that this square is subdivided into N cells of equal area to give a discrete approximation to the continuous distribution, that is, $p(j) = 1/N$ for $j = 1, \ldots, N$. For $j = 1, \ldots, N$, let

$$b(j, z) = 1 - e^{-Nz} \quad \text{for} \quad z \geq 0,$$
$$\delta(j) = 0.$$

One can check that $\mu^* = 1$ for this search.

Consider a Δ-plan that allocates the $(nN + j)$th increment of effort to cell j for $j = 1, 2, \ldots, N$, and $n = 0, 1, \ldots$. Let μ_j be the mean time to find the target given that the target is located in cell j. Consider the first rectangle. The mean time required to find the target given that it is found during the first increment Δ of effort is

$$\frac{N \int_0^\Delta t e^{-Nt} \, dt}{1 - e^{-N\Delta}} = \frac{1 - (1 + N\Delta)e^{-N\Delta}}{N(1 - e^{-N\Delta})}. \tag{7.2.13}$$

If the target is not detected on the first increment, then $N - 1$ additional increments will be placed in the remaining cells until it is again time to place an increment in cell 1. At this point the situation is the same as at the beginning of the search, so that the mean time remaining to detect the target is μ_1,

7.2 Incremental Approximations

the same as the mean time at the beginning of the search. The mean time to detect the target given that it is not found during the first increment is

$$N\Delta + \mu_1. \tag{7.2.14}$$

Since the probability of detecting the target during the first increment is $1 - e^{-N\Delta}$, one can write

$$\mu_1 = \frac{1 - (1 + N\Delta)e^{-N\Delta}}{N} + e^{-N\Delta}(N\Delta + \mu_1).$$

Solving for μ_1, one finds that

$$\mu_1 = \frac{1}{N} + \frac{(N-1)\Delta e^{-N\Delta}}{1 - e^{-N\Delta}}.$$

Since there are $j - 1$ increments of effort exerted before the first increment is placed in cell j, it follows that

$$\mu_j = (j-1)\Delta + \mu_1 \quad \text{for } j = 1, \ldots, N.$$

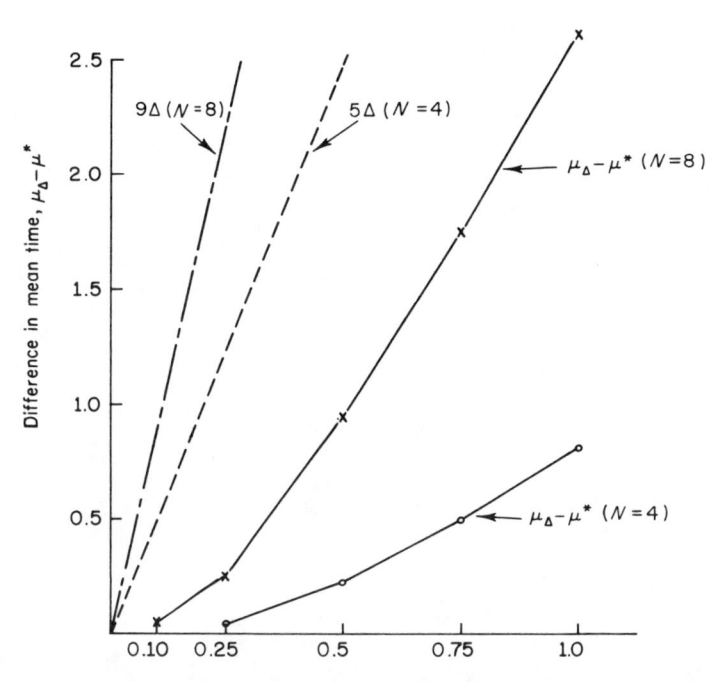

Fig. 7.2.1. Difference in mean time to detection for Δ-plans and optimal plans. [*Note:* - - -, bound on $\mu_\Delta - \mu^*$ obtained from Theorem 7.2.3 for $N = 4$; - - -, bound on $\mu_\Delta - \mu^*$ obtained from Theorem 7.2.3 for $N = 8$.]

Thus,

$$\mu_\Delta = \frac{1}{N}\sum_{j=1}^{N}\mu_j = \frac{(N-1)\Delta}{2} + \mu_1 = \frac{1}{N} + \frac{(N-1)\Delta(1+e^{-N\Delta})}{2(1-e^{-N\Delta})}.$$

Since $\mu^* = 1$,

$$\mu_\Delta = \mu^* + \frac{(N-1)\Delta}{2}\left[\frac{1+e^{-N\Delta}}{1-e^{-N\Delta}} - \frac{2}{N\Delta}\right].$$

For $N = 4$ and $N = 8$ a graph of $\mu_\Delta - \mu^*$ as a function of Δ is given in Fig. 7.2.1. Observe that for small values of Δ, μ_Δ is much closer to μ^* than the inequalities in Theorem 7.2.3. indicate.

NOTES

Bumby (1960) obtained an allocation f that is optimal for cost K within $\hat{F}$ for the search situation that is the one-dimensional analog of that in Example 7.1.3; that is, Bumby took X to be the real line and the target distribution to be one-dimensional normal. He also found necessary conditions for an allocation f to be optimal for cost K within $\hat{F}$ when the search space is X and the detection function is homogeneous and exponential. The conditions are essentially those given in (7.1.7).

The material in Section 7.2 follows that in Stone (1972b). Specific plans for approximating the optimal allocation of effort for a circular normal target distribution that require searching only rectangular regions are given in Koopman (1946) and Reber (1956, 1957). (See Remark 7.1.6.)

Chapter VIII

Conditionally Deterministic Target Motion

This chapter considers search for targets having a rather simple type of motion called conditionally deterministic. The target's motion takes place in Euclidean n-space and depends on an n-dimensional stochastic parameter. If this parameter were known, then the target's position would be known at all times in the future. Thus, the target's motion is deterministic conditioned on knowledge of the parameter.

Conditionally deterministic motion is more restrictive than Markovian motion, which is considered in Chapter IX. However, the class of conditionally deterministic target motions is still rich enough to contain interesting and important search problems. In addition, there are many situations in which optimal search plans can be found explicitly for targets with conditionally deterministic motion. This is in contrast to the situation of Markovian motion, where optimal search plans have been found only in the simplest situations.

Section 8.1 defines the notion of conditionally deterministic motion and states a general class of search problems that involve maximizing the probability of detecting the target subject to certain constraints on total search effort and on the rate at which effort may be applied. While no general method of solving this class of problems is given, optimal plans are found for two interesting subclasses of these problems. Section 8.2 deals with the first class of problems involving factorable target motions. The notion of factorability is defined precisely in Section 8.2, but the crucial requirement is that the Jacobian of the target motion factor into time- and space-dependent parts. Methods of finding uniformly optimal search plans are given in Section 8.2 and illustrated by two examples, the second of which, Example 8.2.5, was motivated by the problem of regaining contact on a moving submarine. For the problems addressed in Section 8.2, there is a constraint on the rate at which search effort may be applied to the whole search space.

Section 8.3 considers a class of problems in which there is a constraint on the rate at which effort density may be applied at a point and a constraint on the total search effort, but no constraint on the rate at which effort may be applied to the whole search space. Theorem 8.3.1 gives a sufficiency result that is useful for finding optimal plans. Example 8.3.2 illustrates the use of Theorem 8.3.1 in finding optimal plans. The optimal plan found in this example has the rather surprising property that it first expands about the mean initial target position, the origin, and then contracts back to the origin despite the fact that the target is continually moving away from the origin.

Section 8.4 presents necessary and sufficient conditions for a search plan to maximize the probability of detecting the target by time t when there is a constraint on the rate at which effort may be applied to the whole search space. In contrast to Section 8.2, the target motion is not assumed to be factorable. The necessary condition is stated but its proof omitted.

8.1. DESCRIPTION OF SEARCH PROBLEM

In this section, conditionally deterministic motion is defined and the problem of optimal search for a target whose motion is conditionally deterministic is stated.

Motivation

To motivate the presentation in this chapter, consider the following example. Let X be the plane. At time 0 the target's location is given by a probability density function p. The target's motion is a simple translation described by a function $\eta \colon X \times [0, \infty) \to X$ such that $\eta(x, t)$ gives the target's position at time t given that it was at point x at time 0; that is,

$$\eta(x, t) = (x_1 + vt, x_2) \quad \text{for} \quad x = (x_1, x_2) \in X, \quad t \geq 0, \quad (8.1.1)$$

where $v > 0$ gives the constant speed of the translation. This situation might occur on the ocean if the target is a life raft known to be drifting with a constant velocity but whose initial position is unknown. In this case an

8.1 Description of Search Problem

optimal search may be obtained by first solving the problem for the stationary case and then "moving" the optimal allocation of effort along with the drift of the target distribution.

To be more precise, let the cumulative effort function M be given and suppose $\varphi \in \Phi(M)$. Let

$$\dot{\varphi}(x, t) = \frac{\partial \varphi(x, t)}{\partial t} \quad \text{for} \quad x \in X, \quad t \geq 0$$

when the partial derivative on the right exists, that is, $\dot{\varphi}$ indicates the derivative of φ with respect to time. Assume $\dot{\varphi}$ exists and

$$\varphi(x, t) = \int_0^t \dot{\varphi}(x, s) \, ds \quad \text{for} \quad x \in X, \quad t \geq 0.$$

Then $\dot{\varphi}(x, t)$ gives the rate at which effort density is accumulating at x at time t. If the target motion is given by (8.1.1) and φ is uniformly optimal for the stationary problem, then at time t we want to apply effort at the rate $\dot{\varphi}(x, t)$ to the point $(x_1 + vt, x_2)$. From this it is clear that if one fixes a point $y \in X$, then $\psi(y, t)$, the rate at which effort density is applied at $y = (y_1, y_2)$ at time t in the uniformly optimal plan is given by

$$\psi(y, t) = \dot{\varphi}((y_1 - vt, y_2), t) = \dot{\varphi}(\eta_t^{-1}(y), t) \quad \text{for} \quad y \in X, \quad t \geq 0,$$

where η_t denotes the transformation $\eta(\cdot, t)$ and η_t^{-1} is the inverse map of η_t.

Observe that if the target starts at point x at time 0, then $\psi(\eta_s(x), s)$ is the rate at which effort density is accumulating on the target at time s and $\int_0^t \psi(\eta_s(x), s) \, ds$ gives the total effort density accumulated by time t. As a result, the probability of detecting the target by time t is

$$\int_X p(x) b\left(x, \int_0^t \psi(\eta_s(x), s) \, ds\right) dx.$$

Definition of Conditionally Deterministic Motion

The target motion takes place in the space Y, which is a copy of Euclidean n-space. The space X, also a copy of Euclidean n-space, is the parameter space. To prescribe the target motion, there is a Borel function

$$\eta: X \times T \to Y,$$

where T is an interval of nonnegative real numbers containing 0 as its left-hand endpoint. The target motion is characterized by a stochastic parameter ξ that takes values in X, that is, if $\xi = x$, then $\eta(x, t)$ gives the position of the target at time t.

The distribution of the parameter ξ is given by the probability density function p, so that

$$\Pr\{\xi \in S\} = \int_S p(x) \, dx$$

for any Borel set $S \subset X$.

Often one takes $X = Y$ and considers ξ as the position of the target at time 0, which has a probability distribution specified by p.

For $t \in T$, let $\eta_t = \eta(\cdot, t)$. Let η_t^i denote the ith component of η_t and $\partial \eta_t^i / \partial x_k$ the partial derivative of η_t^i with respect to the kth component of x. For $t \in T$, it is assumed that η_t has continuous partial derivatives at all $x \in X$. Let $\mathbf{J}(x, t)$ be the absolute value of the Jacobian of the transformation η_t evaluated at x, i.e.,

$$\mathbf{J}(x, t) = \left| \det\left(\frac{\partial \eta_t^i(x)}{\partial x_k} \right) \right| \quad \text{for} \quad x \in X, \; t \in T.$$

The transformations η_t for $t \in T$ are further assumed to have the following properties:

(a) η_t is one-to-one for $t \in T$, and
(b) $\mathbf{J}(x, t)$ is positive for all $x \in X$, $t \in T$.

By (a), one may define η_t^{-1}, the inverse of η_t, on the image of η_t.

Observe that the probability of the target being in a Borel set S at time t is given by

$$\int_{\eta_t^{-1}(S)} p(x) \, dx.$$

Problem Statement

A *moving-target search plan* is a Borel function $\psi: Y \times T \to [0, \infty)$, which may be thought of as specifying a density for search effort in both time and space; that is, $\int_0^t \psi(y, s) \, ds$ gives the effort density that accumulates at point y by time t and $\int_Y \int_0^t \psi(y, s) \, ds \, dy$ gives the total effort expended by time t.

Constraints are imposed on search effort by means of Borel functions $m_1: Y \times T \to [0, \infty]$, $m_2: T \to [0, \infty]$, and $m_3: T \to [0, \infty]$. We say that $\psi \in \Psi(m_1, m_2, m_3)$ if and only if ψ is a moving-target search plan and

(a) $\psi(y, t) \leq m_1(y, t)$ for $(y, t) \in Y \times T$,

(b) $\displaystyle\int_Y \psi(y, t) \, dy \leq m_2(t)$ for $t \in T$,   (8.1.2)

(c) $\displaystyle\int_0^t \int_Y \psi(y, s) \, dy \, ds \leq m_3(t)$ for $t \in T$.

8.1 Description of Search Problem

Condition (a) limits the rate at which effort density may be applied to a point. Intuitively, this bounds the rate at which search effort may be applied to small areas. Condition (b) constrains the rate at which effort may be applied to the whole search space, and condition (c) limits the total amount of search effort that may be applied by time t. If the constraint corresponding to m_j is not imposed, then one writes $m_j = \infty$. For convenience of notation we let

$$\Psi_1(m_1) = \Psi(m_1, \infty, \infty), \quad \Psi_2(m_2) = \Psi(\infty, m_2, \infty), \quad \Psi_3(m_3) = \Psi(\infty, \infty, m_3).$$

As in the case of a stationary target, there is a detection function $b: X \times [0, \infty) \to [0, 1]$ that is Borel measurable. It is assumed that

$$b\left(x, \int_0^t \psi(\eta_s(x), s)\, ds\right)$$

gives the probability of detecting the target by time t with plan ψ given that $\xi = x$. In effect, the target is accumulating search density $\int_0^t \psi(\eta_s(x), s)\, ds$ as it travels long its path during time $[0, t]$. Thus, $b(x, \cdot)$ gives probability of detection as a function of accumulated search density when $\xi = x$. For a search plan ψ, we define

$$P_t[\psi] = \int_X p(x) b\left(x, \int_0^t \psi(\eta_s(x), s)\, ds\right) dx.$$

Then $P_t[\psi]$ gives the probability of detecting the target by time t with plan ψ.

Suppose $\psi^* \in \Psi(m_1, m_2, m_3)$. If, for some $t \in T$,

$$P_t[\psi^*] = \max\{P_t[\psi] : \psi \in \Psi(m_1, m_2, m_3)\},$$

then we say that ψ^* is *t-optimal within* $\Psi(m_1, m_2, m_3)$. If ψ^* is t-optimal within $\Psi(m_1, m_2, m_3)$ for all $t \in T$, then ψ^* is called *uniformly optimal* within $\Psi(m_1, m_2, m_3)$.

When $T = [0, \infty)$, we define for search plans ψ

$$P_\infty[\psi] = \lim_{t \to \infty} P_t[\psi].$$

If $\psi^* \in \Psi(m_1, m_2, m_3)$ and

$$P_\infty[\psi^*] = \max\{P_\infty[\psi] : \psi \in \Psi(m_1, m_2, m_3)\},$$

then ψ^* is simply called *optimal* within $\Psi(m_1, m_2, m_3)$.

For future use it is convenient to note that the problem of finding a t-optimal $\psi^* \in \Psi(m_1, m_2, m_3)$ is equivalent to the following problem:

Find a Borel function $g: X \times T \to [0, \infty)$ to maximize

$$\int_X p(x) b\left(x, \int_0^t g(x, s)\, ds\right) dx \tag{8.1.3}$$

subject to

$$g(x, s) \leq m_1(\eta_s(x), s) \quad \text{for} \quad s \in T, \quad x \in X, \quad (8.1.4)$$

$$\int_X g(x, s) \mathbf{J}(x, s) \, dx \leq m_2(s) \quad \text{for} \quad s \in T, \quad (8.1.5)$$

$$\int_0^s \int_X g(x, u) \mathbf{J}(x, u) \, dx \, du \leq m_3(s) \quad \text{for} \quad s \in T. \quad (8.1.6)$$

To see the equivalence, make the following substitution:

$$\begin{aligned} \psi(\eta_s(x), s) &= g(x, s) \quad \text{for} \quad s \in T, \quad x \in X, \\ \psi(y, s) &= 0 \quad \text{for} \quad y \notin \text{image } \eta_s. \end{aligned} \quad (8.1.7)$$

Then (8.1.3) becomes $P_t[\psi]$. Using the above substitution and setting $y = \eta_s(x)$, (8.1.4) becomes (8.1.2a). Similarly, (8.1.5) becomes (8.1.2b) by using a standard change of variable theorem [e.g., Goffman (1965, p. 104)] as follows:

$$\int_X \psi(\eta_s(x), s) \mathbf{J}(x, s) \, dx = \int_Y \psi(y, s) \, dy \leq m_2(s) \quad \text{for} \quad s \in T,$$

and (8.1.6) is similarly transformed into (8.1.2c).

Thus, from (8.1.7) it is seen that a solution g to (8.1.3)–(8.1.6) yields a ψ that is t-optimal within $\Psi(m_1, m_2, m_3)$, and similarly a plan ψ that is t-optimal within $\Psi(m_1, m_2, m_3)$ yields a solution to the problem of maximizing (8.1.3) subject to (8.1.4)–(8.1.6).

The problem stated in (8.1.3)–(8.1.6) is difficult to solve in general because the objective functional (8.1.3) is not separable in the sense discussed in Section 1.3; that is, in general, one cannot express the objective functional (8.1.3) in the form

$$\int_T \int_X h(x, t, g(x, t)) \, dx \, dt,$$

where $h: X \times T \times [0, \infty] \to (-\infty, \infty)$. In contrast, the constraints (8.1.4)–(8.1.6) are separable. Two cases where this problem may be solved are presented in Sections 8.2 and 8.3. In both of these cases, the nature of the problem allows one to reduce it to a separable problem that may be solved by the Lagrange multiplier methods of Chapter II.

8.2. UNIFORMLY OPTIMAL PLANS WHEN THE TARGET MOTION IS FACTORABLE

The target motion is said to be *factorable* if in addition to satisfying the assumptions of Section 8.1, there exist Borel functions $\mathbf{j}: X \to (0, \infty)$ and

8.2 Optimal Plans for Factorable Motion

$\mathbf{n}: T \to (0, \infty)$ such that

$$\mathbf{J}(x, t) = \mathbf{j}(x)\mathbf{n}(t) \quad \text{for} \quad x \in X, \quad t \in T.$$

One then says that the target motion is factorable with $\mathbf{J} = \mathbf{jn}$.

This section considers search for targets whose motion is factorable and presents a method for finding uniformly optimal moving-target search plans within the class $\Psi_2(m_2)$. Recall that in the class $\Psi_2(m_2)$, the only constraint is on the rate at which search effort may be applied to the whole search space Y. In particular, if $\psi \in \Psi_2(m_2)$, then

$$\int_Y \psi(y, t) \, dy \leq m_2(t) \quad \text{for} \quad t \in T.$$

Theorem 8.2.1 of this section shows that when the target motion is factorable, the problem of finding an optimal plan in $\Psi_2(m_2)$ may be reduced to solving a constrained separable optimization problem. Theorem 8.2.2 gives a method for solving this separable optimization problem and as a result a method of finding uniformly optimal plans within $\Psi_2(m_2)$. Examples 8.2.4 and 8.2.5 illustrate the use of this method.

For this section, let

$$E[f] = \int_X p(x)b(x, f(x)) \, dx,$$

$$C[f] = \int_X \mathbf{j}(x)f(x) \, dx \quad \text{for} \quad f \in F(X).$$

Theorem 8.2.1. *Let the target motion be factorable with $\mathbf{J} = \mathbf{jn}$. Suppose there exists a Borel function $\varphi: X \times T \to [0, \infty)$ such that*

(i) *$\dot{\varphi}(x, t) \geq 0$ and $\varphi(x, t) = \int_0^t \dot{\varphi}(x, s) \, ds$ for $x \in X$, $t \in T$,*
(ii) *for all $t \in T$,*

$$\int_X \varphi(x, t)\mathbf{j}(x) \, dx = M_2(t) \equiv \int_0^t \frac{m_2(s)}{\mathbf{n}(s)} \, ds,$$

and

$$E[\varphi(\cdot, t)] = \max\{E[f] : f \in F(X) \text{ and } C[f] \leq M_2(t)\}. \quad (8.2.1)$$

Then ψ^* defined by

$$\psi^*(y, t) = \begin{cases} \dot{\varphi}(\eta_t^{-1}(y), t) & \text{for } y \in \text{image } \eta_t \\ 0 & \text{otherwise} \end{cases} \quad \text{for} \quad t \in T, \quad (8.2.2)$$

is uniformly optimal within $\Psi_2(m_2)$. Moreover,

$$P_t[\psi^*] = \int_X p(x)b(x, \varphi(x, t)) \, dx \quad \text{for} \quad t \in T. \quad (8.2.3)$$

Proof. First note that (8.2.3) follows from $\dot\varphi(x, t) = \psi^*(\eta_t(x), t)$ for $x \in X$, $t \in T$. To see that $\psi^* \in \Psi_2(m_2)$, define

$$m_2^*(t) = \int_Y \psi(y, t)\, dy \qquad \text{for} \quad t \in T.$$

Using (8.2.2) and the change of variable $x = \eta_t^{-1}(y)$, one obtains

$$m_2^*(t) = \int_Y \dot\varphi(\eta_t^{-1}(y), t)\, dy = \mathbf{n}(t) \int_X \dot\varphi(x, t) \mathbf{j}(x)\, dx \qquad \text{for} \quad t \in T.$$

Then

$$\int_0^t \frac{m_2^*(s)}{\mathbf{n}(s)}\, ds = \int_0^t \int_X \dot\varphi(x, s) \mathbf{j}(x)\, dx\, ds$$

$$= \int_X \varphi(x, t) \mathbf{j}(x)\, dx = \int_0^t \frac{m_2(s)}{\mathbf{n}(s)}\, ds \qquad \text{for} \quad t \in T.$$

Thus $m_2^*(t) = m_2(t)$ for a.e. $t \in T$. For times t when $m_2^*(t) \neq m_2(t)$, one may redefine $\dot\varphi(\cdot, t)$ so that equality holds for all t without altering φ or $P[\varphi(\cdot, t)]$ for any $t \in T$. Thus, one may as well assume that $m_2^*(t) = m_2(t)$ for all $t \in T$, and it follows that $\psi^* \in \Psi_2(m_2)$.

To see that ψ^* is uniformly optimal, we suppose that ψ^* fails to be t-optimal within $\Psi_2(m_2)$ for a fixed $t \in T$. Thus, there is a $\psi \in \Psi_2(m_2)$ such that

$$P_t[\psi] > P_t[\psi^*]. \tag{8.2.4}$$

For $s \in T$ and $x \in X$, let

$$h(x, s) = \int_0^s \psi(\eta_u(x), u)\, du.$$

Then $P_t[\psi] = E[h(\cdot, t)]$, and from (8.2.3) and (8.2.4) it follows that

$$E[h(\cdot, t)] > E[\varphi(\cdot, t)]. \tag{8.2.5}$$

However,

$$C[h(\cdot, t)] = \int_X \int_0^t \psi(\eta_s(x), s)\, ds\, \mathbf{j}(x)\, dx$$

$$= \int_X \int_0^t \frac{1}{\mathbf{n}(s)} \psi(\eta_s(x), s) \mathbf{J}(x, s)\, ds\, dx. \tag{8.2.6}$$

Making the substitution $y = \eta_s(x)$, one obtains

$$\int_X \psi(\eta_s(x), s) \mathbf{J}(x, s)\, dx = \int_Y \psi(y, s)\, dy \leq m_2(s) \qquad \text{for} \quad s \in T. \tag{8.2.7}$$

8.2 Optimal Plans for Factorable Motion

The last inequality follows from $\psi \in \Psi_2(m_2)$. Interchanging the order of integration in (8.2.6) and using (8.2.7), it follows that

$$C[h(\cdot, t)] \leq \int_0^t \frac{m_2(s)}{n(s)}\, ds = M_2(t). \tag{8.2.8}$$

By taking $f = h(\cdot, t)$, one sees that (8.2.5) and (8.2.8) together contradict (8.2.1), which is true by assumption. Thus, ψ^* must be uniformly optimal within $\Psi_2(m_2)$, and the proof is finished.

In effect, Theorem 8.2.1 says that one can find a uniformly optimal plan for a moving target by finding φ to satisfy (i) and (ii). Finding φ to satisfy (ii) is almost equivalent to finding a uniformly optimal search plan for a stationary target having a prior distribution given by the density function p when the detection function is b. The difference is that instead of using the cumulative effort function M given by $M(t) = \int_0^t m_2(s)\, ds$, one must scale the rate m_2 at which effort can be applied by the factor $\mathbf{n}$ of the Jacobian. As a result, the cumulative effort function becomes M_2 as given in the theorem. In addition, the other factor $\mathbf{j}$ of the Jacobian enters into the cost function C of the theorem. Thus, Theorem 8.2.1 says that in the case of a factorable target motion, one may find a uniformly optimal moving-target plan in $\Psi_2(m_2)$ by finding a uniformly optimal plan for a stationary target provided that one appropriately scales the cost functional and the cumulative effort function.

The following theorem gives a method for finding the function φ of Theorem 8.2.1 when the detection function is regular.

Theorem 8.2.2. *Suppose the target motion is factorable with $\mathbf{J} = \mathbf{jn}$ and b is a regular detection function on X. Let*

$$p_x(z) = p(x)b'(x, z)/\mathbf{j}(x) \qquad \text{for } x \in X,\ z \geq 0,$$

$$p_x^{-1}(\lambda) = \begin{cases} \text{inverse of } p_x \text{ evaluated at } \lambda & \text{for } 0 < \lambda \leq p_x(0), \\ 0 & \text{for } \lambda > p_x(0), \end{cases}$$

and let

$$\tilde{U}(\lambda) = \int_X \mathbf{j}(x) p_x^{-1}(\lambda)\, dx.$$

Set

$$M_2(t) = \int_0^t \frac{m_2(s)}{\mathbf{n}(s)}\, ds$$

and define

$$\varphi(x, t) = p_x^{-1}(\tilde{U}^{-1}(M_2(t))) \qquad \text{for } x \in X,\ t \in T. \tag{8.2.9}$$

If

$$\dot{\varphi}(x, t) \geq 0 \quad \text{and} \quad \varphi(x, t) = \int_0^t \dot{\varphi}(x, s)\, ds, \qquad (8.2.10)$$

then ψ^* defined by

$$\psi^*(y, t) = \begin{cases} \dot{\varphi}(\eta_t^{-1}(y), t) & \text{for } y \in \text{image } \eta_t \\ 0 & \text{otherwise} \end{cases} \quad \text{for } t \in T, \qquad (8.2.11)$$

is uniformly optimal within $\Psi_2(m_2)$.

Proof. By Theorem 2.4.3, φ satisfies condition (ii) of Theorem 8.2.1. Since the remaining conditions of Theorem 8.2.1 are assumed to be satisfied, Theorem 8.2.2 follows from Theorem 8.2.1.

Assumption (8.2.10) of Theorem 8.2.2 says that $\varphi(x, \cdot)$ defined by (8.2.9) has a nonnegative derivative and is absolutely continuous. Since $\rho^{-1}(x, \cdot)$ is continuous and decreasing and $\tilde{U}^{-1}$ is continuous and decreasing, $\varphi(x, \cdot)$ is continuous and increasing. Thus, in most cases covered by Theorem 8.2.2, assumption (8.2.10) will hold. However, the following theorem gives conditions on the detection function b that guarantee that (8.2.10) holds. Let $b''(x, \cdot)$ denote the derivative of $b'(x, \cdot)$. Recall that

$$\text{ess sup } p = \inf\{k : k \geq p(x) \text{ for a.e. } x \in X\}.$$

Theorem 8.2.3. *Suppose that the conditions of Theorem 8.2.2 hold and that in addition*

(a) $b'(x, z) < \infty$ *for* $x \in X$, $z \geq 0$,
(b) $b''(x, \cdot)$ *is continuous and* $b''(x, 0) < 0$ *for* $x \in X$,
(c) ess sup $p < \infty$.

Then ψ^* defined by (8.2.11) is uniformly optimal within $\Psi_2(m_2)$.

Proof. All that has to be shown here is that φ defined by (8.2.9) satisfies (8.2.10). Nevertheless, the argument is tedious and is not given here. Instead, the interested reader is referred to Richardson (1971) for the proof.

In the following examples, Theorem 8.2.2 is used to find plans that are uniformly optimal within $\Psi_2(m_2)$.

Example 8.2.4. In this example, an optimal search plan is found for a moving-target problem that may be considered as a generalization of the stationary problem solved in Example 2.2.1, where the uniformly optimal plan is found for the situation of a circular normal target distribution and exponential detection function.

Let $X = Y$ be Euclidean 2-space and the parameter ξ be the target's

8.2 Optimal Plans for Factorable Motion

position at time 0. It is assumed that ξ has a circular normal distribution with parameter $\sigma > 0$ so that

$$p(x) = \frac{1}{2\pi\sigma^2} \exp\left(-\frac{x_1^2 + x_2^2}{2\sigma^2}\right) \quad \text{for} \quad x = (x_1, x_2) \in X.$$

Let $T = [0, \infty)$. The target's velocity is assumed to be constant with its speed proportional to its distance from the origin at time 0 and its heading given by the vector from the origin to its position at time 0. Let u be the proportionality constant for target speed. Then

$$\eta(x, t) = (1 + tu)(x_1, x_2) \quad \text{for} \quad x \in X, \ t \in T,$$

where scalar multiplication of vectors is understood in the usual manner. It is easy to calculate that

$$\mathbf{J}(x, t) = (1 + tu)^2 \quad \text{for} \quad t \in T, \ x \in X,$$

so that the target motion is factorable with $\mathbf{J} = \mathbf{jn}$, and $\mathbf{j}(x) = 1$ for $x \in X$, $\mathbf{n}(t) = (1 + tu)^2$ for $t \in T$.

Recall that $m_2(t)$ specifies the rate at which effort may be exerted at time t and that

$$M_2(t) = \int_0^t \frac{m_2(s)}{\mathbf{n}(s)} ds = \int_0^t \frac{m_2(s)}{(1 + su)^2} ds.$$

The function m_2 is assumed to be continuous so that

$$\frac{dM_2(t)}{dt} = \frac{m_2(t)}{(1 + tu)^2} \quad \text{for} \quad t \in T.$$

The detection function b is assumed to be given by

$$b(x, z) = 1 - e^{-z} \quad \text{for} \quad x \in X, \ z \geq 0.$$

Uniformly Optimal Plan

Theorem 8.2.2 is used to find ψ^*, the uniformly optimal moving-target plan in $\Psi_2(m_2)$. Since $\mathbf{j}(x) = 1$ for $x \in X$, the steps necessary to find ρ^{-1}, $\tilde{U}$, and $\tilde{U}^{-1}$ are identical to those for finding ρ^{-1}, U, and U^{-1} in Example 2.2.1. Thus, one may obtain φ of Eq. (8.2.9) by substituting $M_2(t)$ for Wvt in (2.2.6) to obtain

$$\varphi(x, t) = \begin{cases} \left(\dfrac{M_2(t)}{\pi\sigma^2}\right)^{1/2} - \dfrac{x_1^2 + x_2^2}{2\sigma^2} & \text{for} \quad \dfrac{x_1^2 + x_2^2}{2\sigma^2} \leq \left(\dfrac{M_2(t)}{\pi\sigma^2}\right)^{1/2}, \\ 0 & \text{otherwise.} \end{cases}$$

Observe that

$$\dot{\varphi}(x, t) = \begin{cases} \dfrac{m_2(t)}{2\sigma[\pi M_2(t)]^{1/2}(1 + tu)^2} & \text{for } \dfrac{x_1^2 + x_1^2}{2\sigma^2} \leq \left(\dfrac{M_2(t)}{\pi\sigma^2}\right)^{1/2}, \\ 0 & \text{otherwise,} \end{cases}$$

and that

$$\eta_t^{-1}(x) = x/(1 + tu) \qquad \text{for } x \in X, \quad t \in T.$$

Thus, by Theorem 8.2.2,

$$\psi^*(x, t) = \dot{\varphi}(\eta_t^{-1}(x), t)$$

$$= \begin{cases} \dfrac{m_2(t)}{\pi R^2(t)} & \text{for } x_1^2 + x_2^2 \leq R_2(t) \quad \text{for } t \in T, \\ 0 & \text{otherwise} \end{cases} \qquad (8.2.12)$$

where

$$R^2(t) = 2\sigma^2(1 + tu)^2[M_2(t)/\pi\sigma^2]^{1/2} \qquad \text{for } t \in T. \qquad (8.2.13)$$

Thus, $R(t)$ gives the search radius by time t, and the search plan ψ^* calls for having $\psi^*(\cdot, t)$ constant over the disc of radius $R(t)$ centered at the origin. Because this disc expands with time, cumulative effort tends to be heavier near the origin.

Independent of the distribution of ξ and the dimension of X, one can make the following observation about uniformly optimal plans in $\Psi_2(m_2)$ when the detection function is homogeneous and exponential. Note that

$$\rho_x^{-1}(\lambda) = [\ln(p(x)/\mathbf{j}(x)) - \ln(\lambda)]^+,$$

where

$$[s]^+ = \begin{cases} s & \text{if } s \geq 0, \\ 0 & \text{if } s < 0. \end{cases}$$

Thus the function φ of Theorem 8.2.2 has the form

$$\varphi(x, t) = [\ln(p(x)/\mathbf{j}(x)) - \ln(k(t))]^+ \qquad \text{for } x \in X, \quad t \in T, \qquad (8.2.14)$$

where

$$k(t) = \tilde{U}^{-1}(M_2(t)) \qquad \text{for } t \in T.$$

From (8.2.14) it follows that $\dot{\varphi}(\cdot, t)$ is equal to a constant inside the region $\mathscr{X}_t = \{x : p(x) > \mathbf{j}(x)k(t)\}$ and 0 outside $\mathscr{X}_t$. Since

$$\psi^*(x, t) = \dot{\varphi}(\eta_t^{-1}(x), t),$$

8.2 Optimal Plans for Factorable Motion

one can see that $\psi^*(\cdot, t)$ is constant over the region $\mathcal{R}_t = \eta_t(\mathcal{X}_t)$ and 0 outside $\mathcal{R}_t$. Thus, if one can find $A(t)$, the area of $\mathcal{R}_t$, one can write

$$\psi^*(x, t) = \begin{cases} m_2(t)/A(t) & \text{for } x \in \mathcal{R}_t \\ 0 & \text{otherwise} \end{cases} \quad \text{for } t \in T.$$

Probability of Detection

By Theorem 8.2.1,

$$P_t[\psi^*] = \int_X p(x)b(x, \varphi(x, t))\, dx \quad \text{for } t \in T. \tag{8.2.15}$$

Since φ is equal to the φ^* of Example 2.2.7 with Wvt replaced by $M_2(t)$, it follows from (2.2.21) that

$$\begin{aligned} P_t[\psi^*] &= \int_X p(x)b(x, \varphi(x, t))\, dx \\ &= 1 - \{1 + [M_2(t)/\pi\sigma^2]^{1/2}\} \exp\{-[M_2(t)/\pi\sigma^2]^{1/2}\} \quad \text{for } t \in T. \end{aligned} \tag{8.2.16}$$

It is clear from (8.2.16) that the probability of detecting the target will go to unity if and only if $M_2(t) \to \infty$ as $t \to \infty$. If, for example, $m_2(t) = v$ for $t \geq 0$, then

$$M_2(t) = \int_0^t \frac{v}{(1 + su)^2}\, ds = vt(1 + tu)^{-1}.$$

In this case,

$$\lim_{t \to \infty} P_t[\psi^*] = 1 - [1 + (u\pi\sigma^2/v)^{-1/2}] \exp[-(u\pi\sigma^2/v)^{-1/2}] < 1.$$

The analysis leading to (8.2.12) and (8.2.16) does not depend on the fact that $\mathbf{n}(t) = (1 + tu)^2$ for $t \in T$. The equations remain true for an arbitrary continuous $\mathbf{n} > 0$ provided one takes

$$R^2(t) = 2\sigma^2 \mathbf{n}(t)[M_2(t)/\pi\sigma^2]^{1/2},$$

and

$$M_2(t) = \int_0^t \frac{m_2(s)}{\mathbf{n}(s)}\, ds \quad \text{for } t \in T.$$

In the special case where $u = 0$, the target is stationary and

$$\varphi(x, t) = \int_0^t \psi^*(x, s)\, ds$$

is the effort density accumulated at x by time t. Letting $m_2(t) = Wv$, then $M_2(t) = Wvt$ for $t \in T$ and $\varphi(\cdot, t)$ becomes the optimal allocation of Wvt effort. One can check that $P_t[\psi^*]$ given by (8.2.16) becomes the probability of detecting the target by time t using the uniformly optimal plan for a stationary target that is given in Example 2.2.7.

Example 8.2.5. In this example a search problem is considered that might arise when trying to regain contact with a moving submarine using a field of sensors that are to be distributed in an optimal fashion.

Consider a situation in which the target has previously been detected and then contact is lost. A search for the target commences after a time delay t_l from the loss of contact. The target is assumed to move radially from the last point of contact at a constant speed, which is obtained from a uniform distribution on $[v_1, v_2]$. Similarly, the target's course is assumed to be constant and to be obtained from a uniform distribution on $[\theta_1, \theta_2]$.

Let us take $t = 0$ to be the time at which search commences and choose the origin of the coordinate system to be the target's position at the point of last contact. At $t = 0$, the target is located in the region X_0 shown in Fig. 8.2.1.

Take $r_1 = v_1 t_l$ and $r_2 = v_2 t_l$. The density p for the target location distribution at $t = 0$ is

$$p(x) = \begin{cases} [(r_2 - r_1)(\theta_2 - \theta_1)(x_1^2 + x_2^2)^{1/2}]^{-1} & \text{for } x \in X_0, \\ 0 & \text{otherwise.} \end{cases}$$

The target motion is given by

$$\eta_t(x) = \left(\frac{t + t_l}{t_l}\right)(x_1, x_2) \quad \text{for } x \in X, \quad t \geq 0.$$

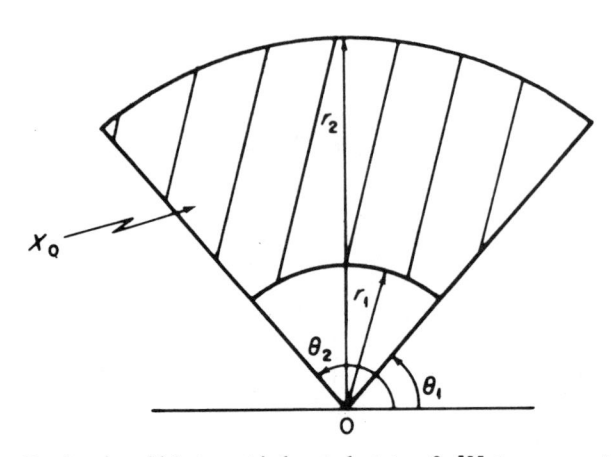

Fig. 8.2.1. Region in which target is located at $t = 0$. [*Note:* $r_1 = v_1 t_l$; $r_2 = v_2 t_l$.]

8.2 Optimal Plans for Factorable Motion

Clearly,

$$J(x, t) = \left(\frac{t + t_l}{t_l}\right)^2 \quad \text{for} \quad x \in X, \ t \geq 0,$$

and the target motion is factorable with $\mathbf{J} = \mathbf{jn}$, and $\mathbf{j}(x) = 1$ for $x \in X$, $\mathbf{n}(t) = [(t + t_l)/t_l]^2$ for $t \in T$.

Let $b(z) = 1 - e^{-z}$ for $z > 0$. This form for b arises from assuming that if one has n sensors uniformly distributed in a region of area A and operating for a length of time S, then the probability of detecting the target given that it is in the region during S is $1 - \exp(-knS/A)$ for some constant $k > 0$. Thus, the effort density that accumulates at points in this region during S is given by $z = knS/A$. If N is the total number of sensors that can be usefully employed at one time, then search effort is applied at the constant rate $m_2(t) = kN$. Thus,

$$M_2(t) = \int_0^t \frac{m_2(s)}{\mathbf{n}(s)} ds = kNt_l\left(\frac{t}{t + t_l}\right) \quad \text{for} \quad t \geq 0.$$

For this example, the detection function is independent of the speed of the target. However, speed could be accounted for by allowing b to depend on x, e.g., $b(x, z) = 1 - \exp(-h(x)z)$.

In order to use Theorem 8.2.2, one calculates

$$p_x(z) = p(x)b'(z) \quad \text{for} \quad x \in X, \ z \geq 0,$$
$$p_x^{-1}(\lambda) = [\ln(p(x)/\lambda)]^+$$
$$= \{-\ln[(r_2 - r_1)(\theta_2 - \theta_1)(x_1^2 + x_2^2)^{1/2}\lambda]\}^+ \quad \text{for} \quad x \in X_0, \ \lambda > 0.$$

Let $H(\lambda) = [r_2(r_2 - r_1)(\theta_2 - \theta_1)\lambda]^{-1}$ for $\lambda > 0$. Then

$$\tilde{U}(\lambda) = \int_{X_0} p_x^{-1}(\lambda) \, dx = \int_{\theta_1}^{\theta_2} \int_{r_1}^{r_2} [\ln(r_2 H(\lambda)/r)]^+ r \, dr \, d\theta$$
$$= \frac{r_1^2(\theta_2 - \theta_1)}{2} \{\tfrac{1}{2}[(r_2 H(\lambda)/r_1)^2 - 1] - \ln(r_2 H(\lambda)/r_1)\}$$

$$\text{for} \quad \frac{r_1}{r_2} \leq H(\lambda) \leq 1,$$

and

$$\tilde{U}(\lambda) = \frac{(\theta_2 - \theta_1)}{2} \{\tfrac{1}{2}(r_2^2 - r_1^2) + r_2^2 \ln(H(\lambda)) - r_1^2 \ln(r_2 H(\lambda)/r_1)\}$$

$$\text{for} \quad H(\lambda) \geq 1.$$

In order to find ψ^* it remains to calculate $\tilde{U}^{-1}$ and to use Theorem 8.2.2. Recalling the discussion that follows (8.2.13), in order to find ψ^*, the uniformly optimal plan within $\Psi_2(m_2)$, we need only determine the region $\mathscr{R}(t)$ and apply effort density at a uniform rate over $\mathscr{R}(t)$; that is, if $A(t)$ is the area of $\mathscr{R}(t)$, then

$$\psi^*(x, t) = \begin{cases} m_2(t)/A(t) & \text{for } x \in \mathscr{R}(t), \\ 0 & \text{otherwise.} \end{cases}$$

In view of this, ψ^* is calculated as follows. Let H be the solution of

$$2M_2(t)/r_1^2(\theta_2 - \theta_1) = \tfrac{1}{2}[(r_2 H/r_1)^2 - 1] - \ln(r_2 H/r_1).$$

If $H \leq 1$, then

$$\psi^*((r, \theta), t) = \begin{cases} \dfrac{kN}{A(t)} & \text{for } \theta_1 \leq \theta \leq \theta_2,\ (t + t_l)v_1 \leq r \leq (t + t_l)Hv_2, \\ 0 & \text{otherwise,} \end{cases}$$

where $A(t) = \tfrac{1}{2}(\theta_2 - \theta_1)(t + t_l)^2[(Hv_2)^2 - v_1^2]$.

If $H \geq 1$, then

$$\psi^*((r, \theta), t) = \begin{cases} \dfrac{kN}{A(t)} & \text{for } \theta_1 \leq \theta \leq \theta_2,\ (t + t_l)v_1 \leq r \leq (t + t_l)v_2, \\ 0 & \text{otherwise,} \end{cases}$$

where $A(t) = \tfrac{1}{2}(\theta_2 - \theta_1)(t + t_l)^2(v_2^2 - v_1^2)$.

The region in which the target may be located expands as time increases, and from the above it is seen that the initial phase of the search covers only part of this region. In fact, if

$$2kN/t_l \leq (\theta_2 - \theta_1)[\tfrac{1}{2}(v_2^2 - v_1^2) - v_1^2 \ln(v_2/v_1)],$$

then the search never expands to cover the entire region in which the target may be located. One may also show that if search effort is applied at a constant rate, then $\lim_{t \to \infty} P_t[\psi^*] < 1$.

The optimal search plan calls for spreading effort or sensors uniformly over a region that is continuously increasing. In practice, one would have to approximate this by choosing a time increment, possibly the lifetime of a sensor, and spreading sensors uniformly over a given region for that increment of time. The given region would, of course, be chosen to approximate the region searched by the optimal plan in that increment.

8.3. OPTIMAL PLANS WITHIN $\psi(m_1, \infty, K)$

Let $T = [0, \infty)$. In this section, a method is presented for finding optimal plans in the class $\Psi(m_1, \infty, K)$. In this class, only constraints (a) and (c) of (8.1.2) are imposed. There is a constraint K on the total amount of search effort available and a constraint $m_1(x, t)$ on the rate at which effort density may be applied to the point x at time t. Theorem 8.3.1 gives sufficient conditions for finding an optimal plan in $\Psi(m_1, \infty, K)$ and Example 8.3.2 shows how Theorem 8.3.1 may be used to find an optimal plan.

By the discussion in Section 8.1, finding ψ^* that is optimal within $\Psi(m_1, \infty, K)$ is equivalent to finding a Borel function $g: X \times T \to [0, \infty)$ that maximizes

$$\int_X p(x) b\left(x, \int_0^\infty g(x, t) \, dt\right) dx \tag{8.3.1}$$

subject to

$$\text{(a)} \quad g(x, t) \leq m_1(\eta_t(x), t) \quad \text{for} \quad t \in T, \quad x \in X,$$

$$\text{(b)} \quad \int_X \int_0^\infty g(x, t) \mathbf{J}(x, t) \, dt \, dx \leq K. \tag{8.3.2}$$

Let $G(m_1)$ be the class of Borel functions $g: X \times T \to [0, \infty)$ such that $g(x, t) \leq m_1(\eta_t(x), t)$ for $t \in T$. Define for $g \in G(m_1)$

$$\tilde{E}[g] = \int_X p(x) b\left(x, \int_0^\infty g(x, t) \, dt\right) dx,$$

$$\tilde{C}[g] = \int_X \int_0^\infty g(x, t) \mathbf{J}(x, t) \, dt \, dx.$$

Then the problem of finding an optimal $\psi^* \in \Psi(m_1, \infty, K)$ is equivalent to finding $g^* \in G(m_1)$ such that $\tilde{C}[g^*] \leq K$ and

$$\tilde{E}[g^*] = \sup\{\tilde{E}[g] : g \in G(m_1) \text{ and } \tilde{C}[g] \leq K\}. \tag{8.3.3}$$

The solution to (8.3.3) may be approached by defining an auxiliary function $\hat{e}$ as follows: For $x \in X$, and $z \geq 0$, let

$$\hat{e}(x, z) = \sup\left\{p(x) b\left(x, \int_0^\infty g(x, s) \, ds\right) : \int_0^\infty g(x, s) \mathbf{J}(x, s) \, ds = z, g \in G(m_1)\right\}.$$

One may think of $\hat{e}(x, z)$ as giving the maximum payoff that can be obtained at x for point cost z. Suppose that for each $x \in X$ and point cost z, one can find a function $h_z(x, \cdot): [0, \infty) \to [0, \infty)$ such that $h_z(x, t) \le m_1(\eta_t(x), t)$ for $t \in T$ and

$$\hat{e}(x, z) = p(x)b\left(x, \int_0^\infty h_z(x, t)\, dt\right),$$

$$z = \int_0^\infty h_z(x, t) J(x, t)\, dt.$$

Then the problem of finding g^* to satisfy (8.3.3) is reduced to the following separable problem: Find $f^* \in F(X)$ such that $\int_X f^*(x)\, dx \le K$ and

$$\int_X \hat{e}(x, f^*(x))\, dx = \max\left\{\int_X \hat{e}(x, f(x))\, dx : f \in F(X) \text{ and } \int_X f(x)\, dx \le K\right\}.$$

For then, $g^*(x, t) = h_{f^*(x)}(x, t)$ for $x \in X$, $t \in T$, satisfies (8.3.3). This is the idea behind the following theorem.

Theorem 8.3.1. Let $g^* \in G(m_1)$ be such that $\tilde{C}[g^*] = K$ and set

$$f^*(x) = \int_0^\infty g^*(x, t) J(x, t)\, dt \quad \text{for} \quad x \in X. \tag{8.3.4}$$

If

$$\hat{e}(x, f^*(x)) = p(x)b\left(x, \int_0^\infty g^*(x, t)\, dt\right) \quad \text{for} \quad x \in X, \tag{8.3.5}$$

and there exists a finite $\lambda \ge 0$ such that

$$\hat{e}(x, f^*(x)) - \lambda f^*(x) = \max\{\hat{e}(x, z) - \lambda z : z \ge 0\} \quad \text{for} \quad x \in X, \tag{8.3.6}$$

then ψ^* defined by

$$\psi^*(y, t) = \begin{cases} g^*(\eta_t^{-1}(y), t) & \text{for } t \in T, \ y \in \text{image } \eta_t, \\ 0 & \text{otherwise,} \end{cases} \tag{8.3.7}$$

is optimal within $\Psi(m_1, \infty, K)$.

Proof. Let $\psi \in \Psi(m_1, \infty, K)$ and define

$$g(x, t) = \psi(\eta_t(x), t) \quad \text{for} \quad x \in X, \ t \in T,$$

$$f(x) = \int_0^\infty g(x, t) J(x, t)\, dt \quad \text{for} \quad x \in X. \tag{8.3.8}$$

Then by (8.3.6) and the definition of $\hat{e}$,

$$\hat{e}(x, f^*(x)) - p(x)b\left(x, \int_0^\infty \psi(\eta_t(x), t)\, dt\right) \ge \hat{e}(x, f^*(x)) - \hat{e}(x, f(x))$$

$$\ge \lambda(f^*(x) - f(x)) \quad \text{for} \quad x \in X.$$

8.3 Optimal Plans within $\psi(m_1, \infty, K)$

Thus

$$\int_X \hat{e}(x, f^*(x))\, dx - \tilde{E}[g] \geq \lambda\left(\int_X f^*(x)\, dx - \int_X f(x)\, dx\right). \qquad (8.3.9)$$

By (8.3.4) and (8.3.5), $\int_X f^*(x)\, dx = \tilde{C}[g^*] = K$ and $\int_X \hat{e}(x, f^*(x))\, dx = \tilde{E}[g^*]$. By (8.3.8) and the fact that $\psi \in \Psi(m_1, \infty, K)$, it follows that

$$\int_X f(x)\, dx = \int_X \int_0^\infty g(x, t)\mathbf{J}(x, t)\, dt\, dx$$

$$= \int_X \int_0^\infty \psi(\eta_t(x), t)\mathbf{J}(x, t)\, dt\, dx$$

$$= \int_Y \int_0^\infty \psi(y, t)\, dt\, dy \leq K.$$

Thus, the inequality in (8.3.9) becomes $\tilde{E}[g^*] \geq \tilde{E}[g]$, i.e., $P_\infty[\psi^*] \geq P_\infty[\psi]$, and the theorem is proved.

Example 8.3.2. In this example an optimal search plan is found by use of Theorem 8.3.1. Let X and Y be copies of Euclidean 2-space. In this case, the parameter space X gives the target's velocity. The target starts at the origin at time 0 and its velocity vector is chosen from a circular normal distribution with density

$$p(x) = \frac{1}{2\pi\sigma^2} \exp\left(-\frac{x_1^2 + x_2^2}{2\sigma^2}\right) \quad \text{for} \quad x = (x_1, x_2) \in X.$$

The target motion function is given by

$$\eta_t(x) = t(x_1, x_2),$$

and the Jacobian by

$$\mathbf{J}(t, x) = t^2 \quad \text{for} \quad x \in X, \quad t \geq 0.$$

Let $m_1 = k$, where k is a positive number, and let

$$b(x, z) = 1 - e^{-z^3} \quad \text{for} \quad x \in X, \quad z \geq 0. \qquad (8.3.10)$$

If for $x \in X$ we let

$$w_t(x, s) = \begin{cases} k & \text{for } 0 \leq s \leq t, \\ 0 & \text{otherwise}, \end{cases}$$

then the resulting cost at $x \in X$ is $\int_0^\infty w_t(x, s)\mathbf{J}(x, s)\, ds = kt^3/3$. Because of the strictly increasing nature of $\mathbf{J}(x, \cdot)$, it is clear that $w_t(x, \cdot)$ gives the least-cost method of applying kt "effort" at $x \in X$. Thus,

$$\hat{e}(x, z) = p(x)b\left(x, \int_0^{(3z/k)^{1/3}} k\, ds\right) = p(x)[1 - \exp(-3k^2 z)].$$

Let

$$h_\lambda(x) = \frac{1}{3k^2}\left[-\ln\left(\frac{2\pi\sigma^2\lambda}{3k^2}\right) - \frac{(x_1^2 + x_2^2)}{2\sigma^2}\right]^+ \quad \text{for } x \in X, \ \lambda > 0.$$

One may check that by taking $f^* = h_\lambda$, (8.3.6) is satisfied for any $\lambda > 0$. By a straightforward calculation, one finds that

$$\int_X h_\lambda(x)\, dx = \frac{\pi\sigma^2}{3k^2} \ln^2\left(\frac{2\pi\sigma^2\lambda}{3k^2}\right).$$

Let $H = \sigma^{-1}k(3K/\pi)^{1/2}$. Then taking λ so that

$$2\pi\sigma^2\lambda = 3k^2 \exp[-H],$$

it follows that $\int_X h_\lambda(x)\, dx = K$, and

$$h_\lambda(x) = \frac{1}{3k^2}\left[H - \frac{(x_1^2 + x_2^2)}{2\sigma^2}\right]^+ \quad \text{for } x \in X.$$

For $x \in X$, let

$$g^*(x, t) = \begin{cases} k & \text{for } 0 \leq t \leq k^{-1}\left\{\left[H - \frac{(x_1^2 + x_2^2)}{2\sigma^2}\right]^+\right\}^{1/3}, \\ 0 & \text{otherwise}. \end{cases}$$

Then $f^*(x) = \int_0^\infty g^*(x, t)\mathbf{J}(x, t)\, dt = h_\lambda(x)$ for $x \in X$, and g^* satisfies the conditions of Theorem 8.3.1. Thus, ψ^* defined by

$$\psi^*(y, t) = \begin{cases} k & \text{for } y_1^2 + y_2^2 \leq 2\sigma^2 t^2(H - k^3 t^3) \\ 0 & \text{otherwise} \end{cases} \quad \text{for } y \in Y, \ t \geq 0,$$

(8.3.11)

is optimal within $\Psi(k, \infty, K)$.

It is interesting to note that ψ^* has the property that search begins at the origin expanding in a circle until a time at which it reaches a maximum distance from the origin. At this time the search circle begins to shrink back to the origin! An examination of (8.3.11) reveals the reason for this behavior. At time t, the optimal plan calls for searching the region of possible target positions that would result from restricting the target's speed to lie between 0 and $v(t)$, where $v(t) = [\sigma(2H - 2k^3 t^3)^{1/2}]^+$ for $t \geq 0$. Note that v is a monotonic decreasing function.

8.4 Necessary and Sufficient Conditions

The choice of b in (8.3.10) is rather unnatural, but it allows one to calculate ψ^* analytically. If the usual negative exponential local detection function is used, one can still find ψ^*, but numerical methods are required. However, without actually calculating the optimal allocation, one can show that it calls for placing some search effort at each point in the plane, the points farthest from the origin receiving search effort for the shortest time. In this case, $\psi^*(0, y) = k$ for $y \in Y$, while for $t > 0$, $\psi^*(t, y)$ equals k inside a circle and zero outside. This circle shrinks down to the origin by a finite time.

8.4. NECESSARY AND SUFFICIENT CONDITIONS FOR t-OPTIMAL SEARCH PLANS

In this section, a necessary condition is stated for a plan to be t-optimal within $\Psi_2(m_2)$. Under an additional concavity assumption, this condition is shown to be sufficient for t-optimality. In contrast to Section 8.2, no assumption of factorability is made here.

Theorem 8.4.1. *Assume that T is a finite interval and that η is a bounded function. Suppose $b'(x, \cdot)$ exists for $x \in X$ and there is a constant $k < \infty$ such that $|b'(x, z)| \le k$ for $x \in X$ and $z > 0$. Then (a) and (b) hold:*

(a) *If ψ^* is bounded and t-optimal in $\Psi_2(m_2)$, there exists $\lambda: T \to [0, \infty)$ such that for $0 \le s \le t$ and $x \in X$*

$$p(x)b'\left(x, \int_0^t \psi^*(\eta_u(x), u)\, du\right) = \lambda(s)J(x, s) \quad \text{for} \quad \psi^*(\eta_s(x), s) > 0,$$
$$\le \lambda(s)J(x, s) \quad \text{otherwise.} \quad (8.4.1)$$

(b) *If $b(x, \cdot)$ is concave for $x \in X$, ψ^* is bounded and a member of $\Psi_2(m_2)$, and there exists a Borel-measurable function $\lambda: T \to [0, \infty)$ satisfying (8.4.1), then ψ^* is t-optimal among all bounded functions in $\Psi_2(m_2)$.*

Proof. The proof of part (a) is not given here, but may be found in Stone (1973b). The proof uses the methods of Dubovitskii and Milyutin, which are presented in Girsanov (1972).

In order to prove part (b), the domain of the function b is extended in the following manner. Let

$$b(x, z) = b(x, 0) + zb'(x, 0) \quad \text{for} \quad x \in X, \quad z < 0.$$

Then $b(x, \cdot)$ is a concave function on $(-\infty, \infty)$ and $|b'(x, z)| \le k$ for $-\infty < z < \infty$ and $x \in X$. The result of extending the definition of b is that $P_t[g]$ is now defined for all $g \in \mathscr{L}_\infty(Y \times T)$, the set of all Borel-measurable, real-valued, bounded functions defined on $Y \times T$.

Let
$$\mathscr{F}[g] = -P_t[g] \quad \text{for} \quad g \in \mathscr{L}_\infty(Y \times T).$$

For $g, h \in \mathscr{L}_\infty(Y \times T)$, define the directional derivative $\mathscr{F}'(g, h)$ of $\mathscr{F}$ at g in the direction of h as

$$\mathscr{F}'(g, h) = \lim_{\varepsilon \to 0^+} \frac{\mathscr{F}(g + \varepsilon h) - \mathscr{F}(g)}{\varepsilon},$$

whenever the limit exists. Using this definition, one has

$$\mathscr{F}'(g, h) = \lim_{\varepsilon \to 0^+} \int_X p(x) \frac{1}{\varepsilon} \bigg\{ b\bigg(x, \int_0^t g(\eta_u(x), u)\, du\bigg) \\ - b\bigg(x, \int_0^t [g(\eta_u(x), u) + \varepsilon h(\eta_u(x), u)]\, du\bigg) \bigg\} dx.$$

Since $|b'(x, z)| \le k$, the integrand is bounded by $p(x)kt\|h\|$, where

$$\|h\| = \sup\{|h(y, s)| : y \in Y, s \in T\}.$$

Thus, one can apply the dominated convergence theorem to obtain

$$\mathscr{F}'(g, h) = -\int_X p(x) b'\bigg(x, \int_0^t g(\eta_u(x), u)\, du\bigg) \int_0^t h(\eta_s(x), s)\, ds\, dx \quad (8.4.2)$$

For $g \in \mathscr{L}_\infty(Y \times T)$, let

$$D_t(g, y, s) = p(\eta_s^{-1}(y)) b'\bigg(\eta_s^{-1}(y), \int_0^t g(\eta_u(\eta_s^{-1}(y)), u)\, du\bigg) \bigg/ \mathbf{J}(\eta_s^{-1}(y), s)$$
$$\text{for } s \in T, \ y \in Y.$$

Applying Fubini's theorem and the change of variable $y = \eta_s(x)$ to (8.4.2), one obtains

$$\mathscr{F}'(g, h) = -\int_0^t \int_Y D_t(g, y, s) h(y, s)\, dy\, ds. \quad (8.4.3)$$

The proof of part (b) now proceeds using an argument by contradiction. Suppose $\psi \in \Psi_2(m_2)$ is bounded and $P_t[\psi] > P_t[\psi^*]$, i.e., $\mathscr{F}(\psi) < \mathscr{F}(\psi^*)$. Let $h = \psi - \psi^*$. Since $b(x, \cdot)$ is concave for $x \in X$, $\mathscr{F}$ is convex, i.e., for $g_1, g_2 \in \mathscr{L}_\infty(Y \times T)$,

$$\mathscr{F}(\theta g_1 + (1 - \theta) g_2) \le \theta \mathscr{F}(g_1) + (1 - \theta) \mathscr{F}(g_2) \quad \text{for} \quad 0 \le \theta \le 1.$$

Thus, for $0 \le \theta \le 1$,

$$\mathscr{F}(\psi^* + \theta h) - \mathscr{F}(\psi^*) = \mathscr{F}((1 - \theta)\psi^* + \theta \psi) - \mathscr{F}(\psi^*) \\ \le (1 - \theta)\mathscr{F}(\psi^*) + \theta \mathscr{F}(\psi) - \mathscr{F}(\psi^*) \\ = \theta[\mathscr{F}(\psi) - \mathscr{F}(\psi^*)].$$

Therefore,
$$\mathcal{F}'(\psi^*, h) \leq \mathcal{F}(\psi) - \mathcal{F}(\psi^*) < 0. \tag{8.4.4}$$

By letting $y = \eta_s(x)$ in (8.4.1), it follows that for $y \in Y$, $0 \leq s \leq t$,
$$\begin{aligned} D_t(\psi^*, y, s) &= \lambda(s) \quad \text{for} \quad \psi^*(y, s) > 0, \\ &\leq \lambda(s) \quad \text{for} \quad \psi^*(y, s) = 0. \end{aligned} \tag{8.4.5}$$

Equations (8.4.3)–(8.4.5) yield

$$0 > \mathcal{F}'(\psi^*, h) = \int_0^t \int_Y - D_t(\psi^*, y, s)[\psi(y, s) - \psi^*(y, s)] \, dy \, ds$$

$$= \int_0^t \int_{\{y : \psi^*(y,s) > 0\}} - D_t(\psi^*, y, s)[\psi(y, s) - \psi^*(y, s)] \, dy \, ds$$

$$+ \int_0^t \int_{\{y : \psi^*(y,s) = 0\}} - D_t(\psi^*, y, s)[\psi(y, s) - \psi^*(y, s)] \, dy \, ds$$

$$\geq \int_0^t - \lambda(s) \int_Y [\psi(y, s) - \psi^*(y, s)] \, dy \, ds = 0.$$

The last equality follows from the fact that

$$\int_Y \psi(y, s) \, dy = m_2(s) = \int_Y \psi^*(y, s) \, dy \quad \text{for} \quad 0 \leq s \leq t.$$

The above contradiction completes the proof.

NOTES

Sections 8.1–8.3 are based on Stone and Richardson (1974). Theorem 8.4.1 is taken from Stone (1973b), and the proof of part (b) of Theorem 8.4.1 subsequent to (8.4.3) is due to D. H. Wagner.

In Stone (1973b), a generalization of conditionally deterministic motion essentially due to Pursiheimo (1973) is considered in which the parameter space is $X \times \Omega$; that is, each choice of $\xi = (x, \omega)$, where $x \in X$, $\omega \in \Omega$, determines a possible target path. Stone (1973b, Theorem 2) expands the result of Theorem 8.4.1 to include this generalization of conditionally deterministic motion. The proof in Stone (1973b) is a completion of the proof in Pursiheimo (1973).

This above-mentioned generalization is needed to consider problems where both the target's initial position and velocity are random. An example of such a problem arises when a moving submarine has been detected but the actual location of the submarine has a circular normal probability distribution centered at the point of detection. Suppose the submarine's speed is assumed known, but its heading is taken to be uniformly distributed on $[0, 2\pi]$. The problem of maximizing the probability of redetecting the submarine by a given time t is then a problem involving a target with this generalized conditionally deterministic motion. In this case, X is the plane and $\Omega = [0, 2\pi]$. This problem is, in fact, considered in Koopman (1946) and some reasonable search plans are developed there. However, a ψ^* that is t-optimal within $\Psi_2(m_2)$ has not been found.

Chapter IX

Markovian Target Motion

This chapter deals with search for targets whose motion is Markovian. With the exception of Section 9.1, results are stated without proof.

In Section 9.1, the simplest type of Markovian motion is considered, namely, a two-cell discrete-time Markov process. The problem of maximizing the probability of detecting the target in n looks is shown to be solvable by the use of dynamic programming. Minimizing the mean number of looks to find the target is formulated as a dynamic programming problem. However, because of the unbounded nature of the problem (i.e., no bound on the number of looks to be taken), its solution is not straightforward. Plans to minimize the mean number of looks are found for two special cases. In one of these cases, detection occurs with probability one if the searcher looks in the cell containing the target.

Section 9.2 considers a continuous-time version of the two-cell problem investigated in Section 9.1. Here the nature of the plan that maximizes the probability of detection by time t is discussed and a method for finding this plan is indicated. Section 9.3 gives necessary conditions for a search plan to maximize the probability of detecting the target by time t when the target's motion is specified by a continuous-time Markov process in Euclidean n-space.

9.1. TWO-CELL DISCRETE-TIME MARKOVIAN MOTION

This section presents a discussion of the problem of minimizing the mean number of looks to detect a target whose motion is Markovian and that can move in discrete time between two cells. In addition, the problem of maximizing the probability of detecting the target within n looks is shown to be a standard dynamic programming problem that may in principle be solved for all cases.

Model and Problem Statement

The target is in one of two cells at any time during the search. At time 0, the start of the search, the target is located in cell 1 with probability $p(1)$ and cell 2 with probability $p(2) = 1 - p(1)$. The target motion is specified by a discrete-time Markov chain with transition matrix

$$A = (a_{ij}),$$

where

$$a_{ij} = \Pr\left\{\begin{matrix}\text{target is in cell} \\ j \text{ at time } n\end{matrix}\middle|\begin{matrix}\text{target in cell} \\ i \text{ at time } n-1\end{matrix}\right\} \quad \text{for} \quad n = 1, 2, \ldots.$$

The transition matrix and prior target location probabilities, $p(j)$, $j = 1, 2$, are assumed to be known to the searcher. The target motion is illustrated schematically by the flow diagram in Fig. 9.1.1.

At each time unit the searcher looks in one of the two boxes. The conditional probability of detecting the target on a look in cell j given that the target is in cell j is assumed to be independent of previous looks. Let

$$\alpha_j = \Pr\left\{\text{target detected}\middle|\begin{matrix}\text{target in cell } j \text{ and a} \\ \text{look is made in cell}\end{matrix}\right\} \quad \text{for} \quad j = 1, 2,$$

Just as in search with discrete effort, a search plan is a sequence

$$\xi = (\xi_1, \xi_2, \ldots).$$

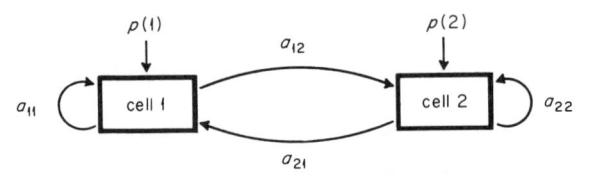

Fig. 9.1.1. Flow diagram for two-cell Markovian motion. [*Note:* $p(j)$ = probability of target being in cell j at time 0; $a_{ij} = \Pr\{\text{target is in cell } j \text{ at time } n \mid \text{target in cell } i \text{ at time } n - 1\}$ for $n = 1, 2, \ldots$.]

9.1 Two-Cell Discrete-Time Markovian Motion

The plan ξ specifies that the first look is to be made in cell ξ_1; if this look fails, then the next look is made in cell ξ_2, and so on. As in Chapter IV, let Ξ be the class of search plans ξ.

Consider the transition matrix A and the conditional detection probabilities α_1 and α_2 to be fixed. Define

$$\hat{P}_p[n, \xi] = \begin{bmatrix} \text{probability of detecting the target with } n \\ \text{looks when using plan } \xi \text{ and the prior target} \\ \text{location probabilities are given by } p \end{bmatrix},$$

$$\mu_p(\xi) = \begin{bmatrix} \text{mean number of looks to detect the target} \\ \text{when using plan } \xi \text{ and the prior target} \\ \text{probabilities are given by } p \end{bmatrix}.$$

Two problems are considered in this section. The first is to find a plan ξ^* to maximize the probability of detecting the target by n looks, i.e.,

$$\hat{P}_p[n, \xi^*] = \max_{\xi \in \Xi} \hat{P}_p[n, \xi]. \tag{9.1.1}$$

The second is to find a plan ξ^* that minimizes the mean number of looks to find the target, i.e.,

$$\mu_p(\xi^*) = \min_{\xi \in \Xi} \mu_p(\xi). \tag{9.1.2}$$

It will be convenient to have a notation for the posterior target distribution after a look has failed to detect the target. For $i = 1, 2$, let

$$\tilde{p}^i(j) = \Pr\begin{Bmatrix} \text{target in cell } j \\ \text{at time 1} \end{Bmatrix} \begin{vmatrix} \text{look in cell } i \text{ at time 0} \\ \text{failed to detect target} \end{vmatrix} \quad \text{for} \quad j = 1, 2, \ldots.$$

Then

$$\tilde{p}^1(1) = \frac{p(1)(1 - \alpha_1)a_{11} + p(2)a_{21}}{1 - \alpha_1 p(1)},$$

$$\tilde{p}^1(2) = 1 - \tilde{p}^1(1),$$

$$\tilde{p}^2(1) = \frac{p(1)a_{11} + p(2)(1 - \alpha_2)a_{21}}{1 - \alpha_2 p(2)},$$

$$\tilde{p}^2(2) = 1 - \tilde{p}^2(1).$$

Maximizing Probability of Detection in n Looks

We now show how dynamic programming may be used to maximize the probability of finding the target in n looks. For each prior target location distribution p, let

$$Q_n(p) = \max_{\xi \in \Xi} \hat{P}_p[n, \xi] \quad \text{for} \quad n = 1, 2, \ldots.$$

By the principle of optimality (see Bellman, 1957, p. 83),

$$\mathbf{Q}_n(p) = \max_{j=1,2} \{\alpha_j p(j) + [1 - \alpha_j p(j)]\mathbf{Q}_{n-1}(\tilde{p}^j)\}. \qquad (9.1.3)$$

A solution for $\mathbf{Q}_n(p)$ may be obtained by using (9.1.3) and an iterative procedure. Note that

$$\mathbf{Q}_1(p) = \begin{cases} \alpha_1 p(1) & \text{if } p(1) \geq \pi_1^*, \\ \alpha_2 p(2) & \text{if } p(1) \leq \pi_1^*, \end{cases}$$

where

$$\pi_1^* = \alpha_2/(\alpha_1 + \alpha_2).$$

Knowing $\mathbf{Q}_1$, one can use (9.1.3) to obtain $\mathbf{Q}_2$, and in a recursive fashion $\mathbf{Q}_n$ may be found for any n. Having obtained $\mathbf{Q}_n$, one may develop the optimal plan ξ^* for a given p by looking in the cell that yields the maximum in (9.1.3) at each stage.

Example 9.1.1 (No learning). There is a situation in which $\mathbf{Q}_n$ and the optimal search plan may be easily found. Suppose $a_{11} = a_{21}$. Then no information is gained about the target's location from looks that have failed to detect the target; that is, a simple computation shows that the posterior target location probabilities are always

$$\tilde{p}(1) = a_{21}, \qquad \tilde{p}(2) = 1 - a_{21},$$

regardless of the prior probabilities or the sequence of looks that have failed to detect the target. Consider two cases.

Case 1: $\alpha_1 a_{21} \geq \alpha_2(1 - a_{21})$. Then it is clear that all looks after the first one will be made in cell 1. Thus, the probability of detecting the target on the remaining $n - 1$ looks given failure on the first is

$$\mathbf{Q}_{n-1}(\tilde{p}) = 1 - (1 - \alpha_1 a_{21})^{n-1}.$$

Hence

$$\mathbf{Q}_n(p) = \begin{cases} 1 - [1 - \alpha_1 p(1)](1 - \alpha_1 a_{21})^{n-1} & \text{for } p(1) \geq \pi^*, \\ 1 - [1 - \alpha_2 p(2)](1 - \alpha_1 a_{21})^{n-1} & \text{for } p(2) \leq \pi^*, \end{cases}$$

where

$$\pi^* = \alpha_2/(\alpha_1 + \alpha_2).$$

Case 2: $\alpha_1 a_{21} < \alpha_2(1 - a_{21})$. By an argument similar to the one used in case 1,

$$\mathbf{Q}_n(p) = \begin{cases} 1 - [1 - \alpha_1 p(1)][1 - \alpha_2(1 - a_{21})]^{n-1} & \text{for } p(1) \geq \pi^*, \\ 1 - [1 - \alpha_2 p(2)][1 - \alpha_2(1 - a_{21})]^{n-1} & \text{for } p(1) \leq \pi^*, \end{cases}$$

where π^* is given as in case 1.

9.1 Two-Cell Discrete-Time Markovian Motion

Remark 9.1.2. Although the recursive procedure described above is straightforward, the computations involved may be complex. It would be desirable for the sake of simplicity in finding optimal plans if there were for each n a threshold π_n^* such that a plan that maximizes the probability of detection in n looks would place its first look in cell 1 if $p(1) \geq \pi_n^*$ and its first look in cell 2 if $p(1) < \pi_n^*$. Although such a result is claimed by Pollock (1970), the supporting argument is incorrect (see Kan, 1972). The existence of such thresholds is still an open question.

Minimizing Mean Number of Looks to Find Target

In order to find a plan ξ^* that minimizes the mean number of looks to find the target, a dynamic programming approach is again employed. Unfortunately, ξ^* has been found for only two special cases. The first is the case where $\alpha_1 = \alpha_2 = 1$ and the second is the "no learning" case, i.e., $a_{11} = a_{21}$. Let

$$M(p) = \inf_{\xi \in \Xi} \mu_p(\xi).$$

Strauch (1966, Theorem 8.2) guarantees that $M(p)$ satisfies

$$M(p) = \min_{j=1,2} \{1 + [1 - p(j)\alpha_j]M(\tilde{p}^j)\}. \tag{9.1.4}$$

By Theorem 9.1 of Strauch (1966) there exists an optimal stationary policy. A stationary policy is one that depends only on the present state of the system. In this case, the state of the system at time n is specified by the posterior target location probabilities at time n. Thus, if there is a unique bounded solution to (9.1.4), that solution must yield $\mu_p(\xi^*)$. The optimal plan ξ^* is recoverable from the recursion (9.1.4) by choosing at each look to search the cell that yields the minimum in (9.1.4).

Example 9.1.3. (Perfect detection). Let us look at the case where $\alpha_1 = \alpha_2 = 1$; that is, if the searcher looks in the cell containing the target, he will detect the target with probability 1. In this case,

$$\tilde{p}^1(1) = a_{21}, \quad \tilde{p}^1(2) = a_{22},$$
$$\tilde{p}^2(1) = a_{11}, \quad \tilde{p}^2(2) = a_{12}, \tag{9.1.5}$$

which are independent of p. Equation (9.1.4) becomes

$$M(p) = \min\{1 + p(2)M(\tilde{p}^1), 1 + p(1)M(\tilde{p}^2)\}. \tag{9.1.6}$$

where $M(\tilde{p}^1)$ and $M(\tilde{p}^2)$ are fixed numbers independent of p.

Suppose **M** is bounded. Then it is clear from (9.1.6) that

$$\mathbf{M}(p) = 1 + p(2)\mathbf{M}(\tilde{p}^1)$$

if and only if

$$1 + p(2)\mathbf{M}(\tilde{p}^1) \leq 1 + p(1)\mathbf{M}(\tilde{p}^2). \tag{9.1.7}$$

Since $p(2) = 1 - p(1)$, (9.1.7) is equivalent to

$$p(1) \geq \frac{\mathbf{M}(\tilde{p}^1)}{\mathbf{M}(\tilde{p}^1) + \mathbf{M}(\tilde{p}^2)} \equiv \pi^*. \tag{9.1.8}$$

Thus, Eq. (9.1.6) may be written as

$$\mathbf{M}(p) = \begin{cases} 1 + p(2)\mathbf{M}(\tilde{p}^1) & \text{if } p(1) \geq \pi^*, \\ 1 + p(1)\mathbf{M}(\tilde{p}^2) & \text{if } p(1) \leq \pi^*. \end{cases} \tag{9.1.9}$$

It is now possible to find $\mathbf{M}(\tilde{p}^1)$ and $\mathbf{M}(\tilde{p}^2)$ and consequently π^*. To do this, we consider 4 cases that are defined in terms of the relation of π^* to a_{21} and a_{11}.

Case 1: $a_{21} \geq \pi^*$, $a_{11} \geq \pi^*$. (This will be shown to correspond to region 1 in Fig. 9.1.2). Let

$$p(1) = a_{21}, \qquad p(2) = 1 - a_{21}.$$

By (9.1.5), $\tilde{p}^1(1) = a_{21}$ so that $p = \tilde{p}^1$. From (9.1.9) and the fact that $a_{21} \geq \pi^*$, we have

$$\mathbf{M}(\tilde{p}^1) = 1 + (1 - a_{21})\mathbf{M}(\tilde{p}^1),$$

i.e.,

$$\mathbf{M}(\tilde{p}^1) = 1/a_{21}.$$

Similarly, if we let $p = \tilde{p}^2$, then since $\tilde{p}^2(1) = a_{11}$, it follows that

$$\mathbf{M}(\tilde{p}^2) = 1 + a_{12}\mathbf{M}(\tilde{p}^1) = 1 + a_{12}/a_{21}.$$

Thus

$$\pi^* = \frac{1}{1 + a_{21} + a_{12}}. \tag{9.1.10}$$

Using this value of π^*, we find that case 1 corresponds to region 1 of Fig. 9.1.2, which is defined by

$$a_{21}^2 + a_{21} + a_{21}a_{12} \geq 1, \qquad a_{21} - a_{21}a_{12} - a_{12}^2 \geq 0. \tag{9.1.11}$$

Thus, when the transition matrix A satisfies (9.1.11),

$$\mathbf{M}(p) = \begin{cases} 1 + \dfrac{p(2)}{a_{21}} & \text{for } p(1) \geq \dfrac{1}{1 + a_{12} + a_{21}}, \\ 1 + p(1)\left(1 + \dfrac{a_{12}}{a_{21}}\right) & \text{for } p(1) \leq \dfrac{1}{1 + a_{12} + a_{21}}. \end{cases}$$

9.1 Two-Cell Discrete-Time Markovian Motion

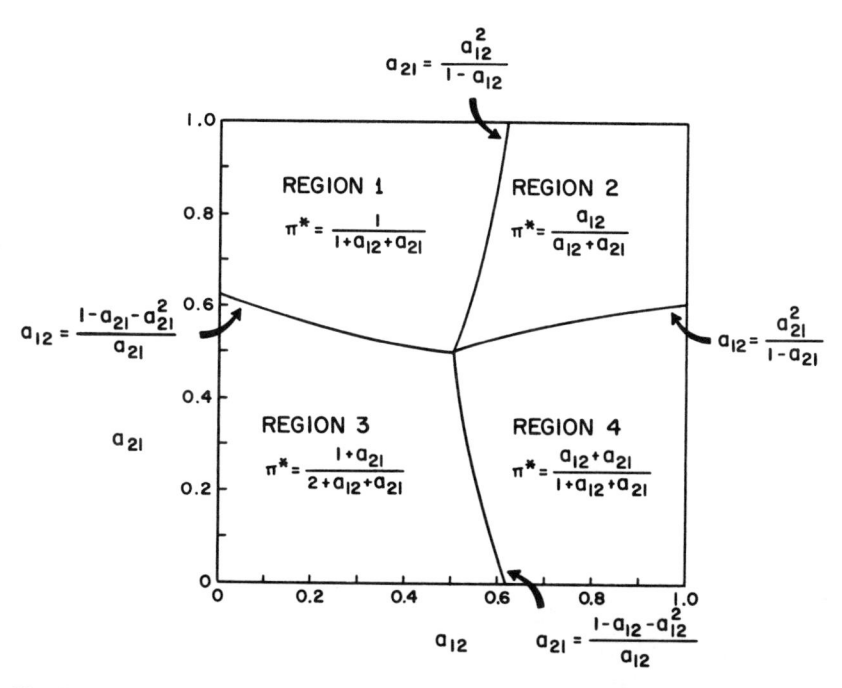

Fig. 9.1.2. The four regions that determine $M(p)$ when detection is perfect ($\alpha_1 = \alpha_2 = 1$).

The value of $M(p)$ and π^* may be found in an analogous way for the remaining three cases.

Case 2: $a_{21} \geq \pi^*, a_{11} < \pi^*$. (This corresponds to $a_{12}a_{21} + a_{21}^2 - a_{12} \geq 0$ and $a_{12}a_{21} + a_{12}^2 - a_{21} \geq 0$, which is region 2 in Fig. 9.1.2.)

$$M(p) = \begin{cases} 1 + p(2)/a_{21} & \text{for } p(1) \geq \pi^*, \\ 1 + p(1)/a_{12} & \text{for } p(1) \leq \pi^*, \end{cases}$$

where

$$\pi^* = a_{12}/(a_{12} + a_{21}).$$

Case 3: $a_{21} < \pi^*, a_{11} \geq \pi^*$. (This corresponds to $a_{12}a_{21} + a_{21} + a_{21}^2 \leq 1$ and $a_{12}a_{21} + a_{12} + a_{12}^2 \leq 1$, which is region 3 in Fig. 9.1.2.)

$$M(p) = \begin{cases} 1 + p(2)\dfrac{1 + a_{21}}{1 - a_{21}a_{12}} & \text{for } p(1) \geq \pi^*, \\ 1 + p(1)\dfrac{1 + a_{12}}{1 - a_{21}a_{12}} & \text{for } p(1) \leq \pi^*, \end{cases}$$

where

$$\pi^* = \frac{1 + a_{21}}{2 + a_{21} + a_{12}}.$$

Case 4: $a_{21} < \pi^*$, $a_{11} < \pi^*$. (This corresponds to $a_{12} - a_{12}a_{21} - a_{21}^2 \geq 0$ and $a_{12}a_{21} + a_{12} + a_{12}^2 \geq 1$, which is region 4 in Fig. 9.1.2.)

$$\mathbf{M}(p) = \begin{cases} 1 + p(2) \dfrac{a_{12} + a_{21}}{a_{12}} & \text{for} \quad p(1) \geq \pi^*, \\ 1 + \dfrac{p(1)}{a_{12}} & \text{for} \quad p(1) \leq \pi^*, \end{cases}$$

where

$$\pi^* = \frac{a_{12} + a_{21}}{1 + a_{12} + a_{21}}.$$

One can check that the above solution to (9.1.6) is the unique bounded one and thus determines the minimum expected number of looks to find the target. The optimal search plan is now readily obtained from Fig. 9.1.2 in the following manner. From the transition matrix A, determine the region of Fig. 9.1.2 that applies. This yields a value of π^*. The first look is placed in cell 1 if $p(1) \geq \pi^*$ and in cell 2 otherwise. After each look, the posterior distribution is calculated. This distribution is simply $\tilde{p}^1$ or $\tilde{p}^2$ as given by (9.1.5), depending on whether the last look was in cell 1 or 2. The next look is placed in cell 1 if the posterior probability in cell 1 is greater than or equal to π^* or in cell 2 otherwise. Further discussion of this example is contained in Pollock (1970).

Example 9.1.4. (No learning). Consider the no-learning situation of Example 9.1.1 (i.e., $a_{11} = a_{21}$). In this case, (9.1.4) may be easily solved by considering two cases.

Case 1: $\alpha_1 a_{21} \geq \alpha_2(1 - a_{21})$

$$\mathbf{M}(p) = \begin{cases} 1 + \dfrac{1 - \alpha_1 p(1)}{\alpha_1 a_{21}} & \text{for} \quad p(1) \geq \pi^*, \\ 1 + \dfrac{1 - \alpha_2 p(2)}{\alpha_1 a_{21}} & \text{for} \quad p(1) \leq \pi^*, \end{cases}$$

where

$$\pi^* = \alpha_2/(\alpha_1 + \alpha_2).$$

9.2 Two-Cell Continuous-Time Markovian Motion

Case 2: $\alpha_1 a_{21} \leq \alpha_2(1 - a_{21})$

$$\mathbf{M}(p) = \begin{cases} 1 + \dfrac{1 - \alpha_1 p(1)}{\alpha_2(1 - a_{21})} & \text{for } p(1) \geq \pi^*, \\ 1 + \dfrac{1 - \alpha_2 p(2)}{\alpha_2(1 - a_{21})} & \text{for } p(1) \leq \pi^*, \end{cases}$$

where again

$$\pi^* = \alpha_2/(\alpha_1 + \alpha_2).$$

Remark 9.1.5. For situations not covered by Examples 9.1.3 and 9.1.4, numerical techniques are available to find $\mathbf{M}(p)$ and the optimal plan ξ^* for any prior probability distribution p, transition matrix A, and conditional detection probabilities α_1 and α_2 (see Pollock, 1970). As with the problem of maximizing the probability of detection by time n, it is still an open question as to whether optimal policies can be characterized by threshold probabilities in the manner discussed in Remark 9.1.2.

In the theory, the approach of this section extends to n-cells. However, the degree of computational difficulty increases dramatically. On the other hand, the solution to the no-learning case, Examples 9.1.1 and 9.1.4, is easily extended to n-cells (see Kan, 1972). In addition Kan (1972) finds the plan that maximizes probability of detection in a fixed number of looks for an n-cell Markovian target motion when the transition matrix is Jordan. The target motion in this case is the discrete time and space analog of conditionally deterministic motion.

The assumption that the conditional detection probability α_j is independent of the number of looks in cell j is analogous to assuming that $b(j, k) = 1 - \alpha_j{}^k$ in the case of a stationary target. It would be desirable to extend the results of this section to more general detection functions.

Two discrete-time, n-cell Markovian search problems are considered in Klein (1968). In both situations, it is assumed that the target is trying to evade the searcher, is aware of the last cell in which the searcher looked, and can base his movements on this knowledge.

9.2. TWO-CELL CONTINUOUS-TIME MARKOVIAN MOTION

The problems of maximizing the probability of detection by time T and minimizing mean time to detection are considered in this section for the case of a continuous-time two-cell Markovian target motion.

Search Model

The target moves in continuous time between two cells. The motion is specified by a continuous-time Markov chain with transition rate matrix $A = (a_{ij})$. The off-diagonal elements of A may be interpreted as follows. Let $h > 0$. Then for $i \neq j$,

$$\text{Pr}\{\text{target in cell } j \text{ at time } t + h \mid \text{target in cell } i \text{ at time } t\}$$
$$= a_{ij}h + o(h) \quad \text{for } t \geq 0, \quad (9.2.1)$$

where o is used to indicate a function such that

$$\lim_{h \to 0^+} o(h)/h = 0.$$

The diagonal elements $a_{ii} = -\sum_{j \neq i} a_{ij}$. The target's position at $t = 0$ is given by the probability distribution p, i.e.,

$$p(j) = \text{Pr}\{\text{target in cell } j \text{ at } t = 0\} \quad \text{for } j = 1, 2. \quad (9.2.2)$$

A *moving-target search plan for J* is a Borel function $\psi: J \times [0, \infty) \to [0, \infty)$. The above definition is the natural analog of the definition of the moving-target search plan given in Chapter VIII for the search space X. In this section we are dealing with a two-cell search space, so $J = \{1, 2\}$.

It is assumed that using search plan ψ the probability of detecting the target in the interval $[t, t + h]$, given that the target is in cell j during $[t, t + h]$, is independent of past detections and is given by

$$1 - \exp\left(-\kappa(j) \int_t^{t+h} \psi(j, s)\, ds\right). \quad (9.2.3)$$

Dobbie (1974) solves the problem of maximizing the probability of detection by time T over all moving-target plans ψ such that

(a) $\psi(1, t) + \psi(2, t) = 1$ for $t \geq 0$,
(b) $\psi(j, \cdot)$ is piecewise continuous for $j = 1, 2$. $\quad (9.2.4)$

The solution is obtained by a change of variable technique that allows one to reduce the problem to an elementary but complex problem involving minimizing a real-valued function of a real variable. Examples of optimal plans are given by Dobbie (1974). By use of Pontryagin's maximum principle, Dobbie is able to show that the optimal plan ψ^* must have the following form. The function $\psi^*(1, \cdot)$ can take on at most three values, namely, 0, r, or 1, where

$$r = \frac{1}{\kappa(1) + \kappa(2)} \left\{ \left(\frac{a_{12}a_{21}}{\kappa(1)\kappa(2)}\right)^{1/2} [\kappa(1) - \kappa(2)] + a_{21} + \kappa(2) - a_{12} \right\}.$$

If $r < 0$ or $r > 1$, then $\psi^*(1, \cdot)$ can take only the values 0 or 1.

9.3 Continuous-Time Markov Motion in Euclidean N-Space

For the problem of finding ψ^* to minimize the mean time to detect the target, subject to (9.2.4) Dobbie (1974) finds the unique ψ^* that satisfies the necessary conditions of Pontryagin's maximum principle. To conclude that ψ^* is the plan that minimizes mean time to find the target, one needs to prove only that a minimizing plan satisfying (9.2.4) exists.

Remark 9.2.1. There are two principal areas in which it would be desirable to extend the results in Dobbie (1974). The first is to find optimal plans for n-cell Markovian motion. The second is to allow a broader class of detection functions. The detection function may be generalized as follows. Suppose that $\{\mathbf{X}_t, t \geq 0\}$ is the Markov process representing the target motion and that Ω is the collection of sample paths of the process. Let $\mathbf{E}$ denote expectation taken over all $\omega \in \Omega$. Assume that

$$\mathbf{E}\left[b\left(\int_0^t \kappa(\mathbf{X}_s(\omega))\psi(\mathbf{X}_s(\omega), s)\, ds\right)\right] \qquad (9.2.5)$$

is the probability of detecting the target by time t when using plan ψ. If $b(z) = 1 - e^{-z}$ for $z \geq 0$, then one obtains the detection function defined by (9.2.3).

The analysis in Dobbie (1974) has been extended in Dobbie (1975a) to the case of an avoiding target, i.e., the transition rate of the target is proportional to the search effort. For n-cells, a discussion of the set of functions ψ that satisfy the Pontryagin maximum principle is given. In Dobbie (1975b) the results of Dobbie (1973, 1974) are extended to a situation in which false targets are detected according to a Poisson process and require an exponentially distributed amount of time to be identified. Contact investigation is assumed to be immediate and conclusive.

*9.3. CONTINUOUS-TIME MARKOVIAN MOTION IN EUCLIDEAN N-SPACE

In this section, necessary conditions are given for a search plan to maximize probability of detection by time T when the target motion is a continuous-time Markov process in the space X.

Search Model

The target's motion in X is given by a Markov process $\{\mathbf{X}_t, t \geq 0\}$. This process is assumed to have a continuous transition density that need not be stationary in time. If ψ is a moving-target search plan, then the probability of failing to detect the target with plan ψ by time t is assumed to be

$$\mathbf{E}\left[\exp\left(-\int_0^t \psi(\mathbf{X}_s(\omega), s)\, ds\right)\right],$$

where **E** denotes expectation over the set Ω of sample paths ω of the Markov process.

As in Chapter VIII, $\Psi_2(m_2)$ is the class of moving-target search plans such that

$$\int_X \psi(x, t) \, dx = m_2(t) \quad \text{for } t \geq 0.$$

Let

$$G(x, s, t, \psi) = \Pr\{\text{target not detected during } [s, t] \text{ using plan } \psi \mid \mathbf{X}_s = x\}.$$

Let $g(x, s, y, t, \psi)$ be the transition density for the process representing undetected targets; that is, for Borel $S \subset X$,

$$\int_S g(x, s, y, t, \psi) \, dy$$
$$= \Pr\{\mathbf{X}_t \in S \text{ and target not detected in } [s, t] \text{ using plan } \psi \mid \mathbf{X}_s = x\}.$$

Necessary Conditions for Optimality

The following theorem gives a necessary condition for a search plan to maximize the probability of detecting the target by time T (i.e., minimize probability of failure to detect).

Theorem 9.3.1. *Let p give the target location distribution at $t = 0$. If $\psi^* \in \Psi_2(m_2)$ is continuous and bounded and satisfies*

$$\int_X G(x, 0, T, \psi^*) p(x) \, dx \leq \int_X G(x, 0, T, \psi) p(x) \, dx \quad \text{for } \psi \in \Psi_2(m_2), \quad (9.3.1)$$

then there exists a function $\lambda : [0, T] \to [0, \infty)$ such that for $0 \leq t \leq T$,

$$\int_X g(x, 0, y, t, \psi^*) G(y, t, T, \psi^*) p(x) \, dx = \lambda(t) \quad \text{for } \psi^*(y, t) > 0,$$
$$\leq \lambda(t) \quad \text{for } \psi^*(y, t) = 0. \quad (9.3.2)$$

Proof. The proof of this theorem is not given here but may be found in Saretsalo (1973).

Remark 9.3.2. While Theorem 9.3.1 gives a necessary condition for a very broad class of Markovian motions, there still remain the questions of when condition (9.3.2) is sufficient and how one uses (9.3.2) to find optimal plans. In the context of Remark 9.2.1, one should observe that Theorem 9.3.1 holds only for the detection function specified by taking $b(z) = 1 - e^{-z}$ for $z \geq 0$ and $\kappa(x) = 1$ for $x \in X$ in (9.2.5).

NOTES

The results in Section 9.1 are based on the work of Pollock (1970) and those in Section 9.2 on the work of Dobbie (1974). The proof of Theorem 9.3.1 given by Saretsalo (1973) is a generalization of the methods used by Hellman (1972), where a necessary condition for optimal plans is obtained for the special case where $\{X_t, t \geq 0\}$ is a diffusion process. Hellman (1970, 1971) also deals with search for a target whose motion is a diffusion process.

Appendix A

Reference Theorems

In this appendix several standard theorems are stated so that readers may conveniently refer to them.

Theorem A.1 is a version of the dominated convergence theorem. This theorem holds for more general measures than Lebesgue measure on Euclidean n-space.

Theorem A.1 (Dominated convergence). *Let $\{h_i : i = 1, 2, \ldots\}$ be a sequence of Lebesgue-integrable real-valued functions defined on Euclidean n-space X such that*

$$\lim_{i \to \infty} h_i(x) \quad \text{exists for a.e.} \quad x \in X.$$

If there exists a finitely integrable function $\bar{h}$ such that for $i = 1, 2, \ldots$

$$|h_i(x)| \leq \bar{h}(x) \quad \text{for a.e.} \quad x \in X$$

then

$$\lim_{i \to \infty} \int_X h_i(x)\, dx = \int_X [\lim_{i \to \infty} h_i(x)]\, dx.$$

Theorem A.2 (Monotone convergence theorem). *If h_i, $i = 1, 2, \ldots$, is* an increasing sequence of nonnegative Lesbesgue-measurable functions defined on X, then

$$\lim_{i \to \infty} \int_X h_i(x)\, dx = \int_X [\lim_{i \to \infty} h_i(x)]\, dx.$$

For $i = 1, 2, \ldots$, let h_i be a real-valued function defined on a real interval $[a_1, a_2]$. Suppose that the series $\sum_{i=1}^{\infty} h_i(s)$ converges for $a_1 \leq s \leq a_2$. Then the series *converges uniformly in* $[a_1, a_2]$ if for $\varepsilon > 0$ there is an integer k independent of $s \in [a_1, a_2]$, such that for $n > k$

$$\left| \sum_{i=1}^{\infty} h_i(s) - \sum_{i=1}^{n} h_i(s) \right| < \varepsilon \quad \text{for} \quad a_1 \leq s \leq a_2.$$

The following theorem is due to Weierstrass.

Theorem A.3. *Suppose there exist real numbers $\bar{h}_i$ for $i = 1, 2, \ldots$ such that*

$$|h_i(s)| \leq \bar{h}_i \quad \text{for} \quad a_1 \leq s \leq a_2$$

and

$$\sum_{i=1}^{\infty} \bar{h}_i < \infty.$$

Then $\sum_{i=1}^{\infty} h_i(s)$ converges uniformly in $[a_1, a_2]$.

The following theorem gives an important consequence of uniform convergence.

Theorem A.4. *Suppose $h_i: [a_1, a_2] \to (-\infty, \infty)$ is continuous for $i = 1, 2, \ldots$. If $\sum_{i=1}^{\infty} h_i(s)$ converges uniformly in $[a_1, a_2]$, then the function H defined by*

$$H(s) = \sum_{i=1}^{\infty} h_i(s) \quad \text{for} \quad a_1 \leq s \leq a_2$$

is continuous on $[a_1, a_2]$.

Appendix B

Necessary Conditions for Constrained Optimization of Separable Functionals

In this appendix it is shown that maximizing the pointwise Lagrangian is a necessary condition for a constrained optimum involving the payoff functional E and vector-valued cost functional C defined in Section 3.4. The first section deals with the discrete space J and the second with the continuous space X.

The notation and definitions of Section 3.4 are followed in this appendix. In particular, recall that e is a nonegative real-valued Borel function and c a vector-valued function with k components such that the ith component c^i is a nonnegative, real-valued, Borel function.

In addition, we let $\mathscr{E}_r$ be Euclidean r-space for $r = 1, 2, \ldots$, and $\mathscr{E}_r^+ = \{y : y \in \mathscr{E}_r \text{ and } y_i \geq 0 \text{ for } i = 1, \ldots, r\}$. Define for $f \in \hat{F}$ and $\lambda \in \mathscr{E}_k$, the *functional Lagrangian L* by

$$L(\lambda, f) = E[f] - \sum_{\lambda^i \neq 0} \lambda^i C^i[f] \quad \text{for } f \text{ such that } \infty - \infty \text{ does not occur.}$$

A pair (λ, f^*) such that $\lambda \in \mathscr{E}_k$ and $f \in \hat{F}$ is said to *maximize the functional Lagrangian with respect to* $\hat{F}$, if $L(\lambda, f^*)$ exists and if

$$L(\lambda, f^*) \geq L(\lambda, f) \quad \text{whenever } f \in \hat{F} \text{ and } L(\lambda, f) \text{ exists.}$$

Let $(c, e) = (c^1, \ldots, c^k, e)$ and $(C, E) = (C^1, \ldots, C^k, E)$. Define
$$\text{im } C = \{C[f] : f \in \hat{F}\}, \quad \text{im}(C, E) = \{(C[f], E[f]) : f \in \hat{F}\}.$$
If S is a set in extended r-space (i.e., infinite coordinates are allowed), then fin S is defined to be the set of points in S having all coordinates finite. Thus fin im(C, E) is the set of points in im(C, E) with all coordinates finite. The interior of S is denoted int S. Thus, int im C is the interior of the image of C.

We say C and $-E$ are *convex functionals* if $\hat{F}$ is convex [i.e., $f_1, f_2 \in \hat{F}$ implies $\theta f_1 + (1 - \theta) f_2 \in \hat{F}$ for $0 \le \theta \le 1$] and whenever $f_1, f_2 \in \hat{F}$ and $0 \le \theta \le 1$, then
$$C[\theta f_1 + (1 - \theta) f_2] \le \theta C[f_1] + (1 - \theta) C[f_2],$$
$$E[\theta f_1 + (1 - \theta) f_2] \ge \theta E[f_1] + (1 - \theta) E[f_2].$$

In this appendix ν denotes Lebesgue measure on $X = \mathscr{E}_n$.

B.1. NECESSARY CONDITIONS FOR THE DISCRETE SPACE J

In this section it is shown that, under appropriate convexity conditions, an allocation f^* that is optimal within $\hat{F}(J)$ for cost $C[f^*]$ must maximize the pointwise Lagrangian with respect to $\hat{F}(J)$.

Theorem B.1.1. *Suppose C and $-E$ are convex functionals and that $f^* \in \hat{F}(J)$ is optimal within $\hat{F}(J)$ for cost $C[f^*]$. If $C[f^*]$ and $E[f^*]$ are finite and there is a $u \in$ im C such that $u^i < C^i[f^*]$ for $i = 1, \ldots, k$, then there is a $\lambda \in \mathscr{E}_k^+$ such that (λ, f^*) maximizes the functional and pointwise Lagrangians.*

Proof. Let
$$S = \text{fin}\{(u, v) : u \ge C[f] \text{ and } v \le E[f] \text{ for some } f \in \hat{F}(J)\}.$$
An example of S is shown in Fig. B.1.1 for the case where $k = 1$. The set S is convex. To see this, let $(u_1, v_1), (u_2, v_1) \in S$ and $0 \le \theta \le 1$. Let $f_1, f_2 \in \hat{F}(J)$ be such that
$$u_1 \ge C[f_1], \quad v_1 \le E[f_1], \quad u_2 \ge C[f_2], \quad v_2 \le E[f_2].$$
By the convexity of C and $-E$, $f = \theta q_1 + (1 - \theta) f_2 \in \hat{F}(J)$ and
$$\theta u_1 + (1 - \theta) u_2 \ge \theta C[f_1] + (1 - \theta) C[f_2] \ge C[f],$$
$$\theta v_1 + (1 - \theta) v_2 \le \theta E[f_1] + (1 - \theta) E[f_2] \le E[f].$$
Thus, $\theta(u_1, v_1) + (1 - \theta)(u_2, v_2) \in S$ and S is convex.

Let
$$R = \text{fin}\{(u, v) : u \le C[f^*] \text{ and } v \ge E[f^*]\}.$$

B.1 Necessary Conditions for the Discrete Space J

Then R is convex, (int R) $\cap$ S is empty (by the optimality of f^*), and $(C[f^*], E[f^*])$ is on the boundary of S (see Fig. B.1.1). By a separating hyperplane theorem (see Luenberger, 1969, Theorem 3, p. 133), there is a nonzero vector $(-\lambda, \lambda^0)$, where $\lambda \in \mathscr{E}_k$, $\lambda^0 \in \mathscr{E}_1$, such that

$$(-\lambda, \lambda^0) \cdot (C[f^*], E[f^*]) \geq (-\lambda, \lambda^0) \cdot (u, v) \quad \text{for} \quad (u, v) \in S. \quad \text{(B.1.1)}$$

From the definition of S it is clear that $\lambda \in \mathscr{E}_k^+$ and $\lambda^0 \geq 0$. Of course, (B.1.1) implies

$$\lambda^0 E[f^*] - \lambda \cdot C[f^*] \geq \lambda^0 E[f] - \lambda \cdot C[f] \quad \text{(B.1.2)}$$

for $f \in \hat{F}(J)$ such that

$$(C[f], E[f]) \in \text{fin im } (C, E).$$

However, it is easy to see that (B.1.2) must hold for all f such that the right-hand side of (B.1.2) is defined. (Note we take $\lambda^i \infty = 0$ whenever $\lambda^i = 0$.)

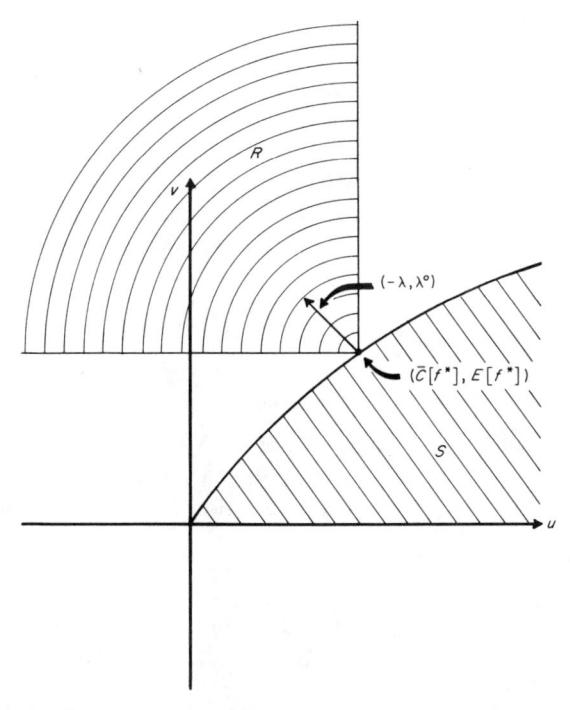

Fig. B.1.1. Example of the sets S and R for $k = 1$. [*Note:* $S = \text{fin}\{(u, v) : u \geq C[f]$ and $v \leq E[f]$ for some $f \in \hat{F}(J)\}$.]

For, if there is a $f \in \hat{F}(J)$ such that

$$\lambda^0 E[f] - \lambda \cdot C[f] > \lambda^0 E[f^*] - \lambda \cdot C[f^*],$$

then there must be $j_0 \in J$ such that

$$(\lambda^0 e(j_0, f(j_0)) - \lambda \cdot c(j_0, f(j_0))) > \lambda^0 e(j_0, f^*(j_0)) - \lambda \cdot c(j_0, f(j_0))). \quad \text{(B.1.3)}$$

Let
$$r(j_0) = q(j_0), \qquad r(j) = f^*(j) \quad \text{for} \quad j \neq j_0. \quad \text{(B.1.4)}$$

Then $E[r]$ and $C[r]$ are finite, but $\lambda^0 E[r] - \lambda \cdot C[r] > \lambda^0 E[f^*] - \lambda \cdot C[f^*]$, which contradicts (B.1.2). Thus,

$$\lambda^0 E[f^*] - \lambda \cdot C[f^*] \geq \lambda^0 E[f] - \lambda \cdot C[f] \quad \text{(B.1.5)}$$

for all $f \in \hat{F}(J)$ such that the right-hand side is defined.

We now show that $\lambda^0 > 0$. Suppose $\lambda^0 = 0$. Then by (B.1.5),

$$\lambda \cdot C[f^*] \leq \lambda \cdot C[f] \quad \text{for all} \quad f \in \hat{F}(J) \quad \text{such that} \quad C[f] \quad \text{is finite} \quad \text{(B.1.6)}$$

By assumption, there is a $u \in \text{im } C$ such that $u^i < C^i[f^*]$ for $i = 1, \ldots, k$. Thus, (B.1.6) must hold for $C[f] = u$. But this is a contradiction, since $u < C[f^*]$, $\lambda \geq 0$, and $\lambda \neq 0$. Thus, $\lambda^0 > 0$, and we may take $\lambda^0 = 1$ in (B.1.5). The result is that (λ, f^*) maximizes the functional Lagrangian with respect to $\hat{F}(J)$. Since this is the case, it is clear that (λ, f^*) must also maximize the pointwise Lagrangian with respect to $\hat{F}(J)$. For if this were not true, then there would exist a $j_0 \in J$ such that (B.1.3) holds with $\lambda^0 = 1$. The function r defined in (B.1.4) would then satisfy $L(\lambda, r) > L(\lambda, f^*)$, which contradicts the fact that (λ, f^*) maximizes the functional Lagrangian. Thus, (λ, f^*) maximizes the pointwise Lagrangian with respect to $\hat{F}(J)$, and the theorem is proved.

Corollary B.1.2. *Suppose $e(j, \cdot)$ is concave, $c^i(j, \cdot)$ is convex for $i = 1, \ldots, k$, and $Z(j)$ is an interval for $j \in J$. If $f^* \in \hat{F}(J)$, $E[f^*] < \infty$, $C[f^*] \in \text{int im } C$, and f^* is optimal within $\hat{F}(J)$ for cost $C[f^*]$, then there is a $\lambda \in \mathscr{E}_k^+$ such that (λ, f^*) maximizes the pointwise Lagrangian.*

Proof. It is easily verified that C and $-E$ are convex functionals. Since $C[f^*] \in \text{int im } C$, $C[f^*]$ is finite and there is a finite $u \in \text{im } C$ such that $u_i < C^i[f^*]$ for $i = 1, \ldots, k$. The lemma now follows from Theorem B.1.1.

Remark B.1.3. Using the proof of Theorem B.1.1, one may easily show that under the conditions of that theorem (λ, f^*) maximizes the pointwise Lagrangian if and only if it maximizes the functional Lagrangian.

Recall that the optimality of f^* for cost $C[f^*]$ is defined in terms of an inequality constraint, i.e.,

$$E[f^*] = \max\{E[f] : C[f] \leq C[f^*]\}. \quad \text{(B.1.7)}$$

B.2 Necessary Conditions for the Continuous Space X

The above theorem and lemma may be extended to the case where the inequality in (B.1.7) is replaced by an equality as follows. The constraint $C[f] = C[f^*]$ may be expressed as two inequality constraints, i.e., $C[f] \leq C[f^*]$ and $-C[f] \leq -C[f^*]$. Suppose f^* is optimal for cost $C[f^*]$ under the equality constraint and the remaining hypotheses of Theorem B.1.1 hold. Then the proof of Theorem B.1.1 yields a $2k$ vector $(\lambda_1, \lambda_2) \in \mathscr{E}_{2k}^+$ ($\lambda_1 \in \mathscr{E}_k^+$, $\lambda_2 \in \mathscr{E}_k^+$) such that

$$E[f^*] - \lambda_1 \cdot C[f^*] + \lambda_2 \cdot C[f^*] \geq E[f] - \lambda_1 \cdot C[f] + \lambda_2 \cdot C[f] \quad \text{(B.1.8)}$$

for $f \in \hat{F}(J)$ such that the right-hand side exists. Letting $\lambda^i = \lambda_1^i - \lambda_2^i$ for $i = 1, \ldots, k$, (B.1.8) becomes $L(\lambda, f^*) \geq L(\lambda, f)$, but it is no longer guaranteed that $\lambda^i \geq 0$ for $i = 1, \ldots, k$. Thus, the conclusions of Theorem B.1.1 and Lemma B.1.2 remain the same, but one must replace $\lambda \in \mathscr{E}_k^+$ by $\lambda \in \mathscr{E}_k$.

B.2. NECESSARY CONDITIONS FOR THE CONTINUOUS SPACE X

In this section, it is shown that an allocation f^* that is optimal within $\hat{F}(X)$ for cost $C[f^*]$ must maximize the pointwise Lagrangian with respect to $\hat{F}(X)$. In contrast to section B.1, no convexity conditions are required here. The reason for this appears in Theorem B.2.1, which shows that fin im(C, E) is convex for the space X.

Theorem B.2.1. (Lyapunov and Blackwell). *Let $\hat{F} = \hat{F}(X)$. Then fin im(C, E) is a convex set in $\mathscr{E}_{k+1}$.*

Proof. If fin im(C, E) is empty, the theorem holds trivially. Thus, we may let f and r be members of $\hat{F}(X)$ such that $(C[f], E[f])$ and $(C[r], E[r])$ are members of fin im(C, E). Let $0 \leq \theta \leq 1$. Define a signed-vector measure g on Borel subsets of $S \subset X$ as follows:

$$g(S) = \int_S (c(x, r(x)) - c(x, f(x)), e(x, r(x)) - e(x, f(x))) \, dx.$$

By Lyapunov's convexity theorem (Halmos, 1948), the image of g is a convex set in $\mathscr{E}_{k+1}$. Thus, there is a Borel set S_0 such that

$$g(S_0) = \theta(C[r] - C[f], E[r] - E[f]).$$

Let

$$f_0(x) = \begin{cases} r(x) & \text{for } x \in S_0, \\ f(x) & \text{for } x \notin S_0. \end{cases}$$

Then $f_0 \in \hat{F}(X)$ and
$$C[f_0] = \int_X c(x, f(x)) \, dx + \int_{S_0} [c(x, r(x)) - c(x, f(x))] \, dx$$
$$= (1 - \theta)C[f] + \theta C[r].$$
Similarly, $E[f_0] = (1 - \theta)E[f] + \theta E[r]$. Thus, $(1 - \theta)(C[f], E[f]) + \theta(C[r], E[r]) \in \text{fin im}(C, E)$, and the theorem is proved.

Theorem B.2.2. *Suppose $f^* \in \hat{F}(X)$, $E[f^*] < \infty$, and $C[f^*] \in \text{int im } C$. If f^* is optimal within $\hat{F}(X)$ for cost $C[f^*]$, then there exists a $\lambda \in \mathscr{E}_k^+$ such that (λ, f^*) maximizes the functional Lagrangian.*

Proof. The allocation f^* satisfies
$$E[f^*] = \max\{E[f] : f \in \hat{F}(X) \text{ and } C[f] \le C[f^*]\}. \tag{B.2.1}$$
Let R be the set in $\mathscr{E}_{k+1}$ defined by
$$R = \text{fin}\{(C[f^*] - C[f], E[f] - E[f^*]) : f \in F(X)\}.$$
By Theorem B.2.1, fin im(C, E) is convex, which implies that R is convex. By (B.2.1), $R \cap \text{int } \mathscr{E}_{k+1}^+$ is empty and $(C[f^*], E[f^*])$ is on the boundary of R, and so by Theorem 3 of Luenberger (1969, p. 133), there is a nonzero vector $(\lambda, \lambda^0) \in \mathscr{E}_{k+1}^+$ (where $\lambda \in \mathscr{E}_k$) such that
$$(\lambda, \lambda^0) \cdot u \le 0 \quad \text{for} \quad u \in R.$$
Thus,
$$\lambda_0 E[f^*] - \lambda \cdot C[f^*] \ge \lambda_0 E[f] - \lambda \cdot C[f] \tag{B.2.2}$$
for $f \in \hat{F}(X)$ such that $E[f]$ and $C[f]$ are finite.

We claim that (B.2.2) implies
$$\lambda^0 e(x, f^*(x)) - \lambda \cdot c(x, f^*(x)) \ge \lambda^0 e(x, f(x)) - \lambda \cdot c(x, f(x)) \tag{B.2.3}$$
for a.e. $x \in X$ whenever $f \in \hat{F}(X)$. To see this, we suppose (B.2.3) does not hold. Then there is a $f \in \hat{F}(X)$ such that the set
$$S = \{x : \lambda^0 e(x, f(x)) - \lambda \cdot c(x, f(x)) > \lambda^0 e(x, f^*(x)) - \lambda \cdot c(x, f^*(x))\}$$
has $\nu(S) > 0$, where ν denotes Lebesgue measure on $X = \mathscr{E}_n$. Since the functions e, c, f, and f^* are Borel, S is Borel. For $r = 1, 2, \ldots$, let
$$S_r = S \cap \{x : |e(x, f(x))| \le r \text{ and } |c^i(x, f(x))| \le r \text{ for } i = 1, \ldots, k\}.$$
Then $\bigcup_{r=1}^{\infty} S_r = S$, and so for some r_0, $\nu(S_{r_0}) > 0$. Let D be a Borel subset of S_{r_0} such that $0 < \nu(D) < \infty$. It follows that
$$\int_D e(x, f(x)) \, dx < \infty, \quad \int_D c^i(x, f(x)) \, dx < \infty \quad \text{for } i = 1, \ldots, k.$$

B.2 Necessary Conditions for the Continuous Space X

Let
$$h(x) = \begin{cases} f(x) & \text{for } x \in D, \\ f^*(x) & \text{for } x \notin D. \end{cases}$$

Then $h \in \hat{F}(X)$ and $E[h]$ and $C[h]$ are finite. However,

$$\lambda^0 E[h] - \lambda \cdot C[h] > \lambda^0 E[f^*] - \lambda \cdot C[f^*],$$

which contradicts (B.2.2). Thus (B.2.3) is true and the claim is proved.

For $i = 0, 1, \ldots, k$, we understand $\lambda^i \infty = 0$ whenever $\lambda^i = 0$. From (B.2.3), it follows that

$$\lambda^0 E[f^*] - \lambda \cdot C[f^*] \geq \lambda^0 E[f] - \lambda \cdot C[f] \tag{B.2.4}$$

for all $f \in \hat{F}(X)$ for which the right-hand side is defined. From (B.2.4) it now follows that $\lambda^0 \neq 0$. For if $\lambda^0 = 0$, then (B.2.4) would yield

$$\lambda \cdot C[f^*] \leq \lambda \cdot C[f] \quad \text{for} \quad f \in \hat{F} \quad \text{such that} \quad \lambda \cdot C[f] \text{ is defined.} \tag{B.2.5}$$

Since (λ, λ^0) is nonzero and $\lambda^0 = 0$, it follows that $\lambda \neq 0$. Thus, (B.2.5) implies that $C[f^*] \notin \text{int im } C$, contrary to assumption. Thus, $\lambda^0 \neq 0$ and we may take $\lambda^0 = 1$ in (B.2.4). The theorem follows.

In order to prove that maximizing a pointwise Lagrangian is a necessary condition for an allocation $f^* \in \hat{F}(X)$ to be optimal for cost $C[f^*]$, the following selection theorem will be used without proof. The theorem guarantees that for any Borel subset of $\mathscr{E}_{n+1}$ there is a Borel function whose graph lies within that subset. This theorem is a special case of Theorem 4.1 (combined with Remark 4.2) of Wagner and Stone (1974), which is a generalization of Von Neumann's (1949) selection theorem.

Let $\Pi(x, z) = x$ for $x \in X = \mathscr{E}_n$ and $z \in \mathscr{E}_1$.

Theorem B.2.3. *Let S be a Borel subset of $\mathscr{E}_{n+1}$ such that $\nu(\Pi(S)) > 0$. Then there is a Borel set $D \subset \Pi(S)$ such that $\nu(D) > 0$ and a Borel function $g: D \to \mathscr{E}_1$ such that*
$$(x, g(x)) \in S \quad \text{for} \quad x \in D.$$

The following is the main necessity theorem for optimal allocations $f^* \in \hat{F}(X)$.

Theorem B.2.4. *Let $\Omega = \{(x, z): x \in X \text{ and } z \in Z(x)\}$ be a Borel set. Suppose $f^* \in \hat{F}(X)$, $E[f^*] < \infty$, and $C[f^*] \in \text{int im } C$. If f^* is optimal within $\hat{F}(X)$ for cost $C[f^*]$, then there exists $\lambda \in \mathscr{E}_k^+$ such that (λ, f^*) maximizes the pointwise Lagrangian with respect to $\hat{F}(X)$.*

Proof. By Theorem B.2.2, there is a $\lambda \in \mathscr{E}_k^+$ such that (λ, f^*) maximizes the functional Lagrangian with respect to $\hat{F}(X)$, that is,

$$L(\lambda, f^*) \geq L(\lambda, f) \quad \text{for} \quad f \in \hat{F}(X) \quad \text{such that} \quad L(\lambda, f) \text{ exists.} \tag{B.2.6}$$

We now show that (λ, f^*) maximizes the pointwise Lagrangian with respect to $\hat{F}(X)$. Suppose that (λ, f^*) fails to maximize the pointwise Lagrangian, and consider the set
$$S = \Omega \cap \{(x, z): \ell(x, \lambda, z) > \ell(x, \lambda, f^*(x))\}.$$
Since Ω, e, c, and f^* are Borel, S is Borel. Thus, $\Pi(S)$ is Lebesgue measurable (see Federer, 1969, 2.2.13), and since (λ, f^*) fails to maximize the pointwise Lagrangian, $\nu(\Pi(S)) > 0$. By Theorem B.2.3, there is a Borel set $D \subset \Pi(S)$ such that $\nu(D) > 0$ and a Borel function $g: D \to \mathscr{E}_1$ such that $(x, g(x)) \in S$ for $x \in D$, that is,
$$\ell(x, \lambda, g(x)) > \ell(x, \lambda, f^*(x)) \qquad \text{for} \quad x \in D.$$
Let
$$r(x) = \begin{cases} g(x) & \text{for } x \in D, \\ f^*(x) & \text{for } x \notin D. \end{cases}$$
Then $r \in \hat{F}(X)$ and $L(\lambda, r) > L(\lambda, f^*)$ in contradiction to (B.2.6). Thus, (λ, f^*) must maximize the pointwise Lagrangian and the theorem is proved.

Remark B.2.5. Using the above proof, one may easily show that under the assumptions of Theorem B.2.4, (λ, f^*) maximizes the pointwise Lagrangian if and only if it maximizes the functional Lagrangian. By referring to Remark B.1.3, it is clear that Theorem B.2.4 can be extended to the case of equality constraints. The conclusion of the theorem remains unchanged, except that $\lambda \in \mathscr{E}_k^+$ must be replaced by $\lambda \in \mathscr{E}_k$.

NOTES

The development of this appendix is based on Wagner and Stone (1974). The use of the set S and separating hyperplane argument in Theorem B.1.1 is modeled on the argument in Theorem 1 of Luenberger (1969, p. 217). Theorem B.2.1 is due to Lyapunov and Blackwell (see Blackwell, 1951). A special case of Theorem B.2.4 is given by Aumann and Perles (1965). Wagner and Stone (1974) give the necessity results of this appendix in a more general form. In particular, they do not require e or c to be nonnegative, and more general measure spaces are considered. In addition, very general existence and sufficiency results are proved.

References

The AD numbers in parentheses at the end of references to unpublished papers may be used to order copies of those papers from the National Technical Information Service.

Arkin, V. I. (1964a). A problem of optimum distribution of search effort. *Theor. Probability Appl.* **9**, 159–160.

Arkin, V. I. (1964b). Uniformly optimal strategies in search problems. *Theor. Probability Appl.* **9**, 674–680.

Aumann, R. J., and Perles, M. (1965). A variational problem arising in economics. *J. Math. Anal. Appl.* **11**, 488–503.

Beck, A. (1964). On the linear search problem. *Israel J. Math.* **2**, 221–228.

Beck, A. (1965). More on the linear search problem. *Israel J. Math.* **3**, 61–70.

Beck, A., and Newman, D. J. (1970). Yet more on the linear search problem. *Israel J. Math.* **8**, 419–429.

Beck, A., and Warren, P. (1973). The return of the linear search problem. *Israel J. Math.* **14**, 169–183.

Belkin, B. (1975). On the rate of expansion of gamma search plans. *Operations Res.* **23**.

Bellman, R. (1957). "Dynamic Programming." Princeton Univ. Press, Princeton, New Jersey.

Black, W. L. (1965). Discrete sequential search. *Information and Control* **8**, 159–162.

Blackwell, D. (1951). The range of certain vector integrals. *Proc. Amer. Math. Soc.* **2**, 159–162.

References

Bumby, R. T. (1960). Prolegomena to a mathematical theory of search. Tech. Rep. Johns Hopkins Univ. Appl. Phys. Lab., Silver Spring, Maryland (AD 654424).

Charnes, A., and Cooper, W. W. (1958). The theory of search: Optimum distribution of search effort. *Management Sci.* **5**, 44–50.

Chew, M. C. (1967). A sequential search procedure. *Ann. Math. Statist.* **38**, 494–502.

Chew, M. C. (1973). Optimal stopping in a discrete search problem. *Operations Res.* **21**, 741–747.

Chow, Y. S., Robbins, H., and Siegmund, D. (1971). "Great Expectations: The Theory of Optimal Stopping." Houghton, Boston, Massachusetts.

Dantzig, G. B., and Wald, A. (1951). On the fundamental lemma of Neyman and Pearson. *Ann. Math. Statist.* **22**, 87–93.

DeGroot, M. H. (1970), "Optimal Statistical Decisions." McGraw-Hill, New York.

DeGuenin, J. (1961). Optimum distribution of effort: An extension of the Koopman basic theory. *Operations Res.* **9**, 1–7.

Dobbie, J. M. (1963). Search theory: A sequential approach. *Naval Res. Logist. Quart.* **4**, 323–334.

Dobbie, J. M. (1968). A survey of search theory. *Operations Res.* **16**, 525–537.

Dobbie, J. M. (1973). Some search problems with false contacts. *Operations Res.* **21**, 907–925.

Dobbie, J. M. (1974). A two cell model of search for a moving target. *Operations Res.* **22**, 79–92.

Dobbie, J. M. (1975a). Search for an avoiding target. *SIAM J. Appl. Math.* **28**, 72–86.

Dobbie, J. M. (1975b). Search for moving targets in the presence of false contacts.

Dreyfus, S. E. (1965). "Dynamic Programming and the Calculus of Variations." Academic Press, New York.

Enslow, P. (1966). A bibliography of search theory and reconnaissance theory literature. *Naval Res. Logist. Quart.* **13**, 177–202.

Everett, H. (1963). Generalized Lagrange multiplier method for solving problems of optimum allocation of resources. *Operations Res.* **11**, 399–417.

Federer, H. (1969). "Geometric Measure Theory." Springer-Verlag, Berlin and New York.

Feller, W. (1957). "An Introduction to Probability Theory and Its Applications," Vol. I. Wiley, New York.

Feller, W. (1966). "An Introduction to Probability Theory and Its Applications," Vol. II. Wiley, New York.

Franck, W. (1965). An optimal search problem. *SIAM Rev.* **7**, 503–512.

Fristedt, B., and Heath, D. (1974). Searching for a particle on the real line. *Advances in Appl. Probability* **6**, 79–102.

Gilbert, E. N. (1959). Optimal search strategies. *J. Soc. Indust. Appl. Math.* **7**, 413–424.

Girsanov, I. V. (1972). "Lectures on Mathematical Theory of Extremum Problems." Springer-Verlag, Berlin and New York.

Goffman, C. (1965). "Calculus of Several Variables." Harper, New York.

Halkin, H. (1964). On the necessary condition for optimal control of non-linear systems. *J. Analyse Math.* **7**, 1–82.

Hall, G. J. (1973). Sequential search with random overlook probabilities. Ph.D. Dissertation, Dept. of Math., Univ. of California, Los Angeles.

Halmos, P. (1948). The range of a vector measure. *Bull. Amer. Math. Soc.* **54**, 416–421.

Hardy, G. H., Littlewood, J. E., and Polya, G. (1964). "Inequalities." Cambridge Univ. Press, London and New York.

Hellman, O. (1970). On the effect of a search upon the probability distribution of a target whose motion is a diffusion process. *Ann. Math. Statist.* **41**, 1717–1724.

References

Hellman, O. (1971). Optimal search for a randomly moving object in a special case. *J. Appl. Probability* **8**, 606–611.

Hellman, O. (1972). On the optimal search for a randomly moving target. *SIAM J. Appl. Math.* **22**, 545–552.

Kadane, J. B. (1968). Discrete search and the Neyman–Pearson lemma. *J. Math. Anal. Appl.* **22**, 156–171.

Kadane, J. B. (1969). Quiz show problems. *J. Math. Anal. Appl.* **27**, 609–623.

Kadane, J. B. (1971). Optimal whereabouts search. *Operations Res.* **19**, 894–904.

Kan, Y. C. (1972). Optimal search models. Rep. No. ORC 72-13. Operations Res. Center, Univ. of California, Berkeley.

Kan, Y. C. (1974). A conterexample for an optimal search and stop model. *Operations Res.* **22**, 889–892.

Karlin, S. (1959). "Mathematical Methods and Theory in Games, Programming, and Economics," Vol. II. Addison-Wesley, Reading, Massachusetts.

Karlin, S. (1968). "A First Course in Stochastic Processes," Academic Press, New York.

Kettelle, J. D. (1962). Least-cost allocations of reliability investment. *Operations Res.* **10**, 249–265.

Kisi, T. (1966). On an optimal searching schedule. *J. Operations Res. Soc. Japan* **8**, 53–65.

Klein, M. (1968). Note on sequential search. *Naval Res. Logist. Quart.* **15**, 469–475.

Koopman, B. O. (1946). Search and screening. Operations Evaluation Group Rep. No. 56 (unclassified). Center for Naval Analysis, Rosslyn, Virginia.

Koopman, B. O. (1956a). The theory of search, Pt. I. Kinematic bases. *Operations Res.* **4**, 324–346.

Koopman, B. O. (1956b). The theory of search, Pt. II. Target detection. *Operations Res.* **4**, 503–531.

Koopman, B. O. (1957). The theory of search, Pt. III. The optimum distribution of searching effort. *Operations Res.* **5**, 613–626.

Larson, R. (1972). "Urban Police Patrol Analysis." MIT Press, Cambridge, Massachusetts.

Lehmann, E. L. (1959). "Testing Statistical Hypotheses." Wiley, New York.

Loane, E. P. (1971). An algorithm to solve finite separable single-constrained optimization problems. *Operations Res.* **19**, 1477–1493.

Loeve, M. (1963). "Probability Theory," 3rd ed. Van Nostrand-Reinhold, Princeton, New Jersey.

Luenberger, D. (1969). "Optimization by Vector Space Methods." Wiley, New York.

McCabe, B. J. (1974). Searching for a one-dimensional random walker. *J. Appl. Probability* **11**, 86–93.

Matula, D. (1964). A periodic optimal search. *Amer. Math. Monthly* **71**, 15–21.

Neyman, J., and Pearson, E. S. (1933). On the problem of the most efficient tests of statistical hypotheses. *Philos. Trans. Roy. Soc. London Ser. A* **231**, 289–337.

Onaga, K. (1971). Optimal search for detecting a hidden object. *SIAM J. Appl. Math.* **20**, 298–318.

Persinger, C. A. (1973). Optimal search using two non-concurrent sensors. *Naval Res. Logist. Quart.* **20**, 277–288.

Pollock, S. M. (1970). A simple model of search for a moving target. *Operations Res.* **18**, 883–903.

Pursiheimo, U. (1973). On the optimal search for a moving target. Unpublished manuscript.

Raiffa, H. (1968). "Decision Analysis: Introductory Lectures on Choices under Uncertainty." Addison-Wesley, Reading, Massachusetts.

Reber, R. K. (1956). A theoretical evaluation of various search/salvage procedures for use with narrow-path locators, Pt. I, Area and channel searching, Bureau of Ships, Minesweeping Branch Technical Rep. No. 117 (AD 881408).

Reber, R. K. (1957). A theoretical evaluation of various search/salvage procedures for use with narrow-path locators, Pt. II, Locating an object whose apparent presence and approximate position are known, Bureau of Ships, Minesweeping Branch Tech. Rep. No. 118 (AD 881410).

Richardson, H. R. (1971). Differentiability of optimal search plans. Daniel H. Wagner, Assoc. Mem. to Office of Naval Res. (AD 785294).

Richardson, H. R. (1973). ASW information processing and optimal surveillance in a false target environment. Daniel H. Wagner, Assoc. Rep. to the Office of Naval Res. (AD A002254).

Richardson, H. R., and Belkin, B. (1972). Optimal search with uncertain sweep width. *Operations Res.* **20**, 764–784.

Richardson, H. R., and Stone, L. D. (1971). Operations analysis during the underwater search for Scorpion. *Naval Res. Logist. Quart.* **18**, 141–157.

Richardson, H. R., Stone, L. D., and Andrews, F. A. (1971). "Manual for the Operations Analysis of Deep Ocean Search." Prepared for the Supervisor of Salvage, Naval Ship Systems Command. (NAVSHIPS 0994-010-7010).

Ross, S. (1969). A problem in optimal search and stop. *Operations Res.* **17**, 984–992.

Rudin, W. (1953). "Principles of Mathematical Analysis." McGraw-Hill, New York.

Saretsalo, L. (1973). On the optimal search for a target whose motion is a Markov process. *J. Appl. Probability* **10**, 847–856.

Savage, L. J. (1954). "The Foundations of Statistics." Wiley, New York.

Savage, L. J. (1971). Elicitations of personal probabilities and expectations. *J. Amer. Statist. Assoc.* **66**, 783–801.

Slater, M. (1950). Lagrange multipliers revisited: A contribution to non-linear programming." Cowles Commission Discussion Paper, *Math. 403*; reissued as "Cowles Foundation Discussion Paper No. 80."

Smith, M. W., and Walsh, J. E. (1971). Optimum sequential search with discrete locations and random acceptance errors. *Naval Res. Logist. Quart.* **18**, 159–168.

Stone, L. D. (1969). "Optimal Search Using a Sensor with Uncertain Sweep Width." Unpublished manuscript (AD 785288).

Stone, L. D. (1972a). Incremental and total optimization of separable functionals with constraints. Unpublished manuscript (AD A007628).

Stone, L. D. (1972b). Incremental approximation of optimal allocations. *Naval Res. Logist. Quart.* **19**, 111–122.

Stone, L. D. (1973a). Total optimality of incrementally optimal allocations. *Naval Res. Logist. Quart.* **20**, 419–430.

Stone, L. D. (1973b). Necessary and sufficient conditions for optimal search plans for targets having randomized conditionally deterministic motion. Daniel H. Wagner, Assoc. Memo. to Office of Naval Res. (AD 785293).

Stone, L. D. (1973c). Semi-adaptive search plans. Daniel H. Wagner, Assoc. Memo. to Office of Naval Res. (AD 785295).

Stone, L. D., and Richardson, H. R. (1974). Search for targets with conditionally deterministic motion. *SIAM J. Appl. Math.* **27**, 239–255.

Stone, L. D., and Rosenberg, J. (1968). The effect of uncertainty in the sweep width on the optimal search plan and the mean time to find the target. Daniel H. Wagner, Assoc. Memo. to Chairman, Techn. Advisory Group, SCORPION Search (AD 785297).

References

Stone, L. D., and Stanshine, J. A. (1971). Optimal search using uninterrupted contact investigation. *SIAM J. Appl. Math.* **20**, 241–263.

Stone, L. D., Stanshine, J. A., and Persinger, C. A. (1972). Optimal search in the presence of Poisson-distributed false targets. *SIAM J. Appl. Math.* **23**, 6–27.

Strauch, R. E. (1966). Negative dynamic programming. *Ann. Math. Statist.* **37**, 871–890.

Sweat, C. (1970). Sequential search with discounted income, the discount a function of cell searched. *Ann. Math. Statist.* **41**, 1446–1455.

Sworder, D. (1966). "Optimal Adaptive Control Systems." Academic Press, New York.

United States Coast Guard (1959). "National Search and Rescue Manual," CG-308.

Von Neumann, J. (1949). On rings of operators: Reduction theory. *Ann. of Math.* **50**, 401–485.

Wagner, D. H. (1969). Non-linear functional versions of the Neyman–Pearson lemma. *SIAM Rev.* **11**, 52–65.

Wagner, D. H., and Stone, L. D. (1974). Necessity and existence results on constrained optimization of separable functionals by a multiplier rule. *SIAM J. Control.* **12**, 356–372.

Zahl, S. (1963). An allocation problem with applications to operations research and statistics. *Operations Res.* **11**, 426–441.

Author Index

Numbers in italics refer to the pages on which the complete references are listed.

A

Andrews, F. A., 18, 44, 63, 188, *248*
Arkin, V. I., 81, *245*
Aumann, R. J., 81, 244, *245*

B

Beck, A., 82, *245*
Belkin, B., 63, 81, 134, 135, *245*, *248*
Bellman, R., 224, *245*
Black, W. L., 116, *245*
Blackwell, D., 116, 244, *245*
Bumby, R. T., 196, *246*

C

Charnes, A., 81, *246*
Chew, M. C., 116, 130, 135, *246*
Chow, Y. S., 126, *246*
Cooper, W. W., 81, *246*

D

Dantzig, G. B., 81, *246*
DeGroot, M. H., 116, *246*
DeGuenin, J., 81, *246*
Dobbie, J. M., 83, 100, 140, 146, 166, 177, 178, 230, 231, 233, *246*
Dreyfus, S. E., 178, *246*
Dubovitskii, A. Ya., 217

E

Enslow, P., 83, *246*
Everett, H., 81, 106, *246*

F

Federer, H., 75, 244, *246*
Feller, W., 51, 158, 175, *246*
Franck, W., 82, *246*
Fristedt, B., 82, *246*

G

Gilbert, E. N., 82, *246*
Girsanov, I. V., 217, *246*
Goffman, C., 202, *246*

H

Halkin, H., 81, *246*
Hall, G. J., 117, *246*
Halmos, P., 241, *246*
Hardy, G. H., 79, *246*
Heath, D., 82, *246*
Hellman, O., 233, *246, 247*

K

Kadane, J. B., 106, 107, 116, 117, *247*
Kan, Y. C., 117, 135, 225, 229, *247*
Karlin, S., 81, 142, 143, *247*
Kettelle, J. D., 106, *247*
Kisi, T., 82, *247*
Klein, M., 229, *247*
Koopman, B. O., 1, 24, 27, 28, 34, 43, 81, 100, 187, 196, 220, *247*

L

Larson, R., 1, *247*
Lehmann, E. L., 81, 107, *247*
Littlewood, J. E., 79, *246*
Loane, E. P., 106, *247*
Loeve, M., 161, *247*
Luenberger, D., 239, 242, 244, *247*

M

McCabe, B. J., 82, *247*
Matula, D., 116, 117, *247*
Milyutin, A. A., 217

N

Newman, D. J., 82, *245*
Neyman, J., 81, 104, *247*

O

Onaga, K., 82, *247*

P

Pearson, E. S., 81, 104, *247*
Perles, M., 81, 244, *245*
Persinger, C. A., 63, 82, 178, *247, 249*
Pollock, S. M., 225, 228, 229, 233, *247*
Polya, G., 79, *246*
Pursiheimo, U., 219, *247*

R

Raiffa, H., 18, *247*
Reber, R. K., 26, 27, 44, 187, 188, 196, *248*
Richardson, H. R., 18, 20, 22, 44, 63, 81, 82, 116, 134, 135, 188, 189, 206, 219, *248*
Robbins, H., 126, *246*
Rosenberg, J., 82, *248*
Ross, S., 122, 129, 130, 135, *248*
Rudin, W., 31, 70, *248*

S

Saretsalo, L., 232, 233, *248*
Savage, L. J., 18, *248*
Siegmund, D., 126, *246*
Slater, M., 81, *248*
Smith, M. W., 178, *248*
Stanshine, J. A., 63, 81, 176, 177, 178, *249*
Stone, L. D., 18, 20, 22, 41, 44, 63, 78, 81, 82, 95, 97, 98, 100, 169, 176, 177, 178, 188, 189, 196, 217, 219, 243, 244, *248, 249*
Strauch, R. E., 121, 225, *249*
Sweat, C., 117, *249*
Sworder, D., 166, *249*

U

United States Coast Guard, 18, 44, 57, *249*

Author Index

V

Von Neumann, J., 243, *249*

W

Wagner, D. H., 41, 81, 97, 106, 107, 219, 243, 244, *249*

Wald, A., 81, *246*
Walsh, J. E., 178, *248*
Warren, P., 82, *245*

Z

Zahl, S., 81, *249*

Subject Index

A

Absolutely continuous function, 70
Allocation, 30
 incremental, 92
 optimal for cost K, 33, 97
Allocation plan, 72

B

Borel measurable, 32
Broad search, 137
Broad-search allocation, 138
Broad-search plan, 138
Broad-search sensor, 137
Broad-search time, 138

C

Comparison of optimal to suboptimal plans, 67, 178
Concave, 38

Conclusive contact investigation, 139
Conditionally deterministic motion, 13, 197–220
 definition of, 199–200
Conditionally optimal pair, 93
Conditionally optimal plan, *see* Optimal plans
Constrained optimization, *see* Optimization
Constrained optimum, 35
Contact, 137
Contact investigation, 137
Convex, 38
Convex functional, 238
Cost function, 30–31, 96
Cumulative effort function, 44

D

Definite-range law, 23
Delay policy with breathers, 152
Δ-plans, 189
Detected, 145

Detection function, 29, 31
  conditional, 59
  for discrete search, 102
  exponential, 4–5, 32, 90
  for false targets, 140
  homogeneous, 32, 90
  posterior, 170
  regular, 31
Detection models, 22–32
Domain, 20
Dominated convergence theorem, 235

F

Factorable motion, 202
False targets, 12, 116, 136–178
  definition of, 137
  density, 140
  detection function, 140
  discrete model, 144
  independent identically distributed model, 139–141
  models for, 137–145
  Poisson model, 141–142
  posterior distribution, 169–173
Found, 145

I

Identified, 145
Image, 20, 38
Immediate and conclusive contact investigation, 148–155
  definition of, 139
Incrementally optimal sequence
  definition of, 98
  see also Optimal plans
Integrably bounded, 71

K

Kuhn–Tucker theorem, 41

L

Lagrange multiplier, 35–41
  definition of, 37

  interpretation of, 84–87
Lagrangian
  functional, 237
  maximization of, 39, 97, 237
  pointwise, 37, 97
Lateral range, 22
Lateral-range function, 23

M

Marginal rate of return, 85
Markovian motion, 14, 221–233
  continuous time and space, 231–232
  discrete time, 222–229
  two cell, 222–231
Mean time to detect, 51–53, 66
  minimization of, 51–52, 82, 110–111, 225–229
Minimizing expected cost, 110–114, 158, 162
Monotone convergence theorem, 235
Moving-target search plan, 200, 230
Multiple constraints, *see* Optimization
Multiple targets, 176

N

Neyman–Pearson lemma, 81, 104, 107
Nonadaptive investigation delay policy, 151

O

Optimal for cost K, 36
Optimal plans
  adaptive, 178
  algorithm for finding, 57
  approximation of, 12, 179–196
  conditionally optimal, 92, 98
  effect of false targets on, 162–164
  incrementally optimal, 10, 91–99
  locally optimal, 9, 86, 108
  necessary conditions for, 38–41, 106, 217, 230, 232
  nonadaptive, 146–155
  properties of, 9, 84–100

Subject Index

Optimal plans—*contd.*
  sufficient conditions for, 36–38, 104–106, 217
  totally optimal, 10, 98
  see also Uniformly optimal plans
Optimal return function, 121
Optimal search and stop plan, 121, 131–132
  existence of, 121
  properties of, 126–130
  sufficient conditions for, 122–125
Optimal within $\Psi(m_1, m_2, m_3)$, 201
Optimization
  constrained, 35–41, 237–244
  incremental, 91–100
  multiple constraints, 41, 96–99, 237–244
  necessary conditions for, 237–244
  sufficient conditions for, 35–38, 70–71, 97–98

P

Pontryagin's maximum principle, 41
Posterior target distribution, 53, 67, 170

R

Random search, 23–27
  definition of, 24
Random-search formula, 24, 34
Range, 20
Rate of return function, 46

S

Search
  bibliographies, 83
  detection, 114
  discrete, 10, 101–117
  linear, 82
  maximum probability, 87–91
  model, 17–34
  model for discrete search, 102–104
  plan, 44, 102
  problem, 2–3, 32–34
  problem in the presence of false targets, 146–148
  problem when target motion is conditionally deterministic, 198–202
  problem when target motion is Markovian, 222–223, 230–232
  space, 32
  with switching cost, 82
  target distribution, 17–22
  whereabouts, 10, 114–116
Search and stop, 11, 118–135, *see also* Optimal search and stop plan
Search and stop plan, 119
Search plan, 102
  adaptive, 146
  for moving targets, 200
  nonadaptive, 146
Semiadaptive plans, 165–177
  comparison with adaptive and nonadaptive plans, 167–169, 176, 178
  definition of, 166
Separable optimization problem, 34
Summably bounded, 72
Sweep width, 23

T

t-optimal, 201
Target density, 21
Target distribution
  bivariate normal, 20
  circular normal, 20
  definition of, 21
Totally optimal pair, 93
Totally optimal plans, *see* Optimal plans

U

Uncertain sweep width, 9, 59–70
Uniformly optimal plans, 7–8, 41–59
  for circular normal target distribution, 52, 63
  computation of, 49–50, 72–73, 203–206
  definition of, 44, 72, 107, 201
  for discrete search, 104–110
  existence of, 77, 80
  for exponential detection function, 52, 55, 63, 86
  for gamma distributed sweep width, 63

Uniformly optimal plans—*contd.*
   lack of, 58, 164–165
   for moving targets, 203–212
   with uncertain sweep width, 59–70
   see also Optimal plans

W

Whereabouts search, 114–116
   definition of, 114
Whereabouts search strategy, 114

Notation Index

Latin Letters

a	140	F	33
a'	150	$\hat{F}$	36
$\mathcal{A}$	166	f	30
B	61	$\mathfrak{F}$	218
B'	61	$\mathfrak{F}'$	218
B_w	60	I	74, 79
B_w'	61	I_ℓ	74, 79
b	33	J	32
b'	31	$\mathbf{J}$	200
C	33, 96	$\mathbf{j}$	202
$\hat{C}$	107	K	33
C^i	97	L	237
c	33, 96	$\mathcal{L}_\infty$	217
c^i	96	ℓ	37, 97
$\mathbf{C}$	130	M	44
$\mathbf{C}'$	132	$m_i\ (i=1,2,3)$	200
$\mathbf{C}_t$	153	$\mathbf{M}$	225
$\mathbf{c}$	153	N	139
d	141	$\mathbf{n}$	203
$\tilde{d}$	143	$\mathfrak{N}$	153, 166
E	70	o	230
e	70	P	33
e'	71	$\hat{P}$	107
$\mathbf{E}$	51	P_t	201
$\mathcal{E}_r$	237	$\mathbf{P}_t$	153
$\mathcal{E}_r^+$	237	$\mathbf{P}_\varphi$	131

259

Latin Letters

p	33	$\tilde{U}$	71, 73
$\mathbf{P}$	153	V	119
Q	170	$\mathbf{V}$	130
q	139	W	23
$\mathbf{Q}_n$	223	$\mathbf{W}$	59
$\mathbf{R}$	120	X	32
$\mathbf{R}_\varphi$	131	Y	199
$\mathbf{S}$	120	Z	36
U	47, 49		

Greek Letters

$\hat{\alpha}$	23	ξ	102
β	102	$\tilde{\xi}$	199
$\tilde{\beta}$	120	ρ	46, 49, 71
Γ	63	ρ_j	49, 73
γ	102	$\rho_j{}^{-1}$	49, 73
$\tilde{\gamma}$	120	ρ_x	46, 72
$\boldsymbol{\Delta}$	141	$\rho_x{}^{-1}$	49, 72
δ	140	τ	141, 143
η	199	Φ	44
λ	37	$\tilde{\Phi}$	72
μ	51, 110	φ	44
$\tilde{\mu}$	148	Ψ	200
Ξ	107	$\Psi_i\,(i = 1, 2, 3)$	201
Ξ_s	119	ψ	200
Ξ_∞	119		

Appendix C

Recent Results in
Optimal Search for a Moving Target

Since the appearance of the first edition of this book in 1975, there have been a number of major developments in the theoretical and computational aspects of finding optimal search plans for moving targets. The purpose of this appendix is to outline some of these developments and to provide the reader with a guide to the literature in this area. The references cited below are listed at the end of this appendix.

A number of books on search problems and theory have appeared since this book was first published. Haley and Stone (1980) contains a series of papers presented at a NATO conference on search theory and applications. Washburn (1981b) discusses a number of interesting problems in search and detection. The book by Ahlswede and Wegener (1987) concentrates on problems in discrete search for stationary targets and demonstrates many interesting connections among search theory, information theory, and coding theory. Koopman (1980) is an update and expansion of the 1946 classic that first formalized the subject of search theory. (See Morse (1982) for a discussion of the origins of search theory.) Chudnovsky and Chudnovsky (1988) is a collection of articles about search for moving targets, particularly pursuit and evasion problems. Hajek (1975) and Gal (1980) are important books on two-sided search problems.

The appendix is divided into three sections. The first section discusses one-sided moving target problems where search effort can be divided as finely as desired and where the application of search effort in a given place at one time does not constrain the location of search effort at any later time. These are called *optimal search density* problems. The second section discusses one-sided problems where the location of search effort at one time constrains the location of effort at a later time. These are called *optimal searcher path* problems. Optimal searcher path problems arise when the searcher's speed is roughly the same as the target's, e.g., one submarine searching for another. Optimal search density problems are reasonable approximations to situations where the searcher's speed is much faster than the target's, e.g., a plane looking for a ship lost at sea. The third section touches very briefly on two-sided search problems.

C.1 OPTIMAL SEARCH DENSITY FOR A MOVING TARGET

Chapter VIII and sections 9.2 and 9.3 deal with a special class of optimal search density problems. The results in Chapter VIII are limited to a class of target motions called conditionally deterministic. In the case where the Jacobian associated with this motion is factorable, the optimal search problem can be reduced to a stationary target problem and solved by the methods of Chapter II. When the Jacobian is not factorable, section 8.4 presents necessary and sufficient conditions for optimal search densities, but gives no indication of how to use these conditions to find optimal plans. Section 9.2 considers a very special two-cell Markovian target motion and for the case of an exponential detection function finds an optimal search density by using Pontryagin's maximum principle from control theory. Section 9.3 gives necessary conditions for the case of Markovian target motion models and exponential detection functions but presents no method for using these conditions to find an optimal search density.

For someone interested in calculating an optimal search density for a moving target problem, these results present a rather limited choice: use a two-cell Markov motion model (and an exponential detection function) or a factorable conditionally deterministic motion model. The situation is not much better for search problems involving discrete looks and Markovian target motion with discrete time and space. The solution method suggested in section 9.1 is to use dynamic programming which quickly becomes impractical when the number of cells is larger than three.

The above discussion describes the state of the theory of optimal search for moving targets in 1975 so that the reader can appreciate the amount of progress that has been made since then.

C.1 Optimal Search Density for a Moving Target

Problem Definition

For the purpose of this discussion, we will use a definition of the search problem very similar to the one given in Chapter VIII. The main difference is in allowing the target motion to be modeled by an arbitrary stochastic process $X = \{X_t, t \geq 0\}$ taking values in Y (Euclidean n-space) with Borel measurable sample paths and in generalizing the detection function to allow it to depend on a weighted search density. The Borel measurability requirement is strictly a technical one. It is always satisfied in applications. For ease of exposition and interpretation, we shall discuss the results on optimal search density for moving targets in the context of discrete time and continuous space. Most of the results stated have analogs for continuous time and for discrete space. The analogs may be found in the articles cited below.

The target's location and motion through Y are specified by the stochastic process $X = \{X_t, t \geq 0\}$ where $X_t \in Y$ gives the target's position at time t. We specify a time horizon $[0, T]$ and seek to maximize the probability of detecting the target by time T. For this discussion time will be discrete so that $t = 0,\ldots,T$.

A search plan ψ specifies the allocation of search effort in space and time. Specifically

$$\psi(y,t) = \text{effort density placed at point } y \text{ at time } t$$
$$\text{for } y \in Y, t = 0,\ldots,T.$$

Search effort is constrained by the rate at which effort can be applied. Specifically there is a function m such that

$$m(t) = \text{effort available for search at time } t \text{ for } t = 0,\ldots,T,$$

and search plans ψ must satisfy

$$\int_Y \psi(y,t) \, dy \leq m(t) \quad \text{for } t = 0,\ldots,T, \tag{C.1.1}$$

$$\psi(y,t) \geq 0 \text{ for } y \in Y, t = 0,\ldots,T. \tag{C.1.2}$$

Let Ψ be the set of search plans satisfying (C.1.1) and (C.1.2).

For each sample path ω of the process X, the probability of detecting the target by time t, given that it follows that path, is a function of the weighted total effort density,

$$\zeta(\psi,\omega,t) = \sum_{s=0}^{t} W(X_s(\omega), s)\, \psi(X_s(\omega), s),$$

which accumulates by time t on the target over the course of the path. The weight $W(y, s)$ represents the relative detectability or sweep width against the target given it is located at point y at time s. There is a detection function $b: [0, \infty] \to [0, 1]$ such that $b(\zeta(\psi, \omega, t))$ is the probability of detecting the target by time t given that it follows sample path ω and that search plan ψ is executed. Letting E denote expectation over the sample paths of X, we have that

$$P[\psi] = E[b(\zeta(\psi, \cdot, T))]$$

is the probability of detecting the target by time T with plan ψ. In the remainder of this discussion, we suppress the variable ω.

The optimal detection problem is to find a plan $\psi^* \in \Psi$ such that $P[\psi^*] \geq P[\psi]$ for all $\psi^* \in \Psi$. Such a plan is called *T-optimal*.

Review of Recent Results

In 1977, Brown applied the Karush-Kuhn-Tucker conditions to the problem of finding optimal plans for targets that move in discrete space and time when the detection function is exponential. By writing these conditions in a suitable form, he observed that the optimal plan for this moving target problem has the interesting property that if one selects a time t and conditions on failure at all times other than t (both before and after t), the optimal plan allocates the effort at time t so as to maximize the detection probability for the *stationary* target problem that one obtains from the conditioning. Since there are efficient methods for finding optimal plans for stationary targets, especially when the detection function is exponential (see Example 2.2.8), Brown was able to take advantage of this fact to devise an iterative algorithm that maximizes the probability of detecting the target in the interval $[0, T]$. This algorithm applies to target motions that are modeled by a mixture of discrete-time-and-space Markov chains and is very efficient (see Brown, 1980). When search effort can be applied only in discrete looks and when detection on each look is independent of detection on any other look, Washburn (1980a) gives an algorithm for finding search plans that satisfy a necessary condition for optimality. This algorithm is a discrete effort analog of Brown's algorithm.

In Stone *et al.* (1978), algorithms were devised for arbitrary discrete-time target motions and exponential detection functions. Stone (1979) generalized the necessary and sufficient conditions of Brown (1980) to target motions that are modeled by an arbitrary stochastic process with any mixture of discrete or

C.1 Optimal Search Density for a Moving Target

continuous space or time. This generalization also applies to regular detection functions. Arnold (1982) has shown that optimal search plans exist for a wide class of search problems in which the search density is bounded.

All of the above results apply to the problem of finding a plan that maximizes the probability of detecting a target by time T, although Stone et al. (1978) also consider a payoff related to mean time to detection. In Stromquist and Stone (1981), the necessary and sufficient conditions for optimal detection search are generalized to a wider class of constrained optimization problems which include problems not related to search as well as numerous search-related ones. Washburn (1983) considers a generalization of the algorithms in Brown (1980) and Washburn (1980a) which applies to finding optimal search allocations for payoffs other than maximizing the probability of detection by time T. Iida (1988) extends the optimal search and stop results of section 5.2 to targets with conditionally deterministic motion.

It is important to emphasize that the foregoing results and developments have two complementary sides, theoretical and practical. On the theoretical side is a set of necessary and often sufficient conditions for optimal plans for detecting moving targets and for more general optimization problems, specifically Brown (1980), Washburn (1980a), Stone (1979), and Stromquist and Stone (1981). On the practical side is a set of efficient algorithms for calculating optimal search plans for moving targets (i.e., Brown (1980); Stone et al. (1978); Discenza and Stone (1981); Washburn (1983)). Typically, these algorithms are developed from the necessary and sufficient conditions mentioned above. Stone (1983) discusses the application of search theory to operational search planning.

In the remainder of this section, we describe an extension of Brown's algorithm to the case of discrete time and continuous space and then present a generalized search optimization technique that encompasses many of the above mentioned results as special cases.

Brown's Algorithm for Continuous Space

For an exponential detection function and a target moving in discrete time and space, Brown's algorithm solves the problem of finding a T-optimal allocation by solving a sequence of stationary target problems. In the following paragraphs, we present an extension of Brown's algorithm to continuous search spaces and relate this extension to the original discrete space algorithm. The continuous space algorithm is based on the following necessary and sufficient condition for a T-optimal search plan proved by Stone (1979).

Define

$$g_t(y) \equiv \Pr\{X_t = y \mid \text{failure to detect at all times other than } t \text{ using plan } \psi\}$$
$$\text{for } y \in Y, \psi \in \Psi, \text{ and } t = 0,\ldots,T.$$

The function g_t is the posterior target location density given failure to detect by the search effort at all times other than t. If the detection function b is exponential, i.e., $b(z) = 1 - \exp(-z)$ for $z \geq 0$, then $g_t(y)$ is proportional to

$$E_{yt}\left\{\exp\left[-\sum_{s \neq t} W(X_s,s)\psi(X_s,s)\right]\right\} p_t(y)$$

where p_t is the probability density function for X_t and E_{yt} denotes expectation conditioned on $X_t = y$.

Necessary and Sufficient Condition for T-Optimality. Assume that the detection function b is exponential. Then a necessary and sufficient condition for $\psi^* \in \Psi$ to be T-optimal is that $\psi^*(\cdot,t)$ maximize the probability of detecting a stationary target with distribution g_t using effort $m(t)$ for $t = 0,\ldots,T$.

Description of Algorithm. For time $t = 0$, the algorithm allocates $m(0)$ effort optimally to the target distribution p_0. For $t = 1,\ldots,T$, the algorithm calculates the posterior target distribution at time t given failure to detect by the effort prior to time t and allocates $m(t)$ effort in a manner that is optimal for the stationary target problem with target distribution given by g_t. The plan that results from this first pass is the incrementally optimal or myopic search plan. It maximizes the increment in detection probability at each time t but does not produce a T-optimal plan.

Subsequent passes proceed as follows for $t = 0,\ldots,T$. The algorithm computes g_t based on the allocation ψ obtained up to that point in the iteration, redistributes the effort $m(t)$ at time t to be optimal for g_t, and changes ψ to reflect the reallocation. The algorithm continues in this iterative fashion until some convergence criterion is met.

To use this algorithm, one must be able to calculate g_t and find the optimal allocation of effort for a stationary target problem when the detection function is exponential. Consider now the case where the search space is discrete rather then continuous. For this case Example 2.2.8 gives an efficient algorithm for finding optimal search allocations for stationary target problems. If one can find an efficient method of computing g_t, then he can implement the above algorithm for the case of a discrete search space. For the case where X is a discrete time and space Markov process, Brown (1980) devised a very efficient algorithm for computing g_t and used it along with the methodology in Example 2.2.8 to produce an algorithm for calculating T-optimal search plans. By making use of the upper bound discovered by Washburn (1981a), one can tell when his solution has come within a specified ε of the detection probability of the T-optimal plan. When using Brown's algorithm, the convergence is usually very rapid.

C.1 Optimal Search Density for a Moving Target

Stone *et al.* (1978) presents a generalization of Brown's algorithm to arbitrary discrete time and space target motions.

Generalized Search Optimization Technique

In the following paragraphs, we state the necessary and sufficient conditions of Stromquist and Stone (1981) and outline how these conditions can be used to produce algorithms to find numerical solutions to a class of constrained optimization problems. The combination of the necessary and sufficient conditions and the algorithms derived from these constitute what we call the *Generalized Search Optimization* (GSO) technique. After presenting the technique, we show that many of the results mentioned above are special cases of the GSO technique.

Necessary and Sufficient Conditions. We now describe the necessary and sufficient conditions that form the basis of the GSO. These conditions apply to a general class of constrained optimization problems. Let

Y be a σ-finite measure space with measure ν
T be a σ-finite measure space with measure τ
$Z = Y \times T$ have the product measure $\mu = \nu \times \tau$.

In search problems the space Y is usually the search space (e.g., Euclidean two-space) and T is the time interval of the search.

Constraints. Let $c: Z \to (0, \infty)$ be μ-measurable, $m: T \to (0, \infty)$ be τ-measurable, and $B \in (0, \infty)$. Define

Ψ = set of real-valued μ-measurable functions $\psi: Z \to [0, B]$

such that

$$\int_Y c(y, t)\, \psi(y, t)\, d\nu(y) \leq m(t) \quad \text{for a.e. } t \in T. \tag{C.1.3}$$

Let Ψ_0 be the subset of Ψ for which equality holds in equation (C.1.3).

Functional. Let P be a real-valued functional on Ψ.

Gateaux Differentials. In order to state the necessary and sufficient conditions of the generalized search optimization technique, we must introduce the notion of a Gateaux differential. Define the Gateaux differential of P at ψ in the direction h as

$$P'[\psi, h] = \lim_{\varepsilon \to 0+} \frac{P[\psi+\varepsilon h] - P[\psi]}{\varepsilon} \qquad (C.1.4)$$

if it exists. If P' exists and is given by

$$P'[\psi, h] = \int_Z d(\psi, y, t)\, h(y, t)\, d\mu(y, t), \qquad (C.1.5)$$

for some function d, then we say that $d(\psi, \cdot, \cdot)$ is the *kernel* of the Gateaux differential at ψ.

Constrained Optimality. We say that $\psi^* \in \Psi$ is optimal within Ψ if $P[\psi^*] \geq P[\psi]$ for $\psi \in \Psi$.

Theorems C.1.1 and C.1.2 below are taken from Stromquist and Stone (1981).

Theorem C.1.1. *If P has a Gateaux differential with kernel d, then a necessary condition for ψ^* to be optimal within Ψ is the existence of a measurable function $\lambda: T \to [-\infty, \infty]$ such that for a.e. $(y, t) \in Z$*

$$\begin{array}{ll} d(\psi^*, y, t) \geq \lambda(t) c(y, t) & \text{if } \psi^*(y, t) = B \\ = \lambda(t) c(y, t) & \text{if } 0 < \psi^*(y, t) < B \\ \leq \lambda(t) c(y, t) & \text{if } \psi^*(y, t) = 0. \end{array} \qquad (C.1.6)$$

Theorem C.1.2. *If the conditions of Theorem C.1.1 are satisfied, P is a concave functional, and $\psi^* \in \Psi_0$, then condition (C.1.6) is sufficient for ψ^* to be optimal within Ψ.*

For concave functionals condition (C.1.6) gives a necessary and sufficient condition for a constrained optimum.

We now outline an algorithm that can be applied to discrete time problems to find optimal allocations ψ^*. The outline is intended to cover a wide range of problems that fit the framework given in this section. As a result, the directions necessarily lack detail. For example, step (v) below requires the developer of the algorithm to find a function that satisfies condition (C.1.6) for one time t and also satisfies the constraint in equation (C.1.7). In the typical detection search problem, step (v) amounts to solving a stationary search problem. Even though this may appear formidable to the reader, it is often the case that when the

C.1 Optimal Search Density for a Moving Target

specific functions are substituted into the algorithm, the way to proceed is clear. In any case, this outline provides a method of attack that has proven successful in a number of problems, some of which are discussed below. Having developed an algorithm along the lines of this outline, one must still prove that the algorithm converges.

GSO Algorithm for Discrete Time. Suppose that time is discrete with $t = 0, 1, \ldots, T$. The algorithm proceeds as follows:

(i) Make an initial guess $\psi_0 \in \Psi$ for an allocation.
(ii) Set $n = 0$.
(iii) Set $s = n \,[\mathrm{mod}(T + 1)]$.
(iv) Compute $d(\psi_n, \cdot, \cdot)$.
(v) Solve for $f^*: Y \to [0,\infty)$ to satisfy

$$\int_Y c(y, s) f^*(y) \, d\nu(y) = m(s) \tag{C.1.7}$$

and so that the function ψ_{n+1} defined by

$$\psi_{n+1}(\cdot,t) = \begin{cases} \psi_n(\cdot,t) & \text{for } t \neq s \\ f^* & \text{for } t = s \end{cases}$$

satisfies condition (C.1.6) for $t = s$.

(vi) Compute the generalized Washburn upper bound (see Washburn (1981a); Stromquist and Stone (1981)). If this shows $P[\psi_{n+1}]$ to be within the desired tolerance of the optimal payoff $P[\psi^*]$, stop. Otherwise, go to step (vii).
(vii) Set $n = n + 1$ and go to step (iii).

Step (v) is often accomplished by choosing a value for $\lambda(s)$ and solving for the function f for which ψ defined by

$$\psi(\cdot,t) = \begin{cases} \psi_n(\cdot,t) & \text{for } t \neq s \\ f & \text{for } t = s \end{cases}$$

satisfies condition (C.1.6) for $t = s$. This is often easy to do but usually leads to functions f that do not satisfy equation (C.1.7). However, since f is often a monotone function (pointwise) of the value of $\lambda(s)$, one can usually perform a binary search to determine the value of $\lambda(s)$ that yields an f that satisfies (C.1.7) to any accuracy desired.

Another method of performing step (v) is to determine an equivalent optimization problem whose solution will yield f^*. This was done by Brown (1980) in his original algorithm for maximizing the probability of detecting a target with discrete-space-and-time Markov motion. In this case step (v) is equivalent to finding an allocation f^* to maximize the probability of detecting a stationary target.

Applications of GSO

We now illustrate how the generalized search optimization technique can be applied to problems in search for moving targets.

Detection Search. Suppose that a target is moving through the plane according to a discrete-time stochastic process $X = \{X_t, t = 0, \ldots, T\}$. We have $m(t)$ effort available at time t, and the function $\psi \in \Psi$ specifies $\psi(y, t)$, the search effort density placed at point y at time t for $y \in Y, t \in T$.

For each sample path ω of the process X, the probability of detecting the target by time t given that it follows that path is a function of the weighted total effort density

$$\zeta(\psi, \omega, t) = \sum_{s=0}^{t} W(X_s(\omega), s) \, \psi(X_s(\omega), s)$$

as discussed above with

$$P[\psi] = E[b(\zeta(\psi, \cdot, T))]$$

giving the probability of detecting the target by time T with plan ψ. In accordance with our convention stated above, we will henceforth suppress the variable ω. Recall that the optimal detection problem is to find a plan $\psi^* \in \Psi$ such that $P[\psi^*] \geq P[\psi]$ for all $\psi \in \Psi$ and that such a plan is called T-optimal.

For this problem the Gateaux differential has a kernel provided that the function b has a bounded nonnegative derivative b'. Let E_{yt} denote expectation conditioned on $X_t = y$, and let p_t be the probability density function for X_t. The Gateaux differential of P is given by

$$P'[\psi, h] = \sum_{t=0}^{T} \int_Y d(\psi, y, t) \, dy$$

C.1 Optimal Search Density for a Moving Target

where

$$d(\psi, y, t) = E_{yt}[b'(\zeta(\psi, T))]W(y,t)p_t(y) \quad y \in Y, \ t \in T.$$

The hypotheses of Theorems C.1.1 and C.1.2 hold so that condition (C.1.6) is necessary and sufficient for $\psi^* \in \Psi_0$ to be a T-optimal search plan.

The conditions of Brown (1980) are the special case of condition (C.1.6) that one obtains by taking $b(z) = 1 - \exp(-z)$ for $z \geq 0$, $Y = \{1, ..., J\}$, and ν to be a counting measure. In this case

$$d(\psi, y, t) = E_{yt}\left\{\exp\left[-\sum_{s \neq t} W(X_s,s)\psi(X_s,s)\right]\right\} \exp[-W(y,t)\psi(y,t)]W(y,t)p_t(y)$$

and one can see that g_t defined by

$$g_t(y) = E_{yt}\left\{\exp\left[-\sum_{s \neq t} W(X_s,s)\psi(X_s,s)\right]\right\}p_t(y) \quad \text{for } y \in Y$$

is proportional to the probability distribution of the target's location at time t given failure to detect the target by the search applied before and after time t but not during time t. Condition (C.1.6) becomes

$$g_t(y)\exp[-W(y,t)\psi^*(y,t)]W(y,t) \quad \begin{aligned} &= \lambda(t) \text{ if } \psi^*(y, t) > 0, \\ &\leq \lambda(t) \text{ if } \psi^*(y, t) = 0, \end{aligned}$$

which is equivalent to the conditions for $\psi^*(\cdot, t)$ to be an optimal allocation for the stationary search problem with target location distribution proportional to g_t.

Multistate Target Search. A generalization of the detection search described above is the multistate target search. In this case the target's motion is given by $\{(X_t, S_t): t = 0, ..., T\}$, where X_t is the target's position at time t and S_t is the target's state at time t. The target may change state as well as location stochastically, and the target's state can affect the target's motion as well as its detectability. As an example, consider a case where there are K states and the sweep width is a function of location, state, and time, so that cumulative effort ζ becomes

$$\zeta(\psi, T) = \sum_{t=0}^{T} W(X_t, S_t, t)\psi(X_t, t)$$

and

$$P[\psi] = E[b(\zeta(\psi, T))]$$

as before. Observe that effort cannot be allocated to states but only to locations. Let E_{ykt} denote expectation conditioned on $(X_t, S_t) = (y, k)$ and let $p_t(y, k) = Pr\{X_t = y, S_t = k\}$. Assuming that b has a bounded derivative b', Discenza and Stone (1981) show that P has a Gateaux differential with kernel

$$d(\psi, y, t) = \sum_{k=1}^{K} E_{ykt}[b'(\zeta(\psi,T))]W(y, k, t)p_t(y, k)$$

$$\text{for } y \in Y, t = 0, ..., T. \quad (C.1.8)$$

Using the definition of d above, one can show that the necessary and sufficient conditions obtained by Discenza and Stone (1981) are also a special case of condition (C.1.6).

Two special cases of multistate search are survivor search and defensive search. In survivor search, the target may be a person missing at sea or lost in the wilderness. The state represents the condition of the survivor (e.g., in a boat, in a life raft, in the water, or dead). By setting the sweep width equal to its appropriate value for each state and to zero if the person is dead, the problem of finding a plan to maximize P becomes that of finding a plan that maximizes the probability of finding the target alive by time T.

In defensive search, as defined by Brown, one is trying to detect an attacker before it launches a weapon. In this case the target has two states, weapon launched or not launched. Once the attacker launches a weapon, the sweep width is set to zero and the target remains in the launched state for the remainder of the problem. In this case, maximizing P is maximizing the probability of detecting the attacker before it launches an attack.

Surveillance. Tierney and Kadane (1983) have developed a technique for solving surveillance problems which builds on the optimal detection search results discussed above. The surveillance problem is to maximize the probability of being in contact (i.e., having a detection on the target) at time T. In contrast to the detection search problem, a detection before time T does not end the problem. It merely helps to obtain a detection at time T. For problems where the target's motion is modeled by a discrete-time-and-space Markov chain, Tierney and Kadane have shown that the optimal surveillance problem can be solved by solving a series of optimal detection search problems. In their method, one starts at time T and works his way backward in time in a fashion similar to dynamic programming. At each time t, one must solve what Tierney

C.1 Optimal Search Density for a Moving Target

and Kadane call a general detection search problem given knowledge of the target's position at time t.

In the general detection problem, the searcher receives a payoff or return $r(j, t)$ if he detects the target in cell j at time t. The search stops the first time the target is detected, and the objective of the general detection problem is to maximize the expected payoff. Suppose that T is the time horizon [i.e., $r(j, t) = 0$ if $t > T$]. Let $\zeta(\psi, -1) = 0$. Then the expected payoff P using plan ψ is given by

$$P[\psi] = E\left[\sum_{t=0}^{T} r(X_t,t) \{b(\zeta(\psi,t)) - b(\zeta(\psi,t-1))\}\right]$$

$$= E\left[\sum_{t=0}^{T} \{r(X_t,t) - r(X_{t+1},t+1)\} b(\zeta(\psi,t))\right] \quad (C.1.9)$$

and the Gateaux differential has the kernel $d(\psi,\cdot,\cdot)$ given by

$$d(\psi,j,t) = \sum_{s=t}^{T} E_{jt}[\{r(X_s,s)-r(X_{s+1},s+1)\}b'(\zeta(\psi,s))\,]\,p_t(j)\,W(j,t) \quad (C.1.10)$$

for $j = 1,\ldots,J$ and $t = 0,\ldots,T$.

Equation (C.1.9) is a special case of equation (16) in Stromquist and Stone (1981). As noted there, conditions (C.1.6) are necessary but not sufficient for optimality in this problem. If one makes the additional assumption that $r(j, \cdot)$ is a decreasing function for $j = 1,\ldots,J$, the conditions become sufficient. In the general detection problems that arise in solving the surveillance problem, however, the functions $r(j, \cdot)$ are likely to be increasing rather than decreasing.

Whereabouts Search. Stone and Kadane (1981) solve the problem of optimal whereabouts search for a moving target. In a whereabouts search one can succeed either by detecting the target or guessing its location. When $Y = \{1,\ldots,J\}$, Stone and Kadane show that solving a whereabouts problem is equivalent to solving J detection search problems.

C.2 OPTIMAL SEARCHER PATH FOR A MOVING TARGET

This problem typically divides into two cases. The first case is the discrete-time, discrete-space problem. The second is the continuous-time, continuous-space problem.

Discrete Time and Space

In the discrete-time-and-space problem, the target motion takes place in discrete cells. The searcher moves among the cells according to constraints. For example, if the searcher is in cell j at time t, he is constrained to be in one of the set $S(j)$ of cells at time $t+1$. The probability of detecting the target at time t depends on the location of the target and the searcher. Detection at one time is independent of detection at all others. The problem is to find a search path that satisfies the constraints and maximizes the probability of detecting the target by time T. Trummel and Weisinger (1986) have shown that the optimal searcher path problem is NP-complete.

A number of heuristic approaches have been taken to produce algorithms for computing optimal searcher paths. See, for example, Stewart (1979 and 1980) and Eagle (1984). Recently Eagle and Yee (1988) have developed an efficient algorithm for this problem based on the work of Stewart (1979).

Continuous Time and Space

In the continuous-time-and-space case, the constraints on the searcher's motion are usually given in terms of a differential equation that the searcher's path must satisfy. The target's motion is modeled as a diffusion process, and the detection function is taken to be exponential. These assumptions on target motion and the detection function are made so that the problem fits into a standard optimal stochastic control framework. One can then find a dynamic programming equation whose satisfaction provides a sufficient condition for optimality. This is the approach taken by Oshumi (1984, 1985). Optimal control theory has also been applied to this problem by Lukka (1977) and Mangel (1981).

Sunahara *et al*. (1982a, 1982b) discuss numerical methods of solving for the optimal searcher path and present an example in a simple case.

Two-sided Search for a Stationary or Moving Target

In this class of problems, the target is trying to avoid detection. In the case of a stationary target, the target's objective is to choose its location to make the search as difficult as possible. This problem is usually modeled as a two-person game with the target wishing to maximize the mean time to detection and the searcher wishing to minimize it. The solutions are typically mixed strategies for searcher and target.

When the target moves, the two-sided game becomes much more complex. The basic references are the books on differential games by Isaacs (1965), pursuit games by Hajek (1975), and search games by Gal (1980). Lalley and Robbins (in Chudnovsky and Chudnovsky (1988)) find simpler and more robust

C.2 Optimal Searcher Path for a Moving Target

versions of Gal's asymptotically minimax plans for search regions that are rectangles, parallelograms, or triangles.

Games of ambush are investigated by Ruckle *et al.* (1976), Ruckle (1981), Ruckle and Reay (1981), and Ruckle (1983). Other articles involving two-sided search include Johnson (1964), Danskin (1968), Dobbie (1975), and Washburn (1980b).

C.3 REFERENCES

Ahlswede, R., and Wegener, I. (1987). "Search Problems." Wiley, New York.

Arnold, L. K. (1982). Existence of search plans when feasible plans are uniformly bounded. Daniel H. Wagner, Associates Technical Report.

Brown, S. S. (1980). Optimal search for a moving target in discrete time and space. *Operations Res.* **28**, 1275-1289.

Chudnovsky D. V. and Chudnovsky G. V. (1988). "Search Theory: Some Recent Developments" Marcel Dekker, New York.

Danskin, J. M. (1968). A helicopter vs. submarine search game. *Operations Res.* **16**, 525-537.

Discenza, J. H., and Stone, L. D. (1981). Optimal survivor search with multiple states. *Operations Res.* **29**, 309-323.

Dobbie, J. M. (1975). Search for an avoiding target. *SIAM J. Appl. Math.* **28**, 72-86.

Eagle, J. N. (1984). Optimal search for a moving target when the search path is constrained. *Operations Res.* **32**, 1107-1115.

Eagle, J. N., and Yee, J. R. (1988). An optimal branch-and-bound procedure for the constrained path, moving target search problem. *Operations Res.* (To appear).

Gal, S. (1980). "Search Games." Academic Press, New York.

Hajek, O. (1975). "Pursuit Games." Academic Press, New York.

Haley, K. B., and Stone, L. D., eds. (1980). "Search Theory and Applications." Plenum Press, New York.

Iida, K. (1988). Optimal search plan minimizing the expected risk of the search for a target with conditionally deterministic motion. *Naval Research Logistics* (To appear).

Isaacs, R. (1965). "Differential Games." Wiley, New York.

Johnson, S. M. (1964). A search game, in "Advances in Game Theory." *Annals of Mathematics Studies* **52**, Princeton University Press, Princeton, NJ.

Koopman, B.O. (1980). "Search and Screening: General Principles with Historical Applications." Pergamon Press, New York.

Lukka, M. (1977). On the optimal searching tracks for a moving target. *SIAM J. Appl. Math.* **32**, 126-132.

Mangel, M. (1981). Search for a randomly moving object. *SIAM J. of Appl. Math.* **40**, 327-338.

Morse, P. M. (1982). In memoriam: Bernard Osgood Koopman, 1900-1981. *Operations Res.* **30**, 417-427.

Oshumi, A. (1984). Stochastic control with searching a randomly moving target. "Proceedings: 1984 American Control Conference", San Diego, Calif., June 1984, pp. 500-504.

Oshumi, A. (1985). Optimal searching for a Markovian target and relation to optimal stochastic control, "Proceedings of MTNS-85: 7th International Symposium on the Mathematical Theory of Networks and Systems", Stockholm, June 10-14, 1985, C. Byrnes and A. Lindquist (eds.). North-Holland, Amsterdam.

Ruckle, W. H., Fennell, R., Holmes, P. T., and Fennemore, C. (1976). Ambushing random walks I: Finite models. *Operations Res.* **24**, 314-324.

Ruckle, W. H. (1981). Ambushing random walks II: Continuous models.*Operations Res.* **29**, 108-120.

Ruckle, W. H., and Reay, J. R. (1981). Ambushing random walks III: More continuous models. *Operations Res.* **29**, 121-129.

Ruckle, W. H. (1983). "Geometric Games and Their Applications." Pitman, Marshfield, Mass.

Stewart, T. J. (1979). Search for a moving target when searcher motion is restricted. *Computers & Operations Res.* **6**, 129 - 140.

Stewart, T. J. (1980). Experience with a branch-and-bound algorithm for constrained searcher motion, in "Search Theory and Applications". K. B. Haley and L. D. Stone (eds.). Plenum Press, New York.

Stone, L. D. (1979). Necessary and sufficient conditions for optimal search plans for moving targets. *Math. Operations Res.* **4**, 431-440.

Stone, L. D. (1983). The process of search planning: Current approaches and continuing problems. *Operations Res.* **31**, 207-223.

Stone, L. D., Brown, S. S., Buemi, R. P., and Hopkins, C. R. (1978). Numerical optimization of search for a moving target. Daniel H. Wagner, Associates Report to Office of Naval Research.

Stone, L. D., and Kadane, J. B. (1981). Optimal whereabouts search for a moving target. *Operations Res.* **29**, 1154-1166.

Stromquist, W. R., and Stone, L. D. (1981). Constrained optimization of functionals with search theory applications. *Math. Operations Res.* **6**, 518-529.

Sunahara, Y., Oshumi, A., and Kobayashi, S. (1982a). On a method for searching a randomly moving target, "Proceedings of the 11th SICE Symposium on Control Theory", Kobe, Japan, May 27-29, 1982.

Sunahara, Y., Oshumi, A., and Kobayashi, S. (1982b). On the optimal search for a randomly moving target, "Proceedings of the 5th SICE Dynamical Systems Theory Symposium", Yokosuka, Japan, December 1982.

C.3 References

Tierney, L., and Kadane, J. B. (1983). Surveillance search for a moving target. *Operations Res.* **31**, 720-738.

Trummel, K. E., and Weisinger, J. R. (1986). The complexity of the optimal searcher path problem. *Operations Res.* **34**, 324-327.

Washburn, A. R. (1980a). On search for a moving target. *Naval Res. Logist. Quart.* **27**, 315-322.

Washburn, A. R. (1980b). Search evasion game in a fixed region. *Operations Res.* **28**, 1290-1298.

Washburn, A. R. (1981a). An upper bound useful in optimizing search for a moving target. *Operations Res.* **29**, 1227-1230.

Washburn, A. R. (1981b). "Search and Detection." Operations Research Society of America c/o Ketron Inc, Arlington VA.

Washburn, A. R. (1983). Search for a moving target: The FAB algorithm. *Operations Res.* **31**, 739-751.

Appendix D

Corrections

This appendix presents corrections to the body of the text. Obvious typos are not listed. The corrections mainly consist of corrected formulas. In this case the correct formula is listed without identifying explicitly the portion of the old formula that is incorrect. The corrections are listed by page number.

Page 39, line 14: "Then a necessary and sufficient condition for f^* to be optimal...".

Page 66, line 15:

$$\frac{\mu(\varphi^*)}{6\pi\sigma^2\alpha/\nu\nu} = \frac{1}{3}\left(\frac{3\nu+1}{\nu-1}\right) \geq 1,$$

Page 70, line 22:

$$\int_a^b g'(y)\,dy = g(b) - g(a) \quad \text{for} \quad a, b \in I.$$

278

Appendix D   Corrections 279

Page 184, below equation (7.1.11):

$$r_n^2 = 2\sigma^2 \left[\frac{K}{2\pi\sigma^2 Nh} + \frac{(N-1)h}{2} - (n-1)h \right] \quad \text{for} \quad n = 1,2,\ldots,N.$$

Page 185, before equation (7.1.13):

$$= e^{-Nh} + N(1 - e^{-h}) \, exp\left(\frac{-K}{2\pi\sigma^2 Nh} - \frac{(N-1)h}{2}\right).$$

Page 185, equation (7.1.13):

$$P[\hat{f}] = 1 - e^{-Nh} - N(1 - e^{-h}) \, exp\left(\frac{-K}{2\pi\sigma^2 Nh} - \frac{(N-1)h}{2}\right), \qquad (7.1.13)$$

Page 204, line 3:

$$m_2^*(t) = \int_Y \psi^*(y,t) \, dy \quad \text{for} \quad t \in T.$$